Information and Communication Technology

Vocational A Level

.... :s are to :

2nd edition

Information and Communication Technology

Vocational A Level

Stephen Doyle

STANLEY THORNES

First published in 1997 by Stanley Thornes (Publishers) Ltd.

Second edition published in 2001 by:
Nelson Thornes Ltd
Delta Place
27 Bath Road
CHELTENHAM
GL53 7TH
UK

01   02   03   04   05   /   10   9   8   7   6   5   4   3   2   1

A catalogue record for this book is available from the British Library

ISBN 0 7487 5317 6

Illustrations by Steve Ballinger
Typeset by Florence Production Ltd, Stoodleigh, Devon
Printed and bound in Spain by Graficas Estella S.A.

Contents

Contents

Introduction

This book is a new edition of *GNVQ Advanced Information Technology* and is written for the Vocational A Level specifications. It is closely mapped to the specifications and is arranged, like the specifications, in six mandatory units.

The Vocational A Level (single award) and the Vocational A Level (double award) both contain the same six mandatory units covered by this book. The double award contains another six optional units; these units vary depending on the examination board issuing the award.

What is ICT?

ICT stands for Information and Communication Technology and it looks at the ways in which the combination of information technology and communication technology can serve people and organisations.

Where does the course lead?

The course can lead straight to employment in many organisations that use ICT as a tool. In these organisations you will use ICT to help you do your main job.

Many of you will want to take your knowledge and expand it by taking a course in Higher Education such as an HND, Foundation Degree, Degree or other similar qualification. You may then want to develop ICT systems for other people to use by taking employment in areas such as programming, systems analysis or networking, or you may just want to use the ICT you have learned as a tool in the course of your job.

The structure of the book

The book is arranged into the six mandatory units which are:

Presenting information
ICT serving organisations
Spreadsheet design
System installation and configuration
Systems analysis
Database design

Although the units can be studied in any order, they will be found in this book in the order that they are presented in the specifications. This is a logical order because the units often build on information in previous units.

Making the most of the features in the book

Here are some of the features of the book and the best way to use them.

What is covered in this unit

This section appears at the start of each unit and lists all the main sections covered in the unit. These headings also match the specifications main headings.

Materials you will need to complete the unit

Different units require different resources and in this section you will find a list of equipment, such as hardware and software, needed to complete the unit.

Quickfire

Quickfire are short questions that require a short answer. They are used to see if you have understood material in the unit. You do not need to write anything down. You can just think about the answer and perhaps discuss it with the rest of your class.

Key Skills

Many of the answers to the activities and assignments in the book can be used to supply evidence for the Key Skills IT, Communication and Application of Number. By building up your portfolio of evidence for your assessor for your GNVQ Advanced course, you will also be building up the evidence for the Key Skills that you are claiming.

Activities

After reading a section within a unit you will come across an Activity. Activities are tasks you complete to build up your knowledge of the subject. By completing them you will be consolidating your knowledge. They require written or computer-generated answers and in answering the activities you will be building up evidence for your Key Skills.

Glossary

The glossary is a list of the technical terms that you will come across when reading this book. If you come across a term that is not described in detail in the chapter, you should look it up in the glossary.

Key terms

Information and Communication Technology (ICT) is a subject full of abbreviations and specialist terms and it important that you understand what they mean and that you are able to use them when writing or talking about ICT. Each unit will introduce new terms and the purpose of the Key terms feature is to enable you to see the terms that are introduced in the chapter and see if you can define them yourself. You can then check that you have the right definition by referring back to the relevant section in the unit, or you could also see if the term is in the glossary at the back of the book.

Review questions

You will find these at the ends of units and are provided for two reasons:

- to test your understanding of the material contained in the unit
- to enable you to gain practice at answering questions for the units which are externally assessed. You can therefore use these questions as a revision tool.

Course assessment

By working through the Activities and Assignments in the textbook you will be producing a portfolio of evidence. It is your responsibility to keep the evidence for your portfolio safe and ready for when the internal or external assessment takes place.

You will receive a mark for each of the units you complete. In each assignment you will be told what evidence you need to produce for each grade and you must read this carefully and make sure that you produce exactly what you are asked for.

The evidence for your portfolio may be in the form of the following:

- documents that you have created (reports, memos, letters, essays, instructions, etc.)
- backup copies of computer files that you have created
- written notes from an assessor to say that they have seen you perform a task correctly (e.g. assembling a computer, giving a presentation on a certain subject)

Acknowledgements

Thanks are due to Dr Martin Wynn of HP Bulmer Ltd for permission to use copyright material.

Screen shots reprinted by permission from Microsoft Corporation. Windows® is a registered trademark of Microsoft Corporation in the United States and the United Kingdom.

Every effort has been made to contact copyright holders and we apologise if any have been overlooked.

Presenting information

What is covered in this unit:

1.1 *Styles of writing and the use of language*
1.2 *Accuracy and readability*
1.3 *Styles of presentation*
1.4 *How organisations gather and present information*
1.5 *Standard ways of working*
1.6 *Managing your work*
1.7 *Keeping information secure*
1.8 *Working safely*

This unit is concerned with the presentation of information. In it you will learn to create original documents in styles that suit the users and improve the accuracy, reliability and presentational quality of the documents you create. You will also be given an understanding of the ways organisations present and gather information and why standard layouts are used for documents. You will learn to choose between standard layouts and understand the need for standard ways of working. Additionally, you will learn to develop good practice in your use of ICT.

At the end of the unit you will use your presentation knowledge and skills to create a portfolio of different documents and an extended report on an investigation you have carried out. You will also evaluate a collection of standard documents used by organisations.

You will be assessed in this unit through your portfolio of evidence and the grade you get for this will be your grade for this unit.

Materials you will need to complete this unit:

- computer hardware including a printer (a colour printer will be needed on an occasional basis)
- fully featured word-processing software (preferably Word 2000 or Office 2000, although earlier versions of Word or Office are not very different from the 2000 versions)
- access to the Internet to help with research tasks
- a range of documents (memos, reports, business letters, invoices, etc.) for inspection.

1.1 Styles of writing and the use of language

Before actually composing a document you need to think carefully about what you want to say. When preparing information for others, there are two main things to bear in mind:

- your reader
- the purpose of your document.

Your reader

The age of your reader is important. Use of gimmicky catch phrases or snazzy presentation of information can work well with the young, but could annoy an older reader.

Although physical age is important, another thing to consider is the reading age of your audience. Not everyone is proficient at reading: the vocabulary of some people is restricted and their knowledge of the English language limited. Reading age is an indication of the reading ability of a proficient schoolchild at that age. For example, if a 23-year-old adult has a reading age of 12 it means that they only have the reading ability of a child of 12 who is competent for their age.

Newspapers make sure that the reading age of the material in the paper is at a level determined by the majority of their readers. Generally, to keep the reading age low, you have to make sure that only well known words are used, that the sentence length is kept short and few punctuation marks are used.

Activity 1

For this activity you can work as a team, to investigate the reading age of a collection of different newspapers. Each newspaper will be aimed at a certain group of readers – for example, the tabloid papers (the small papers) will be aimed at a lower reading age than the broadsheets (the larger papers).

- Look at each paper and produce some statistical analysis of such things as the number of words in an average sentence or the average number of letters in a word.
- Present the information you have found in report format in an interesting and informative manner. Use diagrams/charts to illustrate your results/findings wherever possible.

Before you produce any document, think carefully about the recipient and what they would expect to receive. Using the right kind of language is important. For instance, unusual words (used correctly) might impress the reader of a job application but might annoy a person who wanted something explained simply (such as the directions to your house).

Activity 2

Word has a facility for displaying information about the reading level of a document, which it will display once it has finished checking spelling and grammar.

- Use the help screen to find out how to display information about the reading level of a passage of text. Type in a list of the problems you can have with printers and some simple things to check. Entitle it 'simple printer problems'. Using information from the Help screens, obtain the readability information for your document.

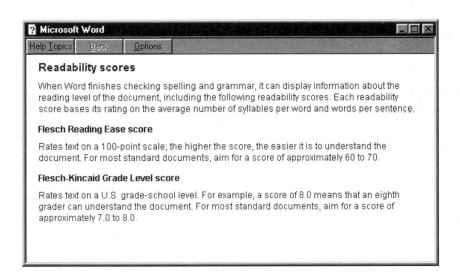

Figure 1.1 You can use the Help facility to obtain information about what the readability statistics mean and how you can check a document

Knowledge of the reader about the subject

If you were writing an article on a subject such as the Data Protection Act 1998 for computer users, you would need to write it differently to an article on the same subject aimed at lawyers, who are able to untangle the intricacies of the law.

Later in the course you will be required to develop a database system for others to use. As part of this you will have to produce documentation aimed at the users of the system and technical documentation aimed at people who are likely to develop the system further in the future. The latter group of people will have specialist knowledge of the subject and will not require every technical term to be described in detail.

The purpose of the document

The purpose of any document produced will determine the style and tone of the document.

- A document used for a formal purpose will demand a formal style.
- A poster for a school/college disco would need to be eye catching and informal.
- A poster for a drink–driving campaign would still need to be eye catching to have impact but would be more formal.

Activity 3

For each of the following documents say whether a formal or an informal approach is needed.

- an invitation to a wedding
- a reminder about an overdue account
- a letter accepting an offer of a job
- a poster outlining the rules of the computer room
- an advert for your computer, to go in the classified section of your local paper
- a letter of complaint to a holiday company regarding the low standard of accommodation
- a letter informing you of a football pools win
- a poster advertising a barbecue your school/college is having, to raise money for charity
- a letter thanking a relative for a birthday present
- a CV for a job
- a letter of complaint to the council about the condition of the pavements along your street
- a request for payment for a job you have completed
- a flyer to be handed out to passers-by about hare coursing.

Not all documents used in an organisation are letters. Documents include business cards, compliments slips, fax cover sheets, invoices, orders and quotations.

Activity 4

What types of documents might the following organisations produce?

- a double-glazing company who fit replacement windows, conservatories, etc.
- a main dealer for cars who sell new and second-hand cars and service and repair them
- a video library
- a doctors' or dentists' surgery.

Writing business letters

Organisations frequently communicate information in the form of business letters. Business letters tend to be formal documents and great care is taken in their construction. In this section we will look at some useful phrases that can be used in business letters. You will be given some practice at putting some of these phrases together and producing some business letters for certain situations.

Reference

In this section you can make reference to something that has happened before (a telephone conversation, advertisement, letter, memo, etc.). Here are some phrases that you could use.

- Following our recent telephone conversation …
- With reference to your letter dated dd/mm/yy …
- With regard to your query about/request for …

- Thank you for drawing attention to ...
- With reference to your recent advertisement in ...
- We were pleased to hear that ...
- I am sorry to hear that ...

Requests

In this section you can ask for something to be done or request further information.

- Could you please ...
- I would like you to ...
- If you would be so kind as to ...
- I would appreciate it if you would ...
- Please let me know if ...
- Please let me know how much ...
- Please let me know when ...

Reminding or drawing the reader's attention to something

- We wish to draw your attention to the fact that ...
- You may not be aware that ...
- I am afraid that I must draw your attention to ...
- We do not appear to have received ...

Offering help or assistance

- We would be happy to ...
- I would like to be able to ...
- Would you like me to ...
- Please do not hesitate to contact us/me if we/I can be of further assistance.

Apologising or expressing regret

- I deeply regret that ...
- We were sorry to hear that ...
- Please accept my apologies ...
- Unfortunately ...
- I am afraid that ...

Giving assurance

- I/we can assure you that ...
- I will make sure that ...
- You can be assured that we will ...

Expressing dissatisfaction/complaining

- We would like to inform you that ...
- I regret to inform you that ...
- We would like this matter to be dealt with to our complete satisfaction immediately.

Politely refusing something

- We fully appreciate your point of view but ...
- I am aware of the problems with ... but am unable to help on this occasion.
- I feel that in the circumstances it would be better if ...

Confirmation

- I wish to confirm the arrangements for …
- Would you please confirm that …
- If you do not inform us otherwise, we will assume that …

An ending for the letter

- I look forward to hearing from you.
- We hope to hear from you shortly.
- A prompt reply to the issues raised in this letter would be greatly appreciated.

Activity 5

Using the phrases outlined above, produce the following word-processed letters:

- to remind a customer that their account is overdue
- to complain about the late delivery of some goods
- to remind a customer of a meeting
- a covering letter (to go with a CV) responding to a job advertisement.

As well as using the phrases outlined above, you will need to supply text to put the letter into context. It is up to you to supply a context for these letters. For example, the letter of complaint might be for some components that actually stopped your production line.

Writing instructions

It is very important for anyone involved in IT to be able to write clear instructions for others to use. You will have to produce user documentation to explain how to use a system that you have developed later on in the course.

Activity 6

a You are giving someone from your school or college directions to your home. Produce a list of instructions (you are not allowed to draw a diagram) for them to use.
b You are now allowed to draw a diagram. Write the instructions that go with your diagram.

Collecting information

The most common method for collecting information is to use a form. Filling in a form is the first stage in many information systems and the completed forms are used as source documents for the system. Such forms include driving licence application forms, passport application forms, application forms for jobs and credit card application forms (Figure 1.2).

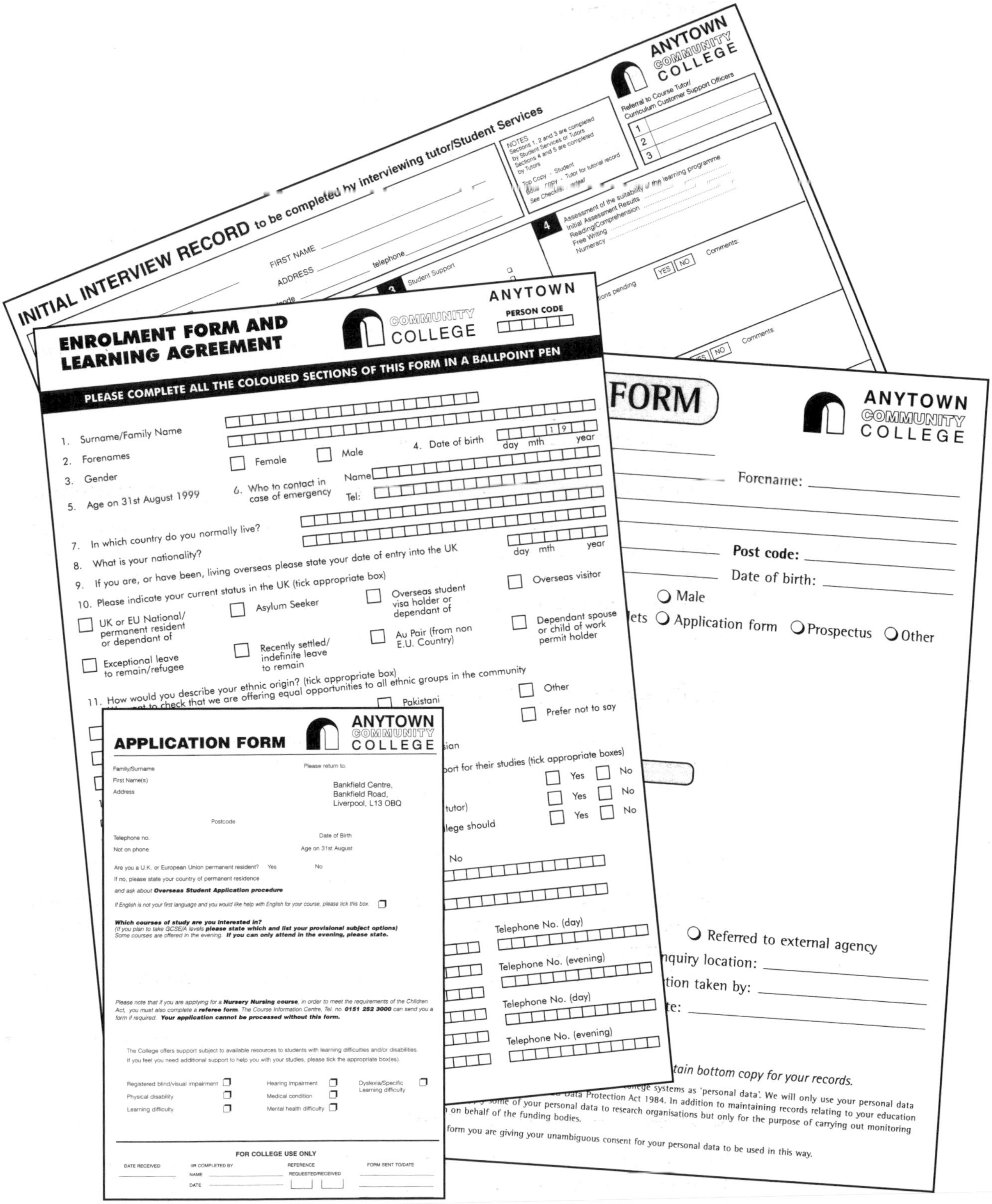

Figure 1.2 A collection of forms used in a college to collect information

CHANGE TO STUDENT RECORD

CSR NUMBER	
EBS	09165

This form is to be used to notify changes to a students course/subject attendance details.
The bottom copy should be retained by the originator when submitting your form to your Centre Office.
PART 1 MUST BE COMPLETED IN ALL CASES PLUS THE RELEVANT SECTION FROM PART 2.

PART 1

Student Details as currently recorded

SURNAME .... D.O.B. ....

FORENAME ....

PERSON CODE

Course Tutor/Originator

NAME .... LOCATION .... EXT NUMBER ....

SIGNATURE .... DATE ..../........./.............

PART 2   Please complete one section only

SECTION A. please enter personal details requiring amendment

SURNAME ....

FORENAME ....

GENDER - MALE/FEMALE .... DATE OF BIRTH ..../........./............

STUDENT ADDRESS (Please tick appropriate box)   ☐ Home   ☐ Term Time

....

PERSON CODE

SECTION B. Student enrolled but never attended

COURSE/SUBJECT REFERENCE   [ ][ ][ ][ ][ ][ ][ ]   CALOCC [ ][ ][ ][ ]

SECTION C. Student withdrawing from Course/Subject (Please see codes on reverse of form)

COURSE/SUBJECT REFERENCE   CALOCC   DATE OF LAST ATTENDANCE

[ ][ ][ ][ ][ ][ ][ ][ ] [ ][ ][ ][ ]   ..../........./.........   [ ][ ] REASON FOR WITHDRAWAL CODE

SECTION D. Cancellation of previous Course/Subject withdrawal (ie Section C CSR)

COURSE/SUBJECT REFERENCE   [ ][ ][ ][ ][ ][ ][ ]   CALOCC [ ][ ][ ][ ]

SECTION E. Student completing Course/Subject early

COURSE/SUBJECT REFERENCE   CALOCC   EARLY COMPLETION DATE

[ ][ ][ ][ ][ ][ ][ ] [ ][ ][ ][ ]   ..../........./.........

	SECTION F Student enrolled on wrong Course/Subject		SECTION G Change to students start date/hours/Fee waiver code Sponsor code		SECTION H Student transferring between Courses/Subjects	
EXISTING COURSE/SUBJECT REF.	[ ][ ][ ][ ][ ][ ][ ][ ]		[ ][ ][ ][ ][ ][ ][ ][ ]		[ ][ ][ ][ ][ ][ ][ ][ ]	
NEW COURSE/SUBJECT REF.	[ ][ ][ ][ ][ ][ ][ ][ ]		███████████		[ ][ ][ ][ ][ ][ ][ ][ ]	
START DATE/TRANSFER DATE						
	EXISTING	NEW	EXISTING	NEW	EXISTING	NEW
WEEKLY HOURS.						
FEE WAIVER CODE						
SPONSOR CODE						

FOR CENTRE OFFICE USE	CSR FORM RECEIVED	PRE-CHANGE REPORTS RUN	CHANGES MADE	POST CHANGE REPORTS RUN	CHANGES VERIFIED	REPORTS ISSUED Finance Originator
NAME						
DATE						

Figure 1.3 A change to student record form

Forms are also used to make amendments to existing systems, for example to update the student records in a college. Over the duration of a course many changes could occur – a female student could get married or divorced and change her surname; a student's address or telephone number could change; a student may leave before completing a course or may change to another course.

Most colleges have forms to enable changes to be made to a student record. Such a form is shown in Figure 1.3.

When creating forms it is important to give clear and precise instructions. The layout of the form should be considered carefully. For example, the user of a form will be annoyed if there is not enough room for their reply.

Once you have created a form you should give it to several people to evaluate (use the form and say if they found the instructions unambiguous and the form easy to complete).

Activity 7

Collect a number of forms that are used to collect information – such as driving licence application forms, UCAS forms, application forms for entry into college, passport application forms.

Evaluate the ease of using these forms. Give each form a mark out of 10 and write a short piece of text justifying the mark you have given.

Suiting the writing style to the needs

The writing style and structure needs to be appropriate to the purpose of the document. Here are some examples.

- *Attracting attention* – Documents such as **flyers**, posters, advertisements, etc. need to attract people's attention and make them read further.
- *Setting out facts clearly* – The readers of your document need to be clear about what you are trying to say so it is essential that facts are set out clearly and succinctly. People who read your letters may be very busy and will not want to have to wade though unnecessary text to get to the main points.
- *Writing to impress* – When applying for a job it is important that you write to impress. Many jobs require applicants to include a covering letter with the application form to explain why they feel they are suited to the job. This gives you the opportunity to sell yourself. Your command and use of English is especially important, as most jobs require such skills.
- *Creating a questionnaire* – Questionnaires are used to collect opinions and need to be constructed in a certain way if they are to be effective. Questionnaires will be looked at later on in this unit.
- *Ordering or invoicing goods* – In many cases when you are placing an order for goods you will use an order form, but you might have to write out an order in the form of a letter. Because you do not have an order form to use you will have to make sure that all the essential parts of your order are present – such as the product numbers/catalogue numbers and descriptions of the goods, the quantities ordered, the total amount, the VAT (if applicable) along with any delivery charges. You should also make sure that the company knows where to send the goods and the invoice, and they should be given a contact name and number.

- An invoice is a request for payment for goods bought or services rendered. It tells the purchaser what they need to pay for. Invoices should have a structure similar to that of an order. In many organisations, the original order is used to produce the invoice.
- *Summarising information* – When you are asked to supply information the people who have asked for it may not want the full details, just a summary. It is therefore important to be able to summarise the main points of a long document, without losing any important parts of the original document.
- *Preparing a draft* – It can be quite daunting to be given a blank piece of paper or a blank screen and be asked to produce a document. **Draft** documents are often produced, which are proofread and passed to other people for comment. The documents are then amended as required, depending on the comments. Important documents often go through several drafts before the final version is obtained.
- *Collecting information from individuals* – There are many instances where information needs to be collected from individuals. The documents that need to be created to do this include time sheets, holiday rotas, attendance monitoring forms and work sheets. These forms do not have a set structure and are produced by the organisation as and when required.
- *Explaining technical details* – If you work for a company involved in the production of equipment or components you could be asked to write technical details about the products the company supplies. Technical information is usually read only by someone who understands it, so you do not need to explain terms that the reader should already know about. As always, you should aim your material at the reader.
- *Writing a reminder* – It is frequently necessary to write reminders – to pay a bill, about an important date or meeting, etc. Writing reminders is quite difficult because if you get the tone of the letter wrong you risk annoying the reader.
- *Preparing a report* – We will look at the construction of reports later in this unit.

Activity 8

Write the following documents. For each document produce at least one draft version in addition to the final version. In writing each of these letters you will need to consider the writing style and the probable reading age of your reader.

a   A letter to a member of a gym/sports club to inform them their membership fee is overdue.

b   A personal statement to explain why you would like to do a certain course at university.

c   A questionnaire to find out about the hardware and software that your fellow students are using at home.

Tools to help with style

There are software tools to help with style. For example, if you are an author writing a story for 10-year-old children you need to be sure that children of this age will be able to read and understand what you have written. This is where the reading age comes in. You could produce a story using word-processing software and then get the package to assess the **reading age**. If it is too high for the intended readership you could reduce the number of long words and shorten the sentences.

Software tools are available with most word-processing packages that allow you to choose a particular writing style – such as business, letter, memo, advertising or report. You are guided through writing each of these by the software. If you select a certain style the thesaurus will suggest only words that are appropriate to the style adopted.

Using Word to create documents

As soon as you decide to create a new document in Word, a screen appears from which you can choose the type of document (Figure 1.4).

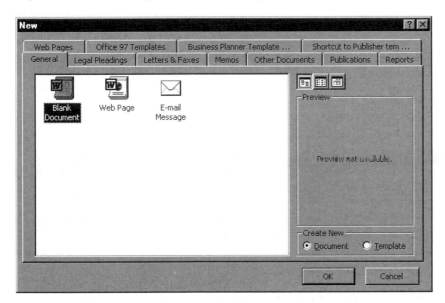

Figure 1.4 If you create a new document you can choose a wizard or template to help you, using this screen

Click on Letters & Faxes. The screen in Figure 1.5 appears.

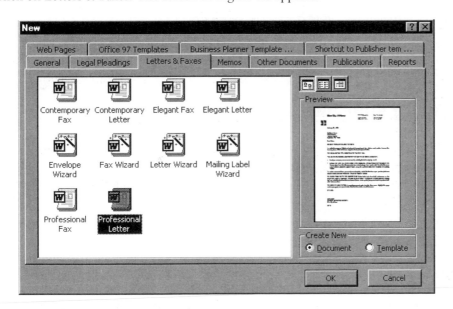

Figure 1.5 The Letters & Faxes templates and wizards in Word

If you single-click on Professional letter, a small version of the style of the letter will appear in the preview section. Click OK to create a professional letter.

The document now appears (Figure 1.6). All you now have to do is insert your own details into the spaces.

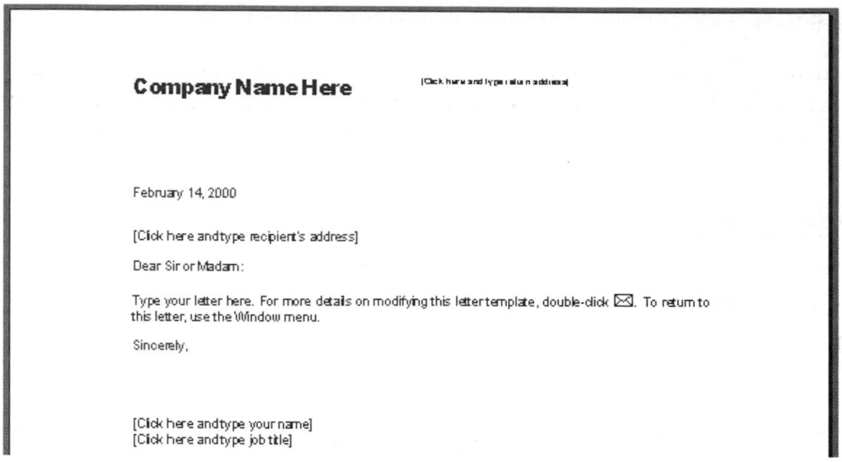

Company Name Here

February 14, 2000

[Click here and type recipient's address]

Dear Sir or Madam:

Type your letter here. For more details on modifying this letter template, double-click ✉. To return to this letter, use the Window menu.

Sincerely,

[Click here and type your name]
[Click here and type job title]

Figure 1.6 The Professional Letter template. Notice that today's date is automatically inserted. All you have to do is insert the variable information for your letter into this template

Activity 9

Assess the style or styles from the bulleted list on pages 9 and 10 that are most suited to the following documents:

- a form for an opinion poll
- a league table of results for a sporting activity
- a draft outline of the results of a survey
- a report to colleagues of a meeting that you attended on their behalf
- a questionnaire
- a CV
- a formal letter responding to a job advertisement
- the minutes of a meeting in a sports and social club
- a newspaper advertisement to sell something second hand
- a formal invitation to a party or other special event
- a glossy single-page advertisement for some new cosmetics
- an e-mail to a company asking for information on a product.

Writing your own advert

Here are some tips when producing an advertisement.

Headline

Get the reader's attention with a headline or an eye-catching phrase. Keep this short and simple and avoid the use of controversial phrases or slang.

Length of the advert

Use a suitable length to get the message across. Although long messages are good for technical or business products, a short message leaves space for graphics to brighten up the advert.

Comparison

Use comparative advertising phrases such as 'You've tried the others. Now try us!' if the product or service you are trying to sell has obvious advantages over its competitors.

Benefits

The main body of the advertisement should list the benefits or reasons why the customers should buy your product or service. It is best to emphasise the customer by using the word 'you' rather than 'we'. Also use bulleted text to highlight the key points.

Closing section

The closing part of the advert should mention the contact name, telephone, address and any other important details for finding out more information or for placing an order.

Activity 10

a Write adverts for any two of the following:
  * a disco/barbecue at your school/college in aid of a charity of your choice
  * a service of your choice (e.g. plumbing, electrician, mobile hairdresser, beautician) to appear in a local free newspaper
  * a computer system you want to sell
  * a holiday.

b The advertisement in Figure 1.7 appeared in a local free magazine. Although the text has been placed in a text box no other techniques have been used to make the advertisement stand out. Improve the appearance of this advertisement. You are not allowed to alter the written details in any way but feel free to use your artistic and computing talents.

Man with a Van
Don't Skip It. I'll Tip IT

Household Rubbish Removed
Single Items Delivered from £10
Small removals
3 Piece Suites moved or removed

House Clearances
Tel. Bill 0331–132–9800
Mobile 0998 009 0912
Locally Based

Figure 1.7

1.2 Accuracy and readability

Information presented in any manner *must* be accurate. The most common mistakes are incorrect spellings and missing or incorrect punctuation.

Spellcheckers

Spell checking is a facility offered by word-processing, DTP and graphics packages. A **spellchecker** checks that the words in the document are correctly spelt (if they are in the dictionary) but not whether the words are correct for the context. Therefore it would not detect 'care' typed instead of 'core' or 'which' instead of 'witch' because neither is spelt incorrectly. When it recognises a spelling mistake the spellchecker will suggest possible corrections.

One common mistake is to type the same word twice – a spellchecker will spot where this occurs.

Spellcheckers can check only those words that are included in the dictionary and will query any words not in the dictionary. Such words will include specialist terms (e.g. legal or medical terms or computer jargon) along with proper names such as 'Jones' or 'GNVQ'. To accommodate such words it is necessary to create a custom dictionary that collects these specialist words over a period of time. The spellchecker will then recognise them when they next appear.

Spellcheckers do have their limitations. For example, they cannot spot if you have used a word incorrectly – they will not be able to correct 'their' for 'there' or 'to' for 'too'. However, they can spot if you have omitted a capital letter at the start of a sentence.

Activity 11

Try to locate the mistakes in the text below that a spellchecker would detect and those it wouldn't.

The Internet can be described as a network of networks. All most any type of information can be found. It is possible to access anything from a database on space science to train timetables. The Internet is accessed though Internet Service Providers. Through use of the Internet, it is possible for data to be rapidly transfered around the world alowing people all over the world to read it.

Grammar checkers

Grammar checkers (Figure 1.8) are used to check the grammar of a document against set rules, although they do not check the meaning. Mistakes such as the subject and verb disagreeing, a sentence that is too long or full stops not followed by capital letters will be detected. However, they do have their limitations due to the complexity of the English language.

Grammar checkers can be used to check that:

- sentences end with only one full stop
- there is a capital letter at the beginning of a sentence
- the subject and verb of a sentence agree

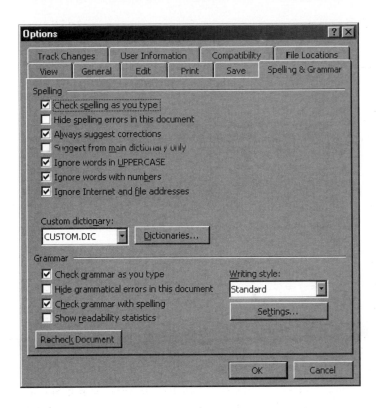

Figure 1.8 You can tailor the way the spellchecker and grammar checker work to suit your requirements

- common errors like writing 'you and I' rather than 'you and me' have been avoided
- for direct effect you write in the active voice rather than the passive voice
- the readability statistics meet the needs of your readers.

Activity 12

Locate the mistakes that would be detected by a grammar checker in the following text.

The World Wide Web (WWW) is an part of the Internet where graphics, sound, video and animation are used, as well as text. the word used for this mix of media is 'hypertext'. Special hypertext links are built into the World Wide Web that allow the user to move around by clicking with a mouse on words or graphics on the screen.. special browser software is needed to take full advantage of the World Wide Web, such as Netscape, Internet Explorer and Mosaic.

Even using spell and grammar checkers your text may still contain mistakes, so there is no substitute for slowly **proofreading** a document. Remember, the last thing you want is for other people to spot any mistakes. When a document is proofread it is marked by hand to indicate the changes that are needed. Standard **proofreading marks** are used to indicate corrections, such as:

- deletion
- start a new paragraph
- insertion of letters, words, spaces, numbers, other characters, etc.

Instruction	Mark in text	Mark in margin
Correction is concluded	None	/
Leave unchanged	----- under characters to remain	(✓ encircled)
Insert in text the matter indicated in the margin	ʌ	New matter followed by ʌ
Insert additional matter identified by a letter in a diamond	ʌ	Followed by for example ◇A
Delete	/ through character(s) or ⌐——⌐ through words to be deleted	⟨⟩
Delete and close up	through character or through characters	⟨⟩ (close up)
Substitute character or substitute part of one or more words	/ through character or ⌐——⌐ through word(s)	New character or new words(s)
Set in or change to italic	___ under character(s) to be set or changed	⊔
Set in or change to capital letters	≡ under character(s) to be set or changed	≡
Set in or change to bold type	~~~ under character(s) to be set or changed	~~~

Instruction	Mark in text	Mark in margin
Change capital to lower case letters	Encircle character(s) to be changed	╪
Substitute or insert full stop or decimal point	/ through character or ʌ where required	⊙
Substitute or insert colon	/ through character or ʌ where required	⊙
Substitute or insert semi-colon	/ through character or ʌ where required	;
Substitute or insert comma	/ through character or ʌ where required	,
Start new paragraph	⌐	⌐
Run on (no new paragraph)	⌒	⌒
Transpose characters or words	⌐	⌐
Close up. Delete space between characters or words	linking characters ⌒	( )
Insert space between words	Y	Y
Reduce space between words	T	T

Figure 1.9 The common proofreading symbols

- transpose (i.e. swap around)
- change case (i.e. from small letters (lower case) to capital letters (upper case) and vice versa).

The main proofreading symbols are shown in Figure 1.9.

Editing and proofreading documents

The terms editing and proofreading are often taken to mean the same thing but there are many differences between them. In computing:

- editing involves looking at the overall document to check its content and organisation
- proofreading looks more closely at the actual sentences and word combinations to make sure that they make sense and are grammatically correct.

What is the best way to edit a document?

After you have produced a document, it is a good idea to wait a couple of days before editing it because, with a fresh mind, you are likely to spot more errors. It is common to miss words if you just read in your head. It is better to read aloud (make sure there is no one around!) because you can't skip words.

You need to remember that you are writing for others and they will need to understand what you have written.

What is the best way to proofread a document?

The first thing is to read slowly. If you read too fast you may read a word that should be there but isn't. Every word must be read before moving to the next. Again by reading aloud you are more likely to read every word. Some people like to read with a cover such as a piece of paper covering the sentences below. Doing this allows the reader to concentrate on a single line at a time.

Improving the readability of your work

Try to keep sentence lengths short and make sure that only easy to understand words are used if you need the document to be understood by all readers. The idea is to get the message of the document across in as easy and friendly way as possible.

An average sentence length should ideally be around 15–20 words. Use the readability information feature in your word-processing software to monitor the reading age of your documents.

Use the active rather than the passive voice

To understand what the active and passive voice means you need to understand what a verb is. A verb is a *doing* word, such as typing, talking, running, writing, making. A verb may be classed as an active verb (e.g. 'I did this') or a passive verb (e.g. 'this was done'). Using lots of passive verbs in a document makes it dull to read but by using active verbs you can make a sentence crisp and more interesting. The passive voice takes more words to say the same thing, and it is not always clear who does what to whom.

- *Paul surfed the Internet all night.* Paul is the doer and the Internet is the thing. The verb is 'surfed'. If you look at the order in which these occur you will see that the doer is first, then the verb and finally the thing.

Name the checks that you can make on a document before sending it out.

ICT facilities for checking the accuracy of your documents do not guarantee that there are no errors. Why?

- *As it was Friday and there was nothing interesting on the television, Paul decided to surf the Internet all night.* Again the order is doer (Paul), verb (surf) and thing (the Internet). When the verb appears with the doer and the thing in this order it is called an active verb. If we swap the sentence around and make the thing come first the sentence becomes more awkward.
- *It was Friday and the Internet was surfed by Paul all night.* Because this sentence has a passive verb it is written in the passive rather than the active voice.
- *A complaint was filed by the union* is written in the passive voice, whereas 'The union filed a complaint' is written in the active voice.

Activity 13

Proofread the following passage, marking up any alterations using the standard proofreading symbols.

Notice that this text has been typed in double line spacing to allow room for any alterations.

Personal Data is a very valuable comodity and if one company collects personal data, then it can be sold to varoius other companies and organisations. some catalogue companies can even make as much money through the selling of the personel data they collect about their customers as they do selling their own goods to them! Although you may think that under the Data Protection act companies are not allowed to trade personal data in this way it is allowed provided that the data subject has given them permission. If you look at order order forms, application forms for innsurance or loans then there is usually a box to tick if you do not wish for your data to be passed to other organisations who may send you details of goods or services.

Most people do not think to much about this and do not tick this box and then they wonder why other organisations send them junk mail or seem to know quite alot about them

Activity 14

The following sentences are written in the passive voice. Rewrite each one in the active voice.

- A refund was given to her by the supermarket.
- A good score was achieved by the quiz team.
- A box of chocolate and a dozen red roses were presented to the girl by her boyfriend.
- A virus was introduced onto the computer by the student.
- I was given the scanner by my father.
- He was persuaded to leave by the human resources department.
- The car was repaired by the mechanic.

Making sure that your sentences have a subject and a verb that agree

It is important to make the subject and the verb agree with each other, not with a word that comes between them. Take the following sentence:

The Internet, along with other developments in IT, were responsible for the communications revolution.

The middle part of the sentence refers to other developments in IT but *the Internet* is the main subject of the sentence and so it should be 'was responsible' rather than 'were responsible'.

Activity 15

The following sentences are grammatically incorrect because they contain a subject and a verb that do not agree. Correct each of them.

- The laptop computer, as well as all the disks in the case, were stolen.
- The footballer, along with his team members, were at the bar.
- A box of blank disks were left next to the computer.
- A new system, to make sure that parts needed by production never run out, were introduced.

Do not use a long sentence when a short one will do

If you can say the same thing using fewer words then, from the point of view of readability, you should. Here some examples of sentences saying the same thing.

The order will be dispatched in the least possible time. (10 words)

or

The order will be sent quickly. (6 words)

One of our pressing needs is to sell our surplus vehicles. (10 words)

or

We must sell our surplus vehicles. (6 words)

Avoid pompous phrases

There are a number of phrases that have crept into common usage, mainly among older letter writers. These phrases appear pompous and old fashioned and should be avoided. Here are some of them:

- It is clear that …
- It is interesting to note that …
- It is obvious that …
- There is little doubt that …
- Actually …

Use tact when dealing with sensitive issues

It is extremely important to use tact when dealing with sensitive issues such as letters of complaint or letters asking for a late bill to be paid.

Both of the sentences below appeared in a letter sent by a company that manufactures self-assembly furniture in reply to a customer who is having difficulty assembling the furniture.

- *Lack of tact* – 'Obviously, if you had read the instructions carefully before trying to assemble your furniture, you would be able to answer these questions yourself.'
- *More tact* – 'We are sorry to hear that you are having difficulty assembling the furniture. We hope that these further questions will help you in your quest.'

What is the difference between 'accuracy' and 'readability'?

1.3 Styles of presentation

Information should be presented clearly. It is very annoying if the headings are inconsistent and the font and font size (measured in points) are changed too often.

Always think about what you want to achieve with your document and what will appeal to the readers.

When thinking about the **page layout** you will need to consider:

- position of common items
- page layout
- textual styles
- paragraph formats
- special features.

Position of common items

Common items are those things that appear on many (and sometimes all) of the pages in a multi-page document. These items need to be in the same position on each page to ensure consistency. The date, page number, your name, the name of the document can all appear on each page. Setting up such information in the headers and footers on the first page of the document will ensure that all the other pages will have the same information in the same positions.

Page layout

What items on the pages of a document would you make consistent with all the other pages?

You can create an effective page layout by using suitable:

- margins
- headers and footers
- page orientation
- pagination
- paper size
- gutters.

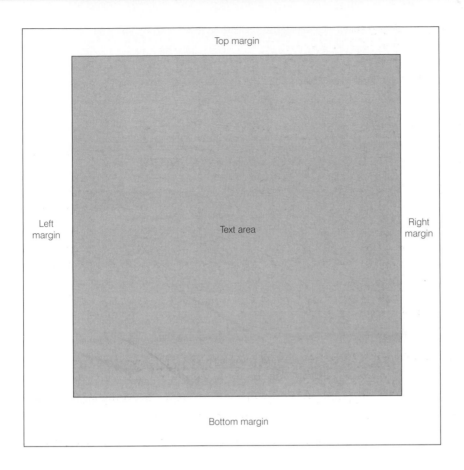

Figure 1.10 *The margins may be altered or kept at the default settings*

Margins

Margins mark the boundary of the text itself and therefore determine the amount of blank space left at the sides, top and bottom of the page (Figure 1.10). It is important to leave plenty of space at the sides in case the printed pages are to be bound in some way – the binding may occupy considerable space and could encroach upon the text. Also, if the text is set too near the sides of the paper, parts could be cut off when pages are photocopied.

Headers and footers

Headers and footers are text that is repeated at the top of every page (**headers**) and the bottom of every page (**footers**). Headers and footers are often used for easy reference to the title of the document, section in which the page falls, author's name, date or version of document. The headers at the top of this and the facing page show common usage in books.

Page orientation

There are two ways in which a page can be printed onto paper. With portrait orientation the height is greater than the width; with landscape orientation the page is turned sideways so that its width is greater than its height. Portrait orientation is more common and is used for most business documents such as letters, memos and reports; this book is portrait format. However, landscape format can be useful for charts, spreadsheets and notices. Figure 1.11 shows the page set-up selection menu from Microsoft Word.

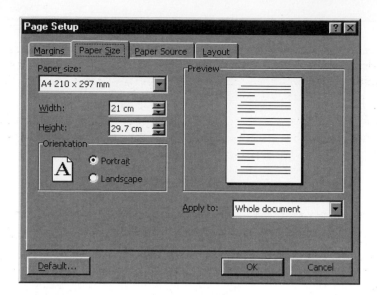

Figure 1.11 Word's page set-up selection menu

Pagination

The **pagination** determines when a page finishes and a new page starts. You could of course leave the computer to sort all this out but there are problems in doing so. For example, you may find that a heading appears at the bottom of a page, where the reader is least likely to look. It would be even worse if the heading appeared at the bottom of the page on its own with the text to which it refers over the page. It is therefore important to look out for this when proofreading and correct it. You can make the computer start a new page by inserting what is called a page break.

Paper size

Most word processors have a default paper size, usually A4 (297 × 210 mm), and this is the size assumed unless you change it in the 'page set-up' menu of the program. Other paper sizes may be used, such as 'legal' (which has the same width as A4 but is slightly longer) and most word processors will allow you to adjust the page size to any value (within limits). Before you decide on the paper size it is important to consider the use to which the final document will be put.

Unless you are using the default size (the size for which the printer and software are set up), you need to alter either the paper size setting using the word-processing software or the setting on the printer. Sometimes both are necessary.

Gutters

In either word processing or desktop publishing a **gutter** is the space between two columns.

Textual styles

You can create suitable textual styles by careful selection and use of:

- fonts
- heading and title styles
- bold, italic and underline
- text orientation
- superscript and subscript
- text animation (on screen).

Fonts

Fonts are collections of letters, numbers, symbols and punctuation marks of common design. Fonts and their sizes can be selected using the applications software but they will only be printed out if the printer is enabled to print them. Otherwise a default font will be used.

Changing fonts can make the design more interesting, but care should be taken not to use too many fonts. It is probably best to use no more than two, making the documents more interesting and readable by using different sizes, bold, underlining or italics.

Activity 16

Experiment with the font and font size facilities offered by your word processor. See if your printer supports all the fonts, and produce a small booklet that you can refer to in the future, showing the different fonts and font sizes.

Heading and title styles

Headings and titles separate related sections of text. You will need to decide where to place titles and headings, what size and style to use and whether subheadings are necessary. Generally the size of a heading will be related to its importance, and, as with the text, it is best not to use too many fonts.

Bold, italic and underline

Bold (text in heavy type), *italic* (text slanting to the right) and underlining can be used to highlight certain items in order to draw attention to them.

Text orientation

Special text effects can be obtained by setting text at an angle. This is particularly useful for the column headings in spreadsheets where the heading is too wide for the column width and making the column wider would restrict the number of columns that can be seen on the screen at one time. By inclining the text, we can get around this problem (Figure 1.12).

	A	B	C	D	E	F	G
1			HOME				
2		Played	Win	Drawn	Lost	Goals For	Goals Against
3	Man United	38	14	4	1	45	18
4	Arsenal	38	14	5	0	34	5
5	Chelsea	38	12	6	1	29	13
6	Leeds	38	12	5	2	32	9

Figure 1.12

Superscript and subscript

Suppose you want to type x^2 into a word processor. The power of two needs to be smaller and higher up than normal text. This textual style is called superscript.

In a similar way you may want a smaller letter below the line, called a subscript. For example, in the chemical formula for water (H_2O) the number two is a subscript.

In Microsoft Word you can simply highlight the text that is to have either the superscript or subscript and then click on the relevant button (Figure 1.13).

Figure 1.13 Superscript Subscript

Text animation

Some word-processing packages such as Word or presentation managers such as Powerpoint have special moving text effects called text animation. Obviously, such effects are only suitable for eye-catching screen displays.

Paragraph formats

You can create a variety of presentation styles using different paragraph formats, including:

- tabs and indents
- justification
- bullet points
- paragraph numbering
- line spacing
- widows and orphans
- tables
- hyphenation.

Tabs and indents

Pressing the **tab** key on your keyboard (it is the one labelled Tab, or with two arrows pointing in different directions on it) moves the cursor to a new position to allow the entry of text. The place where the cursor moves to when the Tab key is pressed will depend on the tab stops. There are default settings for the tab stops (usually every half inch) but you can also alter them using the screen shown in Figure 1.14.

Figure 1.14 The screen used to alter the tab stop settings in Word

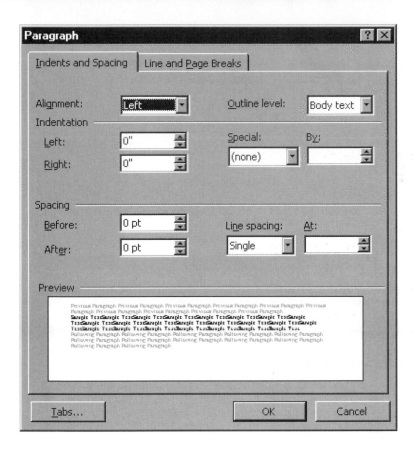

Figure 1.15 The screen used to alter the paragraph settings in Word

In Word, you can alter the presentational style of a paragraph using the paragraph screen (Figure 1.15). Notice that you can set the indentation of a paragraph and alter the spacing of the text in the paragraph.

Justification

Most word processors are pre-set to align the text flush with the left margin while leaving the right margin ragged (left justified). There may also be the option of centring the text (centre justified) or having only the right margin aligned (right justified – this is useful for addresses). If both the left and right margins are lined up, the text is said to be fully justified.

Figure 1.16 By selecting the text and clicking one of these buttons you can select the justification

Bullet points

Bullet points are used to draw the reader's attention to a list of points you would like to make. Rather than just write an ordinary list, you can put a bullet point (a dot, arrow, diamond, square or even a small picture) at the start of the point. The text for your points is then indented a small amount from the bullet. Figure 1.17 shows the general format for a bulleted list.

Figure 1.17 The general format for a bulleted list

To produce a bulleted list, select the text and then either click on the Bullets button or call up the format menu and select bullets and numbering (Figure 1.18).

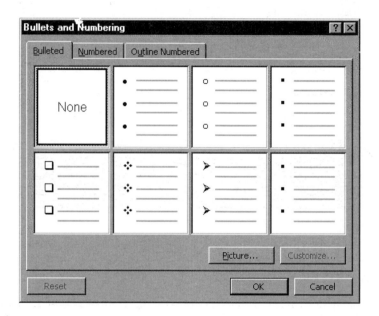

Figure 1.18 Using the Bullets and Numbering screen you can select the appearance of your bullets

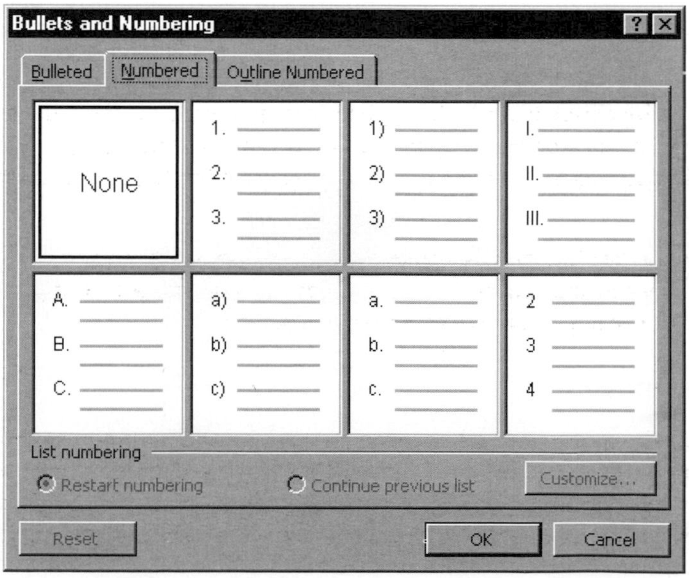

Figure 1.19 Use the Numbered screen to produce a numbered list

Paragraph numbering

It is often useful to number a series of points, which may done automatically using numbering (Figure 1.19). You can use numbering to number whole paragraphs or just a couple of important points.

If you want to add a hierarchical structure to your document, you can use the screen shown in Figure 1.20.

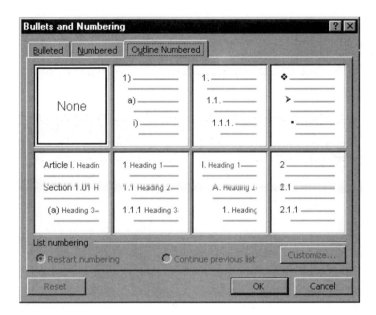

Figure 1.20 You can structure your documents in Word by layering them using selections from the Outline Numbered screen

Line spacing

The line spacing determines the amount of white space between the lines of text. Normally single line spacing is used but sometimes it is useful to print draft copies of documents using one and a half or even double line spacing so that readers can insert their comments between the lines.

Widows and orphans

An **orphan** is the first line of a paragraph that becomes separated from the rest of the paragraph by a page break – the orphan line appears as the last line at the end of one page, and the rest of the paragraph is at the top of the next page. This looks untidy, and many modern word processors have a facility to prevent this from happening.

A **widow** is the last line of a paragraph that appears by itself at the top of a page, which can ruin the appearance (and readability) of a document. Again, modern word processors can prevent widows from occurring.

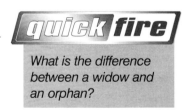

What is the difference between a widow and an orphan?

Tables

Tables are made up of rows and columns that can be filled with text. They are primarily used to organise and present information but are also used to align numbers in columns so that you can perform calculations on them. There are many different ways of creating a table. One way in Word is to use the Insert Table menu (Figure 1.21).

27

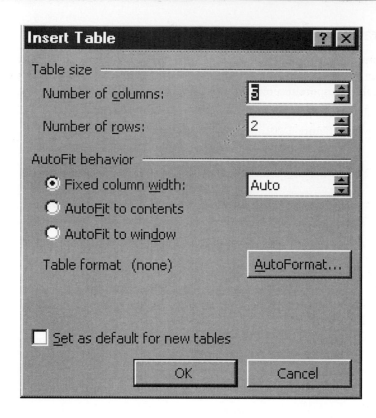

Figure 1.21 The Insert Table screen is used to place a table in a document. You can select the number of rows and columns in the table

To make creating tables easy, a number of formats have been created for you to select the one that suits your needs (Figure 1.22).

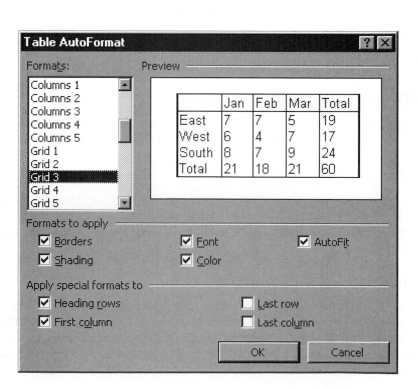

Figure 1.22 In Word, you can see all the pre-set table formats available using the Table AutoFormat screen

Hyphenation

If you **justify** text it means that either one or both of the margins are kept straight. In order to do this the word processor has to adjust the amount of white space between the words and sometimes large gaps are caused by this adjustment. To get around this problem you can choose to **hyphenate** words so that they are broken – part of the word appears on one line and the other part on the next. Hyphenation is particularly useful if you need to maintain even line lengths in narrow columns.

What is meant by hyphenation?

Special features

You can make use of special features to develop special presentation styles, including:

- borders
- shading
- a bibliography
- sounds
- background/text colour
- text/picture boxes
- a contents page
- an index
- an appendix.

Borders

A border can be used to add more emphasis to a word, paragraph, section of text or table. A border can be just a line, or you can have a picture border (such as a row of Christmas trees) around the edge of a page. The selection of clip art on your computer usually includes a good selection of picture borders.

Shading

Shading can be used for the background to a table or a section of text. It can be in colour, with a pattern, or both – but you need to be careful that you can still read the text under the shading. This is particularly important if you choose to print out the document in black and white.

A bibliography

A bibliography is a list of books, articles etc., and their authors, that you have used in your research. The bibliography is normally included at the end of a document.

Sounds

Many sales presentations have moved from a static slide display to a full multimedia presentation making use of the full facilities of the computer – such as text, graphics, animation and sound. Multimedia can also be used to present training material making use of sound, which can be voices or music.

Background/text colour

The background or the text (or both) can be coloured to make parts of a document stand out. However, if you do not have a colour printer, then don't bother with colour because the contrast between the text and the background can make the text hard to see. It is important to experiment with colours – some combinations work well while others do not.

Text/picture boxes

If you wanted to put some text that needs to stand out in the middle of a page you should use a text box. You can then place a border around it, give it a coloured background or shade the background.

A simple text box

If you want to create an effect like you see with newspapers, where the text starts on one page and finishes on another, you should put the text into two different text boxes and link them.

Once you have created a text box you can alter the text size and font, make the text italic, bold, etc. and add borders, etc. using the screen shown in Figure 1.23.

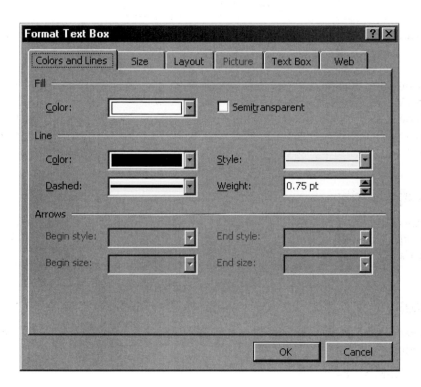

Figure 1.23 You can format the text box using this screen in Word

A contents page

A contents page lists the main headings in a document so that the reader can quickly get an overview of the document and navigate the document easily. You can produce a contents page by telling your word-processing software that every heading needs to be placed in the contents page (to learn how to produce a contents page search the Help facility in your software). Always remember to include a contents page in any extended documents (i.e. running over several pages) that you produce.

An index

To understand what an index is and how it is used, take a look at the index at the back of this book. If you require information on, say, spellcheckers then you would

Activity 17

a Produce a text box and insert a piece of clip art into it (like this).

Try to size the image and reposition it on the screen.

b Produce a text box, type in the text as shown in the box below and alter the colours of the text and the background. Produce a table with the combinations of text and background along with a comment about the result for future reference.

> This is a text box that has text of one colour and a background of another.

look up 'spellchecker' in the index and turn to the page or pages listed against the word. You can create an index by marking the words or phrases that you want to include in the index. Once you have done so you build the index, listing all the words in alphabetical order along with their page numbers. If the word is duplicated on any page the word-processing software will include the page number only once.

An appendix

An appendix is something that is added to a document. It could be a supplement, containing material that is extra, but not essential, to the original document. The material contained in an appendix should be self-explanatory.

Using graphics to improve presentation style

You can use a variety of different types of graphics to improve presentation style, including:

- graphs or charts
- pictures or drawings
- clip art or scanned images
- lines or borders.

Some examples are shown in Figure 1.24.

Activity 18

Screenshots are useful when explaining how to use an application that you have developed when producing user documentation. Explain briefly how you would perform a screendump and incorporate it into a word-processed document.

(a)

(b)

(c)

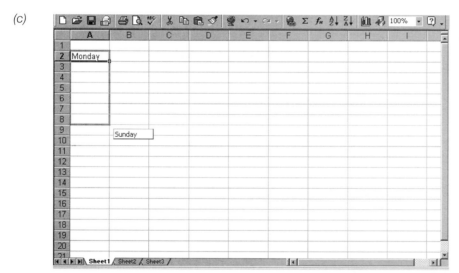

(d)

(e)

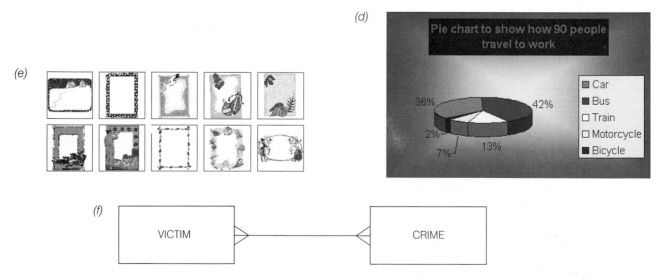

(f)

Figure 1.24 *You can use many types of graphics to improve the presentational style of a document: (a) clip art or scanned images; (b) images from a digital camera (c) screenshots (d) graphs or charts; (e) borders; (f) pictures or drawings*

Positioning important items in a document

You will need to understand how to position the following important items on a document:

- references
- dates
- addressee names
- signatures
- logos
- headings.

Activity 19

Look at the templates in Microsoft Word (use the Help facility if you don't know how to do this). Produce three possible outline designs for letters, showing the position of the important items in the above list.

The many different types of document used in organisations all have their own standard layout. The structure of a typical letter on non-headed paper is shown in Figure 1.25.

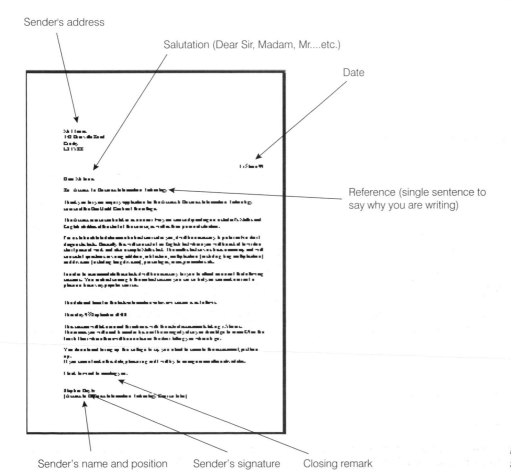

Sender's address

Salutation (Dear Sir, Madam, Mr....etc.)

Date

Reference (single sentence to say why you are writing)

Sender's name and position   Sender's signature   Closing remark

Figure 1.25 The basic structure of a letter

As part of your evidence for the Advanced GNVQ in ICT, you will need to find out about the purpose and structure of the following documents used in business:

- memos
- agendas
- minutes
- publicity flyers
- invoices
- questionnaires
- business letters
- newsletters
- itineraries
- draft documents
- reports
- CVs
- forms for collecting information
- fax cover sheets
- e-mail
- technical specifications
- purchase orders
- sales orders
- Web pages.

Memos

Basically, **memos** are used to tell the reader about new information, such as the release of a new product or price increase, or to persuade the reader to take an action – such as attend a meeting, reduce the amount of paper used or keep the area around the photocopier tidy.

Memo structure

The first part is a *heading* consisting of:

TO: (your reader's names and job titles)
FROM: (your name and job title)
DATE: (current date)
SUBJECT: (what the memo is about)
CC: (A list of the people who need copies of the memo)

The next part is the *opening segment*, consisting of three parts:

- The *context*, which is the background to the problem you are solving.
- The *task statement*, which explains what you are doing to help solve the problem.
- A *statement of the purpose* of the memo.

If the memo is more than a page long it should contain a *summary segment*, where you outline the recommendations you have reached so that your reader will understand the key points of the memo immediately.

The next segment is the *discussion segment*, where you include all the details that support your ideas. Start with the most important things and, to make for easy reading, put the important points or details into lists.

The final segment, called the *closing segment*, needs to contain a courteous ending stating what action you want your reader to take.

Attachments

Sometimes it is necessary to provide further detail (e.g. newspaper articles, lists, graphs and tables) by attaching another document to your memo.

Agendas

An agenda is a document that is prepared before a meeting, outlining what is to be covered and thus enabling the people who will be attending the meeting to prepare for it in advance. An agenda usually deals with the topics in the order in which they will be discussed. It usually contains the following items, though not necessarily in the order shown.

- *Apologies for absence:* members unable to attend should have given their apologies in advance along with reasons for their non-attendance.
- *Minutes of the last meeting:* this is to refresh members' memories of what was said and decided at the last meeting and usually consists of just a summary. Provided that there are no objections members will agree that the minutes are a correct reflection of the previous meeting.
- *Matters arising from the minutes of the last meeting:* all the items arising from the last meeting that need to be discussed at this meeting.
- *New business:* a list of all the items of new business that the members need to discuss.
- *Any other business (AOB):* usually matters arising at the last minute or after the agenda was set, and which need discussing.
- *Date of the next meeting:* this is important because agreement can be reached while everyone is present.

Minutes

Minutes are a detailed account of who said what during a meeting and what was eventually agreed. A secretary will often take the minutes, summarise the main points and present them at the next meeting.

Publicity flyers

Flyers are small leaflets that are usually distributed by handing them directly to people as they walk past or by posting them through letterboxes. Usually they advertise goods or services – anything from a car boot sale, a computer fair, a new restaurant or club to the services of a gardener.

Invoices

When a supplier sends out goods they also send an **invoice**, either with the goods or separately by post. The invoice is a demand for payment. Invoices should be checked against the original order on receipt, because they can contain mistakes (often to the purchaser's disadvantage!). Many larger companies now deal with invoices electronically, using a system called EDI (electronic data interchange).

Activity 20

Investigate what should be included on a typical invoice. Look at as many invoices as you can find. Are any clearer than others? What factors enhance their appearance? Produce a brief list of hints for someone who is about to design an invoice. You can choose any organisation you like.

Produce your final invoice and incorporate clip art into the design. Show your design to your tutor, who will make sure there are no important omissions.

Questionnaires

Questionnaires are useful for collecting opinions from a large number of people in a short time. They are used extensively during systems analysis to find out facts about the existing system and to elicit the users' opinions on parts of the new system that is being developed. Questionnaires should be used whenever feedback is needed; they are a useful way of ascertaining that what the developer thinks the user wants is *actually* what they want.

When designing questionnaires, several types of question should be avoided.

Ambiguous questions
Questions need to be self-explanatory, especially if the questionnaire is to be filled in by the user on their own.

Leading questions
Leading questions point to a certain answer. For example 'Do you agree that the new report layout is better than the previous one'? is a leading question – the respondent is being asked to agree with the person asking the question.

Burdensome questions
Burdensome questions involve the respondent in a lot of work and time.

Complex questions
Complex questions are too difficult for the user to answer (e.g. 'How many hours of television were watched in your house last year?').

Open and closed questions
In most cases it is best to avoid open-ended questions. These are questions that do not have answers for the respondent to choose from. A question such as 'What sort of music do you like?' is open ended; 'Do you like chocolate?' is not open ended because it can elicit only a certain set of answers. In systems analysis open-ended questions can be particularly useful for finding out about a system – for example 'What are the different ways in which a customer can make an order?'

Questionnaires are particularly useful for finding out the users' opinions about a part of a system that you have just designed – perhaps about a form designed for entering the data into a database. Figure 1.26 lists a few questions that might be found on such a questionnaire.

Business letters

Business letters are usually formal and, like most letters, can be divided into three parts:

- the *introduction*, which explains why you are writing
- the *main points* of the letter (i.e. the detail)
- the *end/close*.

Figure 1.25 (on page 33) shows the layout of a standard business letter. Although the layout will change slightly, depending on the preferences of the letter writer, the actual components remain the same.

Letters usually consist of:

- letterhead or typed heading
- date
- the name and address of the recipient
- the body of the text
- salutation

Usability questionnaire

1 How comfortable did you feel using the system?

1	2	3	4	5

Uncomfortable   Very comfortable

2 How easy was the new interface to use compared with the old interface?

1	2	3	4	5

Much less easy   The same   Much easier

3 How did you like the appearance of the new interface?

1	2	3	4	5

Did not like it   It was fine   Really liked it

4 How satisfied were you with the positioning of the items in the interface?

1	2	3	4	5

Dissatisfied   Very satisfied

5 How efficiently did you feel the new system allowed you to enter data?

1	2	3	4	5

Quite inefficiently   Very efficiently

Figure 1.26 Usability questionnaire

- complimentary closing (Yours sincerely, Yours faithfully, etc.)
- signature
- typed name and position.

Newsletters

Some firms, usually the larger ones, produce their own **newsletters**, which they use to foster an involvement in the company. New developments/investments, promotions, sports fixtures and results, results of exams taken by staff, births, marriages and deaths feature in most newsletters, such as the one shown in Figure 1.27.

Itineraries

An **itinerary** is a list of things to do and places to visit during a trip, such as a holiday, or it could be a list of things to look at during a visit to an organisation. The itinerary should include the day, date or time for the activity or visit and an outline of what or where the activity or place is. Figure 1.28 shows an itinerary for a transatlantic cruise.

Draft documents

A **draft** document is a rough version of a document that may be passed to others for comment. It is always wise to get other people's opinions on documents that you have produced, especially if you intend to put them into wide circulation.

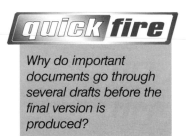

Why do important documents go through several drafts before the final version is produced?

37

newsline

BULMERS GROUP NEWSPAPER

Bulmer shapes up for the 21st Century

HP BULMER has again enjoyed a successful trading year, followed within days by the announcement of a major restructuring and completion of the first phase of its prestigious £20-million Moorfields development.

On the trading front, any comparison between the progress made in 1993-94 and that of the previous 12 months, is complicated by two major factors.

Firstly, the year just ended contained 52 trading weeks; the previous year had 53.

Secondly, this year's con-

Another sound performance in a difficult market

was 13 per cent above that achieved in the previous 53-week year.

Interest costs at £3.3 million were substantially lower than in the previous year, due to effective control on working capital, lower rates and the sale of Pectin.

Pre-tax profit before exceptional items increased by 10 per cent at £21.5 million and compares

Also, following a wide-ranging review of its business strategy, Bulmer announced in May that is would now be focussing its activities on its core business of cider, both at home and abroad, while at the same time continuing to develop its role as a distributor of premium beers.

As a result, the company is to withdraw from the soft drinks sector in which it currently acts as UK agent for Orangina and

Fenella Tyler, leading the way on training

Winning ways!

SOME companies might be content to sit back and bask in

time into personal development and, of course, I've also has

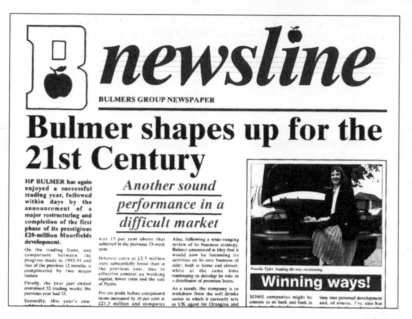

Figure 1.27 A newsletter from the cider producer HP Bulmer

Itinerary details: transatlantic Miami/Barcelona

Day	Port of call	Arrives	Departs
1	Miami, Florida		5:00 pm
2	At sea		
3	At sea		
4	Charlotte Amalie, St. Thomas	7:00 am	6:00 pm
5	At sea		
6	At sea		
7	At sea		
8	At sea		
9	At sea		
10	Tenerife, Canary Islands	8:00 am	5:00 pm
11	Funchal (Madeira), Portugal	8:00 am	5:00 pm
12	At sea		
13	Malaga, Spain	8:00 am	5:00 pm
14	At sea		
15	Barcelona	8:00 am	

Figure 1.28 Itinerary for a transatlantic cruise

They may spot obvious omissions, which will prevent embarrassing and sometimes costly mistakes. Draft documents are improved by taking into account the comments made by the people asked to evaluate the document and can undergo several revisions before the final document is produced.

Reports

Reports contain feedback from a 'finding out' exercise. For instance, a manager might ask you to look at the buildings and office equipment to make sure that your company is complying with the Health and Safety at Work Act. After completing your research you would provide your boss with a summary of the main points arising from your investigation, and it is this that would constitute a report.

There are many different types of report and the structure of one type, called a formal report, is outlined in Figure 1.29.

TO:	The name and address of the person who has asked for the report
FROM:	The name of the person who is writing the report
Terms of reference:	Outline what it is you have been asked to look at.
Procedure:	Explain how the information was collected.
Findings:	Outline the main findings of the report as a series of numbered points.
Conclusion:	Sum up the main findings of your report.
Recommendations:	Make recommendations based on your findings.

Figure 1.29 Structure of a formal report

CVs

CV is short for *curriculum vitae*, which means 'course of life'. Your CV is a factual account of your qualifications and work experience. A CV is used to give a potential employer information about you.

Although there is no real set format for a CV, some things are essential to include.

- *Personal details* – These would usually include your name, address, telephone numbers (mobile and land lines), e-mail address and driving licence.
- *Personal profile and career history* – You can start off this section with a couple of sentences saying why you feel you are suitable for the post. This is a good opportunity to sell yourself.
- *Your achievements* – Mention your achievements – things you have done on your course or in your past job that are relevant to the post you are applying for.
- *Work history* – In this section you should outline the jobs you have held, starting with the most recent. Make sure you do not have any gaps. As well as full-time work, include part-time posts and voluntary work. You need to include details about the type of job but keep it brief.
- *Training and qualifications* – Include qualifications gained whilst at school, college or university, starting at the most recent qualification. You only need to outline your education details from secondary school onwards.
- *Interests and spare-time activities* – Employers are always interested to see what you enjoy doing in your spare time so detail sporting activities, hobbies and membership of any clubs or societies here. Never put down anything that you do not do – you could have the world's leading expert as your interviewer!
- *Additional information* – You can put in this section any information that you want to clarify or which does not fall into any of the other sections. If there are any gaps in your employment or education, you can use this section to explain why.

- *References* – Give the names and addresses of people who have known you well for a long time, who are not relatives and who have consented to give you a reference. If you have worked, then one person should be your previous employer; if you are at college you could use the course tutor or another member of the lecturing staff.

Forms for collecting information

It is a very important part of systems analysis to be able to design forms for collecting information from people. These can be on paper or on screen. More and more data is being collected directly, meaning that the information is keyed straight into the computer without any paperwork being completed first. As more business is conducted using the Internet, being able to design forms to collect information about customers and the goods/services they order is a very important part of computing.

Fax cover sheets

A fax (facsimile) is a document copy that is sent via a telephone line either through a fax machine or from a computer file using a modem. It is essential that the receiving computer or fax machine is switched on, a feature unnecessary for e-mail because messages can be stored until the user decides to read them. It is normal to have an accompanying **fax cover sheet**, which will include information about the organisation/company sending the document.

e-mail

Electronic mail (**e-mail**) is a very useful way for organisations to communicate with each other. It is easy, quick and cheap. The only disadvantage is that many people do not have the facilities to send or receive e-mail. However, e-mail is set to become a very popular form of correspondence.

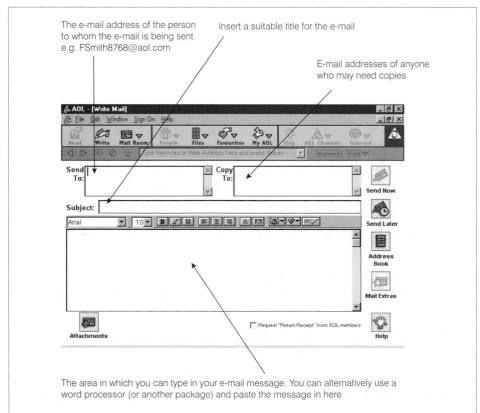

Figure 1.30 How to compose a typical e-mail message

Sending an e-mail

1 Obtain the e-mail address of the person you are sending the e-mail to. Notice the address book (Figure 1.30), where you can store all your e-mail addresses.
2 Put in the cc window the e-mail addresses of anyone else who may need to see the e-mail. They will be sent a copy automatically.
3 Give your e-mail a suitable title that indicates to the receiver what it is about.
4 Type in your message or import it from another package.
5 If you want, you can attach files to your e-mail. You could send a many-page document, a spreadsheet, a drawing prepared using CAD, some sound clips, a photographic image as a file. However, the receiver would need to have the applications software to use the data files you have transferred to them.
6 Log on to your Internet service provider (ISP).
7 Click on the 'Send' button' to send your e-mail.

File attachment

To attach a file, complete your e-mail message and briefly explain what types of file you are sending. Click on the **file attachment** button and select the file you want to send. A box will appear to allow you to select the drive, folder and eventually the file you want to send.

If you want to send more than one file simply repeat the process. If there is more than one file to send, the files will be compressed. By clicking on Send Now the e-mail will be sent. The time it takes to send and receive information will depend on the size of the attached files. The file attachment menu for the ISP America Online is shown in Figure 1.31.

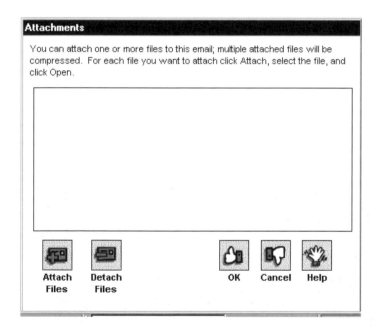

Figure 1.31 The file attachment menu for America Online

Tips when making up e-mail messages

Although there are no hard and fast rules for constructing an e-mail message some unofficial rules have evolved among e-mail users.

• *Reply promptly* – E-mail is only faster than regular post if people check their e-mail regularly. The whole point of e-mail is its speed and a prompt reply is important.
• *Think before you send a message* – Other forms of communication (letters, memos, etc.) require a lot of thought. Just because an e-mail reaches you

quickly it does not mean that you should not give its contents careful consideration. The message that you send represents you to the recipient and you should consider carefully what you want to say.

- *Formal or informal?* – E-mail messages normally fall in a category between a short note and a memo so you do not need the formal structure of a business letter. When sending an e-mail, you may only know the recipient's name and e-mail address so you can treat everyone with similar informal respect. Many e-mails are composed very quickly so people do not bother about the odd spelling mistake (although you can spellcheck them).
- *Make it clear who you are* – The recipient of your e-mail will have only your e-mail address from which to work out who has sent it. It therefore makes sense to say exactly who you are and the organisation you are from and maybe your job title.
- *Do not use abuse* – E-mails can be intercepted by your organisation or someone else. The laws of libel apply to e-mails in the same way as they apply to written correspondence so be careful about what you say about other people. Abusive e-mails (called flames) should not be sent, and if you receive one, don't bother to reply to it.
- *Don't shout* – Avoid typing an e-mail in capital letters because this is considered to be shouting. However, capital letters *can* be used sparingly to add emphasis to a word – e.g. 'What is YOUR problem?' emphasises the word 'your'.

Technical specifications

Technical specifications are documents that go into great detail about a particular product and are aimed at a person who is a specialist in the area described in the document. Any specialist terms used will not have to be explained because the document will interest only people who know about the subject.

Purchase orders

When companies place orders for goods or services they often do so using their own documents, called purchase orders. This allows them a standard way of ordering, which makes it easy for them to check the goods (when they arrive) against the original order.

Activity 21

A new publishing company is designing an order form to be included at the back of their catalogues. The first draft is shown in Figure 1.32.

a Help to improve the draft design of the order form. Collect at least five order forms (any order forms will do) and write down a list of common items on them.

b Using appropriate software, design a simple company logo that can be used in the form and save it in a suitable file format so that it can be imported into word-processed documents.

c Take the order form shown in Figure 1.32 and improve the design, incorporating your logo into your design. You should produce several draft versions and at least three final versions so that the directors of the company can choose the one that they prefer.

d Look at the table opposite and match up the most appropriate description with the type of document.

```
                    Nelson Thornes Ltd
                       Delta Place
                       27 Bath Rd
                  Cheltenham, GL53 7TH, UK


                        ORDER FORM

    To:    Clover Hill High School
           Clover Hill Lane
           Glanbury
           G08 5JD


    Date:  06/04/00
```

ISBN	Title	Qty	Price	Amount

You must quote the order number when referring to this order.

Figure 1.32 A first draft of an order form for a book publishing company

Document Type	Description
Fax cover sheet	An electronic message sent between computers
Newsletter	An organised list of events during the a holiday
Business letter	A collection of opinions/answers used for analysis
E-mail	An account of what has been discussed and agreed at a meeting
Memo	A page supplementing a document copied via a telephone system
Itinerary	A short document/message sent usually to solve problems/inform readers of new information
Agenda	Small leaflet, usually used for advertising
Questionnaire	A document outlining what is to be covered in a meeting
Minutes	Formal letter to or from an organisation
Publicity flyer	An account of new information of interest to a particular group of people

What is the main difference between a sales order and a purchase order?

Sales orders

If an organisation is involved in the sale of goods and services they will have a standard order form that constitutes an agreement to buy. It is important to understand what is required on a typical order form.

Web pages

A Web page is a document on the World Wide Web (the Internet). Each page is located by a unique reference called its uniform resource locator (URL). Web pages are becoming increasingly important as more and more people and organisations are setting them up to advertise products and services.

1.4 How organisations gather and present information

Organisations, no matter how big or small, from corner shops to multinationals, all need information in order to function. In organisations people work together as a group to make something or provide a service, and data needs to flow between them and be managed. In this section we will look at:

- the types of information that organisations need to use
- how organisations collect information
- the need to acknowledge sources of information
- the flow of information to and from outsiders such as suppliers and clients
- presenting information both within and outside the organisation
- typical uses of illustrations, technical drawings, pictures and artwork
- commonly accepted standards for the layout of formal documents
- essential information that appears on formal documents
- methods of presenting a corporate image
- how templates might be used to enforce corporate standards.

The types of information organisations need to use

The types of information that organisations need to use can be put into three groups: commercial, financial and legal information.

Commercial information

Commercial organisations hold information in order to trade and, hopefully, make a profit. Organisations can be divided into those that provide a service and those that sell a product.

Commercial organisations hold information about their suppliers, customers, costings, sales, orders, budgets, production, etc. Some of this information would be of benefit to a competitor. Take, for example, costings. Suppose company A was in the process of tendering to resurface a large stretch of motorway for the Highways Department. It would be of benefit to company B, who was also tendering, to know what price company A intended to charge, because company B could then bid slightly lower. Most commercial organisations hold information that would be of great interest to their competitors, so it is not surprising that they take precautions to keep this valuable data secure.

/ **Activity 22** /

Classify each of the following commercial organisations according to whether they provide a service or make or supply a product:

- computer dating agency
- bank
- oil company
- shop
- building society.

Financial information

Organisations need to keep financial details relating to sales, purchases and general ledgers, cashbooks, cashflow predictions, payroll and final accounts. Again, all of these details need to be kept secure and there should be stringent security procedures to prevent anyone from tampering with them. Financial records are also required to ensure that the correct income tax and VAT are paid.

Legal information

A variety of legal information is held by most organisations, and is often confidential. The sort of legal data held includes documents for the registration of the business (if it is a limited company), contracts of employment for employees, contracts made between suppliers and purchasers, copies of all the acts pertaining to the business.

How organisations collect information

All organisations need to manage information in some way, but before they can do this they need to *collect* the information. Some information is originated from within the organisation – projected sales figures, department budgets, details of late payers, price lists, brochures, memos are information that normally comes from within the organisation. Other information – such as details on income tax, bank accounts, council tax bills, utilities (gas, electric, water, etc.) – all originate from outside the organisation. The details of any person who deals with an organisation will be stored and processed in some way by that organisation.

When information enters the organisation it has to be managed. For example, it has to be directed to the person or department allocated to deal with it (all requests for payment will need to be sent to the accounts department, all orders to the sales department, applications for jobs to the human resources department, etc.). Some information will need to be passed to several departments and information flows must be managed so that the information is received by everyone who needs it at the right time. An example of this is when an order arrives from a customer: the accounts department needs a copy of the order so that an invoice can be created and sent to the customer; the warehouse staff will need to know the details of the order so that they can make it up and dispatch the goods to the customer. Some orders are specially made for a customer. For example, conservatories are often tailor-made to customer requirements so the production department will need to be kept informed of the order.

The need to acknowledge sources of information

The source of any information that is used within an organisation and then subsequently used to present information should be known because it may be under copyright. One organisation may have spent money to collect facts and figures, and therefore they might want compensation if these are copied illegally by another. Some organisations are happy that the information they have collected is used – provided the source of the information is acknowledged.

Presenting information internally and externally

Information presented externally will need to be carefully worded because the reader might not be familiar with the terminology used. This is especially true if a customer is buying a product that has certain technical specifications. For example, a quote for a new central heating system could contain a lot of technical information concerning the size of the boiler, the heat losses in each room, etc. All the customer really needs to know is if it will heat the house to a suitable temperature when it is very cold outside and whether it will give enough hot water.

Internal documents are intended for people within the organisation, most of whom will be familiar with the specialist terminology used and the procedures that need to be applied to the information.

When you are designing a form, make sure that the instructions for completing it are clear. With some complex forms, such as forms for applying to university, a tax return or forms for applying for a passport, an accompanying booklet or leaflet explains clearly how to fill it in.

Typical uses of illustrations, technical drawings, pictures and artwork

There is a saying that 'a picture is worth a thousand words', and it is hard to imagine trying to fix a car engine or explain how to put together self-assembly furniture without the use of pictures. Whether is it for putting ideas across using presentation graphics or producing technical diagrams from which engineers can construct components, the production of a diagram using a computer is an important use of IT.

Graphics can be *presentational* or *technical*. These two types serve very different purposes and tend to be produced with different types of software. **Presentational graphics** are used to enhance the printed page, making the content more eye-catching in some way. They are used to add interest to word-processed documents, and presentation slides are often employed in commerce for marketing and promotional leaflets. Presentational graphics software is specialist software used to help design, produce and manipulate all the components of presentational graphics.

Technical graphics include maps, designs and technical drawings. These last are often produced with computer-aided design (CAD) packages. The emphasis with technical graphics is on accuracy and clarity, rather than necessarily on an attractive appearance. More generally, designs and technical illustrations are produced with specialised drawing packages, such as Adobe Illustrator, which was used to produce many of the diagrams in this book.

Graphic design software

Graphic design software broadly generates two types of image: bit-mapped images and vector images. Presentation graphics programs such as Microsoft PowerPoint and painting and image manipulation programs such as Adobe Photoshop store their images in bit-mapped form. Drawing programs such as CorelDRAW! and Macromedia FreeHand save their files in vector format. If you do not have specialist graphic design software do not worry – it is possible to produce good diagrams using the drawing features in ordinary word-processing software.

Bit-map graphics

With black and white bit-mapped graphics, each pixel is stored as a bit of information. Images are not, of course, always made up of black and white elements: many images have shades of grey in them (grey-scale images), while others consist of many colours. For grey-scale and coloured images, each grey or coloured pixel is made up from several bits. The position of each pixel is mapped out and stored as a series of bits in a file: hence the name 'bit-map'. The main disadvantage of bit-mapped graphics is that the number of pixels is set when the file is created. Therefore if the image is enlarged the pixels must move further apart, so the image loses clarity. In a similar way, you may have noticed that a television with a large screen will generally have a less sharp picture than one with a small screen. This is because the number of pixels used to make up the picture is the same, but they are further apart on the larger screen. Photographs or drawings are often scanned into computers. These images are stored as bit-map files and may be manipulated with graphics, DTP or word-processing software.

Vector graphics

With **vector graphics**, the information defining an image is stored as a series of equations within the computer. These graphics are defined in vector format – for example, an individual straight line or a triangle would be defined by reference to starting points, lengths and directions. The user does not need to be aware of how these graphics are constructed within the computer, only how to create and use such images. A vector graphic might be created out of dozens, hundreds or even thousands of individual lines, boxes and curves. Any text which is added to the graphic can be stored either in text format or in vector graphic format. The quality of a vector graphic is independent of its size: it can be enlarged or reduced and the image quality will depend on the quality of the output device, not upon how much the image has been enlarged.

Components of graphics and drawings

The components of the various types of graphics include:

- *Lines* – Most of the simpler graphics images consist of combinations of lines which can be straight, curved, freehand or part of a circle (called an arc).
- *Shapes* – All packages provide a range of shapes that can be manipulated and sized, using a combination of mouse and keyboard. Shapes available include rectangles, circles, ovals and polygons. Both lines and shapes can also be drawn freehand.
- *Colour and shade* – Lines and shading can be coloured to make a design or diagram more appealing, although the use of colour greatly increases the size of the file needed to store the graphic. Colour ink-jet printers are quite cheap, and posters, leaflets and other promotional material certainly look much better in colour.
- *Text* – There are three main elements to consider with text: font, size and colour. The way in which fonts are used must be thought through carefully. For example, if you are preparing an illustration for use on a poster, you

should ensure that the text is large enough and resist the temptation to have too much text on the illustration.

- *Attributes* – The attributes of some of these components can be changed. For example, lines can be dotted, dashed, have arrows on them or be of different thicknesses. Shapes can be 'flood-filled' (the inside of the shape is coloured or patterned). A 'paint spray' can be used to provide shadings in a similar way to that produced by an artist working with an airbrush. You can alter the shape of a 'paintbrush' to produce different painting effects. Many packages now have huge numbers of features that are worth exploring in detail.

Activity 23

Produce a short set of notes that will help a person new to graphics packages use the package you have chosen. Describe the main components of the package and show how it can be used to construct a diagram of your choice. Illustrate your notes with examples.

Technical drawings

Technical drawings are produced using vector-based software. Such drawings range from simple technical diagrams, such as flow charts, through to complex three-dimensional drawings of, say, cars and buildings. CAD software is used to produce engineering drawings and architectural drawings.

Technical drawings can be categorised as follows.

Layout drawings

These drawings are produced using vector-based software to show the plan view of something – that is, as if you were looking down on it from directly above.

Figure 1.33 shows a series of possible plans for a bathroom, which have been produced using a CAD package called AutoCAD. Producing drawings or plans on a computer is efficient because you can quickly modify drawings or plans to produce a range of alternatives. Some of the designs in Figure 1.33 are much better than others.

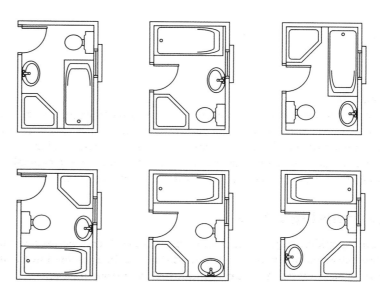

Figure 1.33 Bathroom plans produced using AutoCAD

Product design

Technical drawings are used in product design to show how a particular object should appear when manufactured. Such drawings usually define the dimensions of the product, the materials from which it is to be constructed and other detailed specifications.

Figure 1.34 shows the nozzle of a fire hose, which has been designed using a CAD package. Notice how the software can generate isometric or perspective drawings from the original two-dimensional drawing.

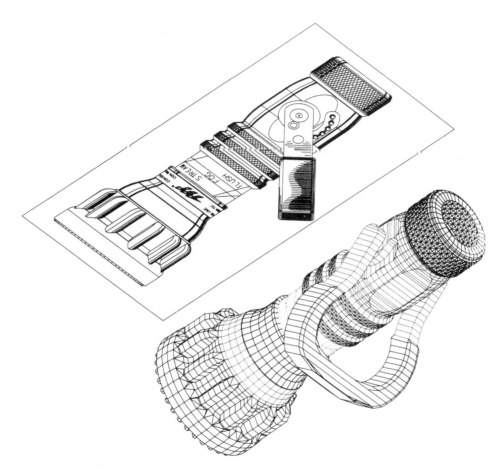

Figure 1.34 Using CAD to design the nozzle of a fire hose

Block schematic

Block schematic drawings are created using vector-based drawing software. They represent in simple terms any system flow or organisation. Such diagrams often consist of various types of labelled box, with arrows showing flows of information or activity. Such diagrams can be used to describe the operation of a control system the functions of an organisation or the operation of a microprocessor.

Presentational graphics

Presentational graphics packages are used to create artwork and pictures, perhaps for use in a slide presentation or for an illustration that is to be presented in electronic format. Such material is mainly created and stored as bit-map images.

Charts

A variety of charts can be produced. They broadly fit into the following categories:

- *Text charts* – These make useful handouts or slides in a presentation when the concepts can be conveyed better in words than in pictures. Text charts are ideal for making comparisons or for presenting the benefits and drawbacks of a particular issue. When text is used in this way it is often a good idea to present information as a bulleted list.
- *Graph charts* – These are used mainly for numerical information. Rather than just listing series of numbers it is better to make them visually appealing by presenting the information as bar charts, pie charts, etc. You can make comparisons by placing charts alongside each other: you might, for example, compare the monthly sales figures over one year with those for the next year (see Figure 1.35).

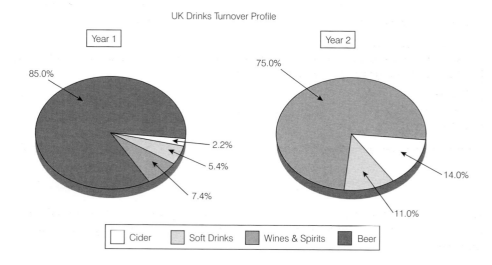

Figure 1.35 Pie charts that illustrate the turnover profile over two years

Many spreadsheet packages have a graphics facility. The user simply enters the numbers and selects the appropriate type of chart; the graph is then produced automatically.

Slide shows

Using presentation graphics it is possible to develop a **slide show** that consists of a series of slides presented on a computer screen one after the other. The software also enables you to print charts onto transparencies or slide film (the latter needs special equipment) and then make a presentation using an overhead or slide projector. You can, of course, also print the slides onto paper for use in handouts. With a colour printer you can make the slides very eye catching.

Slides developed using presentation graphics packages are often images, such as photographs that have been scanned in, clipart, company logos, graphs or charts.

Guidelines for slide presentation are shown in Figure 1.36.

HP Bulmer sells itself: a case study

HP Bulmer is the UK's largest cider producer. Company presentations are regularly given to the City, potential and existing customers and to the company's own staff. The reputation of a company rests on how others see it, so good presentations are vital. To meet the wide range of presentation demands, Bulmer has developed a

---

Guidelines for overhead projector presentation

Simplicity

Keep the information on the slide simple. Research has shown that there should be no more than six words per line and six lines per slide. The tendency is to put too much detail on the slides and use them as notes rather than as a presentation of the main points.

Brevity

In a presentation most listeners can retain only between three and five brief points. Any more could bore or confuse your audience.

Language

Keep the language simple and avoid the use of any specialist jargon which might be known only to a few of the audience.

Notes

Do not use the graphics for notes. People do not want to see reams of text and if notes are needed it is best if you give them copies after the presentation so that they can refer back to them.

Graphics charts

Try to present numerical information in a more interesting way than just a list. Use bar, pie, line and other charts to help get the figures across to your audience.

Fonts

Don't mix too many fonts on the same diagram and do not overuse upper case (capital letters). Use both upper and lower case letters.

Figure 1.36 Guidelines for slide presentation

system for producing presentations. Because of the high costs involved in sending presentation material to outside agencies the company is now able to produce the material itself.

Bulmer now has a presentation graphics unit, with equipment for scanning in photographs, usually of the company's products but sometimes of the various processes and equipment used in cider-making. A high-specification PC is used to prepare material on a 37-inch screen and this is then printed on a colour laser printer. Staff in all departments are able to produce work on their own PCs using a package called Freelance and transfer their work over the network to the presentation graphics unit for colour laser printing. As well as providing colour output on paper, the unit produces acetates for overhead projectors and 35-mm slides. The unit can also produce slides for use on a liquid crystal display (LCD) projector, which is used in larger presentations to project a computer screen onto a large white screen.

The presentation graphics unit is staffed by two graphics analysts who work on their own projects as well as checking the material from other departments to ensure that it is of a high standard.

The annual cost of sending the material to an outside agency used to be around £200 000: doing the task within the company has reduced it to about £70 000.

Commonly accepted standards for the layout of formal documents

There is a set way in which documents (letters, memos, agendas, etc.) are normally laid out. These documents all have a certain structure (the structure has been looked at in previous sections). When constructing one of the these documents you will need to use a standard format.

Essential information that appears on formal documents

Certain essential information appears on all formal documents, such as:

* name and address of the sender
* name and address of the intended recipient
* telephone, fax and mobile numbers and e-mail address of the sender
* the date the document was produced
* a reference to what the document is about (this may refer back to another document or telephone conversation)
* a signature, underneath with the person's name is typed.

Methods of presenting a corporate image

Many organisations have a corporate style specification for commercial documents that ensures all documents look the same, even when produced in different areas by different people. Such a specification also ensures that all documents have a professional appearance.

Figure 1.37 Templates like this are included as part of most word-processing software

The corporate or 'house' style is designed to ensure that the information is presented clearly and is easy to assimilate. Consideration is also given to the use of company logos, creating an image of the company which improves awareness and provides positive publicity.

Use of templates

It is common for members of an organisation to be given templates, one for each type of document, to ensure a common or corporate style. A template for a document is a blueprint for the text, graphics and formatting, providing a framework on which to build and ensuring that all documents of a particular type have the same overall look. The template is an electronic file which holds the basic outline of the document. The user needs only to input the variable data – the formatting, font size, etc. will already be included. The main advantage of using templates is consistency of style, irrespective of the author.

Most word-processing software has template facilities (Figure 1.37) so all you need do is 'fill in the blanks'. Each time you want a similar document, simply open the template, fill in the blanks, save the document (under a different name to that of the template) and then print it out.

Wizards are used for creating templates. What is a wizard and why are they so useful?

/ **Activity 24** /

Investigate the availability of templates (sometimes called style sheets) within your word-processing package. Print out the most useful templates, try putting in the variable data and print out some examples of your work. How easy are they to use and adapt? Produce a brief report outlining your findings.

1.5 Standard ways of working

Most organisations have rules and guidelines to help people work effectively and avoid problems. These ways of working, called 'standard ways of working', are particularly important to people working with ICT because there are many ways that hardware, software and data can be lost or misused. The loss of even a small amount of work can be frustrating and large losses can be catastrophic.

If everyone who used ICT in an organisation could use it in their own way, a number of problems could arise.

- The look of letters would be different because they would depend on who constructed them.
- Other documents (quotes, memos, reports, questionnaires, etc.) would look different.
- Staff would be free to store their files wherever they wanted.
- It would be up to the staff to keep regular backups.
- Staff could load software illegally onto the computer.
- Staff could start to use the computer to play games.
- By loading programs and data from elsewhere viruses could be introduced onto the system.
- Users could leave confidential information displayed on their screen whilst not at their desks.
- Disks containing important data could be copied without anyone knowing.

- Unsavoury material could be accessed using the Internet.
- Suspect files could be downloaded from the Internet.
- Data and programs could be deliberately corrupted.
- People could copy work (e.g. a project such as a game they were working on) and present it as their own.
- Staff could damage computer hardware.

Activity 25

There are many ways in which data can be damaged or misused by employees, many more than the ones shown above. Identify ten more examples of misuse or damage to data.

Most organisations have rules and regulations as to what their employees can and cannot do whilst working on computers or IT equipment, which are intended to help people work more effectively and avoid problems. Rules and regulations are very important for staff employed in ICT because there are many sources of potential problems in this area.

Activity 26

You are working for a medium-sized business who use a lot of IT and telecommunications equipment. The business, called Perfect Partners, is a computer dating agency which arranges dates for their clients. They have a local area network and several of their computers are connected to the Internet.

Perfect Partners has had a few problems with some members of staff regarding the use of the computer equipment, software and data. They would like to make sure that all the staff who are using the computers are aware of what is not allowed during the course of their employment.

They are particularly concerned about the following unacceptable activities:

- unauthorised access to confidential client information by staff who do not need this information in the course of their work
- staff taking copies of the client data and selling it to a rival dating agency who is undercutting Perfect Partners on price
- staff copying the software (an integrated suite) and using it at home
- one member of staff actually copied the software, used a CD writer to produce multiple copies of the software and then sold them at a car boot sale
- staff not keeping regular backups, resulting in a large amount of data being lost and then having to be keyed back in
- staff deliberately altering client details.

Write a memo using word-processing software to outline what staff should not do with the computer equipment and explaining why they should not do it.

Data files may be lost, corrupted by a virus or damaged in other ways

When you work on lots of different computers there is a real danger of transferring a virus from one of them to the others. Not all computers have up-to-date virus checkers and some have no virus checker at all. Others may have the software but it may not have been properly configured.

Although to avoid viruses you should not use 'strange' machines such as the ones in your computer rooms or in the learning resource centre, in practice this is impossible. You *do* need to use other computers, but you must be careful that, when you take any work home or to use with an important computer, you scan the disk using the latest virus-checking software.

Computers may be damaged so that the data stored in them cannot be recovered

There are occasions when an equipment malfunction occurs, making it impossible to recover the files off a disk. This could happen to either the hard or floppy disk and necessitates re-creating or copying the data from the backup copies.

Copying work and presenting it as your own

If you obtain material that you intend to publish by scanning it, copying it or downloading it from the Internet, you need to be careful about copyright. People reading a newsletter, your Internet home page or anything that others may use will think it's your own original work. It is therefore vital that you know the source of the information and have explicit permission to use it. This applies to text, photographs, diagrams, pieces of clipart, etc.

Remember that authors and professional photographers rely on the royalties and any copyright payments to make a living and that illegal copying of such material deprives them of this revenue. Unauthorised copying is theft. The law takes this very seriously – it is a criminal offence.

1.6 Managing your work

The way you manage your ICT work is important. You need to learn to:

- plan your work to produce what is required to given deadlines
- use spaces, tabs and indents correctly to ensure consistent layout and easy editing
- use file names that are sensible and remind you of the contents
- store files where you can easily find them in the directory/folder structure
- keep a log of any ICT problems you meet and how you solve them.

Plan your work to produce what is required to given deadlines

You are seldom given a task to do without a deadline and it is extremely important to keep to deadlines in an IT environment. The tendency to sit back in the early stages of a project should be resisted, as it is impossible to foresee problems that might cause you to lose time. Identifying the milestones in a large project is important – you can work towards them in the short term, and passing them will give you a sense of fulfilment.

You will be given assignments throughout your GNVQ course. At some times you may feel pressured by the number of assignments you have to complete. It is best not to try to do bits of all of them – it is better to concentrate on one at a time. At the start of each week you need to set yourself a task or a series of tasks for the week. Get in the habit of writing 'to do' lists for each day. Do not waste time wondering or worrying about what you have to do. Make sure that, when you go into any practical workshop, you know exactly what you have to do and have all the notes, disks, etc. you need to do it.

Make sure that you produce any printouts ahead of the deadline as computer equipment has a habit of 'playing up' when you are under pressure to complete some work. Do not leave anything to the last minute.

Use spaces, tabs and indents correctly to ensure consistent layout and easy editing

All the work you produce for your assignments will need to be word processed to a fairly high standard. You must use spaces, tabs and indents correctly to ensure consistent layout and easy editing.

Use file names that are sensible and remind you of their contents

By default, word processors will give documents names (such as doc1, doc2), but trying to find a file that you stored six months ago using this naming system would mean having to look at the contents of each document. It is therefore important to consider the file names carefully. To make it easy for you to find documents, it is best to use long descriptive file names. For example a letter could be called 'letter inviting applicants in for interview'.

1.7 Keeping information secure

Protecting information from loss or misuse is essential in an ICT environment. You therefore need to understand the importance of

- keeping information secure (for example, protecting it from theft, loss, viruses, fire, etc.)
- protecting confidentiality (for example, preventing illegal access to medical or criminal records)
- making sure that work kept on an ICT system is not lost
- respecting copyright.

Security checks

Computer security is concerned with taking care of the hardware, software and – most importantly – the data. The cost of recreating the data from scratch can often far outweigh the cost of any hardware or programs lost. Apart from theft, probably the most important threat to data comes from viruses, which are programs designed to damage data, introduce annoying messages, or sometimes both. In the past viruses infected only programs, but there are now many viruses that can damage your data. The risk of viruses must be taken seriously and anti-virus software should be installed on all computers to scan the computer's memory and disks for viruses and destroy them.

Protecting confidentiality

Many documents are confidential to the originator and the recipient (examples are references for employees, medical records in a hospital and social worker reports). Other documents, such as monthly sales figures or new brand promotions, are of use to competitors and so need to remain within an organisation. In this section we look at the confidentiality issues when processing commercial documents.

Many of the documents produced by organisations are confidential and it is important that only certain employees are allowed to see them. Some documents may be of interest to outside bodies and must be kept securely within an organisation. It is worth remembering that documents are vulnerable on and off the computer and there is little point in ensuring data security on the computer if printed copies of confidential documents are left on a desk for anyone to read or steal.

Passwords

Passwords are used to prevent unauthorised access to documents held as computer files, and should be changed on a regular basis. More details on passwords are given on pages 179–182 and 267.

Non-disclosure

Although passwords prevent unauthorised access, some staff need to read confidential data and could be tempted to pass on the information they have seen. The only real way of reducing the risk of this is to make people aware of the consequences of such action. Most organisations put non-disclosure agreements in their contracts of employment, which prevent staff from disclosing any information gained in the course of their employment to any outside organisation. Public service employees are often made to sign the Official Secrets Act as part of their conditions of employment. This makes it a criminal offence to pass on information, with penalties as severe as a prison sentence.

Making sure that work kept on an ICT system is not lost

If work stored on an ICT system is lost, it is important that there is another file that can be used in its place. There are two ways to make this possible:

- by keeping dated copies of files on another disk and in another location
- by saving work regularly, and using different file names.

Some software saves data automatically after a certain period. The time between the saves can often be altered to suit the user. Regular file saving ensures that work is not lost if a problem occurs with the computer or its power supply.

Backing up

Most word processors can automatically produce a backup file in addition to the original document. They usually have a different file extension – in Microsoft Word, for example, backup files have a .bk extension (the original documents have the .doc extension).

Some word processors automatically create a temporary backup file while you are working, allowing the last version of the document to be recovered in the event of a system crash.

It is important that you get into the habit of making regular backup copies of your files and keeping the backup disks away from the computer. It is best to keep a series of disks and rotate them as the backup copies for your most important files. The reason for this is that some viruses now attack data and you do not always know if your disk is infected, so any backup copy could also be infected. If your original disk and first backup become infected, you could still have uncorrupted files on a second backup disk, although you will lose any work done between the first and second backups.

Retain source documentation

A source document is the original document used when preparing a new document on a computer. For example, if you are writing up an assignment, you may have decided to first write out a rough version on a piece of paper, from which you type the final version into the computer. In this case the rough version would be the source document.

If the document held on the computer is lost or corrupted, as long as you kept the rough, you would have the source document from which to work again.

What is a 'source document'?

Backup procedures

Backing up is the creation of copies of programs and data so that if they are lost they can be recreated. Although there are many software utilities (e.g. Norton Utilities) that will aid recovery of lost data, these don't work in every situation – there is no substitute for making a backup copy.

Backup copies should be kept on a separate disk or tape and stored away from the computer system. If you keep a backup copy on the same disk as the original, then if this disk is stolen or destroyed by fire you will also lose your backup.

Rules for backing up

What backup procedures can you adopt to make sure that you do not lose your work?

- Never keep backup disks near the computer. If the computer is stolen, the thieves will probably take the disks as well (unless they are ignorant about computers). Never keep the disks in the drawer of your desk – this is one of the first places thieves will look.
- If you hold a lot of data that would be very expensive to recreate, you should invest in a fire-proof safe to protect your backups against theft and fire.
- Keep at least one set of backup disks at a different site.

Equipment theft

Theft of computer equipment can obviously result in the loss of any documents the equipment contains and possibly the loss of backups if these were stored close by.

Software/data theft

Software is owned by the developer. To use it without a proper licence is theft. Organisations have to make sure that they are not party to software theft by checking that their employees' hard disks do not contain unauthorised software. If an organisation uses networks, it must have the correct number of licences for the number of users to whom the software is available.

Copyright

If work is prepared by an organisation's staff in work time, copyright to any new software or data they prepare belongs to that organisation. It is illegal to copy data or software without the owner's permission. Copyright is covered by the Copyright, Designs and Patents Act 1989.

You must be careful about the way you use a scanner. If you are scanning material that you intend to circulate widely – for example in a newsletter or magazine or maybe your Web page – it is vital that you know the source of the material and have explicit permission to use it. This does not just apply to text that you might have copied – it also applies to diagrams, artwork and photographs. It is important to realise why this is so important. Professional writers, artists and photographers make a living from royalties and selling their work, and in copying it freely you are depriving them of money that is rightfully theirs.

1.8 Working safely

Several health problems can occur when working with computers for long periods, but if you are aware of them you can take steps to prevent them. As more and more computers are used in the workplace, health problems have increased. In this section we will look at the main problems and how to prevent them.

Eye strain

Most people who use computers have experienced eye strain. It is caused by several factors – reflections from lights on the screen, concentration on the screen for long periods and shifting focus between the screen and paper (both keeping focused on the screen and refocusing the eyes lead to eye strain). The early symptom of eye strain is hazy vision, which is usually followed by a headache. When this happens the person needs to rest and lie down if possible.

Preventing and relieving eye strain

Give your eyes a break
Take regular breaks from the computer screen. Experts recommend that a 15-minute break should be taken for every hour of intensive work on the computer. During this break you should try to relax but if there is so much work to do that this is impossible, then you should do some non-computer work in this period.

Refocus your eyes every so often
This should be done every 10 minutes. It involves looking up from the computer and focusing your eyes on a distant object.

Use suitable lighting
You need to make sure that there are no concentrated sources of light, such as pendant lights, near your screen. These will produce reflections on the screen. Fluorescent tubes that have plastic covers, called diffusers, which disperse the light and illuminate the work area evenly, should be used.

Use a copyholder
Shifting your vision between paper and the screen can cause eye strain, and can also give rise to neck strain. It is best to use a copyholder, which will keep the paper you

are working from at the same height as the computer screen , so your eyes will not continually have to refocus.

Use adjustable blinds on the windows

A common problem is the glare of sunlight on the screen. To get round this, use adjustable blinds on the windows. Try to avoid using curtains – they cannot be adjusted easily and are not as easy to clean as blinds.

Have frequent eye tests

Employers are required by law to pay for their employees who work with computer screens to have their eyes tested regularly. If employees need to wear glasses because of their work the employer should pay for them.

Radiation

There have been many scare stories in the press about the dangers of electromagnetic radiation given out from computer screens. Many other devices also give out this radiation (most of the research on it is centred around radar installations and electricity power lines, which give out quite strong emissions). Some women using computer screens for long periods have had abnormal pregnancies, but there is little evidence to support a connection between the two. Nevertheless, many employees encourage pregnant women to work in an area that does not use display screens. There are also many devices that can be placed over the screen to reduce radiation. Several new European and international regulations have been introduced to reduce the amount of radiation emitted by computer screens, and all new computer screens must conform to these standards.

Repetitive strain injury

People who spend long hours working at a keyboard can develop a condition called **repetitive strain injury** (RSI). This condition is caused by the constant pounding that the joints in the fingers, hands and wrists take during keying. When an operator is keying in at high speed, he or she presses the keys quite hard. When the key reaches the end of its depression it stops, and a shock wave travels up through the bones in the hand, causing damage to the muscles and tendons in the fingers, wrists, arms and neck. Symptoms of RSI include soreness or tenderness of the fingers, wrists, elbows, arms or neck. If left without treatment it can give rise to a painful condition, similar to arthritis, that causes long-term disability.

Good keyboard design, a well positioned keyboard, good typing technique and frequent breaks can help prevent RSI. Special keyboard supports and wrist-guards are available to ease fatigue, and these are worth using.

Physical stress

Physical stress is a general condition brought about by the body working in the wrong environment or trying to do a task for which it is not really designed. Physical stress causes direct damage to the body and a doctor can see the damage that has been done. Examples of physical stress include RSI, eye strain and backache. Many employers are aware of these stresses and have tried to improve conditions in the workplace by using **ergonomic** workstation designs.

Psychological stress

Psychological stress can be caused by using inappropriately designed software. An example might be if you change over to a new word-processing package and are unable to do a task that before was quite simple. You feel frustrated because you are wasting time and this gets worse if the job is urgent. Losing files or finding a virus on your machine are among the most stressful situations.

What can be done about these problems?

Because of the various health problems that can occur with incorrect computer use, the government has laid down certain laws which require employers to provide the following:

- *Inspections* – Employers should periodically inspect the workplace environment and the equipment being used to check that it complies with the law. Any shortcomings in the conditions, working practices or equipment should be reported and corrected. Desks, chairs, computers, etc. should be inspected for possible risks to workers' eyesight and to their physical and mental health.
- *Training* – All employees should be trained in the health and safety aspects of their job. They need to be told about the correct posture when using a keyboard, etc.
- *Job design* – Each employee's job should be designed to allow them periodic breaks or changes of activity when using computers.
- *Eye tests* – Computer users should undergo regular eye tests, and be provided with glasses if necessary, at the employer's expense.

The law also lays down certain minimum requirements for computer systems and furniture. All new and existing equipment should meet the following requirements:

- *Display screens* must be easy to read, have no 'flicker' and be very stable. Brightness, contrast, tilt and swivel must all be adjustable and there must be no reflection from the screen.
- *Keyboards* must be separate from the screen and tiltable. Their layout should be easy to use and the surface should be matt in order to avoid glare. There must be sufficient desk space to provide arm and hand support.
- *Desks* must be large enough to accommodate the computer and any paperwork and must not reflect too much light. An adjustable document holder should be provided so as to avoid uncomfortable head movements.
- *Chairs* must be adjustable and comfortable, and allow easy freedom of movement. A footrest must be available on request.
- *Lighting* – There should be suitable contrast between the computer screen and the background. There must be no glare or reflections on the computer screen, so point sources of light should be avoided and the windows should have adjustable coverings such as blinds to eliminate reflections caused by sunlight.
- *Noise* should not be so loud as to distract attention and disturb speech.
- *Software* must be easy to use and should be appropriate to the user's needs and experience. Although software can be used to monitor an employee's performance, this is not allowed without the knowledge of the employee.
- *Heat, humidity and radiation* – Heat and humidity should be kept to optimum levels while radiation emissions should be kept to a minimum.

Other hazards in the workplace

Fire hazards

- *Overloaded power sockets* – Never overload power sockets. This is frequently a source of fire. If there are insufficient power sockets the room should be rewired to cope with the level of power consumption.

- *Large quantities of paper lying around* – Fire frequently starts from cigarette ends disposed of in wastepaper bins. There should be a 'no smoking' policy in office areas and bins should be emptied regularly. Paper stores should be separate from the working area.

Obstructions

- *Trailing wires* – These are dangerous, so make sure that wires are long enough to go around the walls of the room or are placed in plastic or rubber trunking.
- *Boxes of paper* – Boxes of paper are frequently left around the work area. As well as constituting a fire hazard, they are often the cause of people tripping.

Electrical

- *The wrong size fuse being used* – If a fault occurs and the electrical current starts to rise, the thin piece of wire inside the fuse will melt. This cuts off the electricity and prevents damage to the computer equipment. As computers, scanners, printers, etc. all use different amounts of electricity, you should make sure that the correct fuse is placed in each plug.
- *Bare wires showing* – There should be no bare wires showing from any plugs or sockets.
- *Tampering* – When taking the casing off any piece of equipment, make sure that the equipment is unplugged first.

Other safety aspects

- *Lifting* – Any lifting of computers, printers, heavy boxes of printer paper, etc. should always be performed with your knees bent and your back straight. There is a legal obligation on a company to show its employees how to lift properly if that is a requirement of a person's job.
- *First aid* – A qualified first aider available at all times.
- *Fire precautions and emergency evacuation procedures* – These should all be in place and should be practised at regular intervals. Employees must be told not to use lifts during emergencies.

Activity 27

You are employed as the data processing manager of a mail order company involved in the sale of the latest fashions direct to the public via a catalogue. Your main duties involve the day-to-day running of the data processing department and making sure everything runs smoothly. You have in your charge 40 staff, most of whom use computers all the time or periodically in the course of their work. The majority of these staff enter details of orders, invoices, etc. using a keyboard. Much of this work necessitates them having to work at high speed from source documents or customer telephone calls and this means some of them spend long periods at the computer.

A television programme is broadcast one evening which draws attention, in an alarmist way, to the problems that could occur when working for long periods with VDUs and keyboards. Many of the staff watched it and are worried that their work may be affecting their health. The safety and union representative has approached you to allay some of the fears the workforce might have. She has only a superficial knowledge of computers and has come to you for further information, and she would like to know what steps the company is taking to reduce the risks. ▶▶

The Managing Director also saw the programme and has heard that the staff are voicing their concern.

a   The safety and union representative is concerned about the following problems raised in the programme:

- eye fatigue leading to poor eyesight caused by continually looking at a screen
- backache caused by incorrect posture
- the dangerous radiation given out by the screen (particularly the dangers to pregnant women)
- skin rashes
- stress caused by a large workload.

   i   Research each of these problems, making use of information from reliable sources. Plan a presentation outlining the facts relating to each problem and what the company intends to do to reduce the risk of each problem. You should also mention any other problems that you come across during your research.

   ii The Managing Director suggests that you find out if there is any equipment that could be bought to help with the problems outlined above. He would also like you to see if a change in working practices would help. Carry out research into the points raised by the Managing Director, and plan a structured presentation.

b   You have recently been looking at the contents of the hard drives used by the department and have found some unlicensed software as well as some games. The company intends to bring in a terminal connected to the Internet on which it intends, eventually, to put the entire catalogue so that customers will be able to place their orders direct using a computer.

   You are concerned about the following issues and need to mention each one to the staff:

- copyright infringement
- responsible attitudes to uncensored materials available on the Internet
- the confidentiality of data.

   Plan a presentation to cover each of these issues.

c   Staff have lost important and valuable files due to viruses, operator error, etc. Write a brief report about the security methods that could be used routinely by staff to reduce these problems.

Key terms

After reading this unit you should be able to understand the following words and phrases. If you do not, go back through the unit and find out, or look them up in the glossary.

Agenda	*Border*
Appendix	*Bullet points*
Bibliography	*Contents page*
Bit map graphics	*CV*

Draft
e-mail
Ergonomics
Fax cover sheet
File attachment
Flyer
Font
Footer
Grammar check
Graphic design software
Gutter
Header
Heading and title style
Hyphenation
Indent
Index
Invoice
Itinerary
Justification
Line spacing
Margin
Memo
Minutes
Newsletter

Orphan
Page layout
Page orientation
Pagination
Paragraph numbering
Presentational graphics
Proofread
Proofreading marks
Readability
Reading age
Report
RSI (repetitive strain injury)
Shading
Slide show
Spellcheck
Subscript
Superscript
Tab
Table
Text orientation
Text/picture box
Vector graphics
Widow
Writing style

Review questions

1 Organisations often use templates for setting out documents.
   a What is a template?
   b Give two examples where using a template would be beneficial.

2 The boss of a large company decides that everyone in the organisation who uses a word-processing package must use predefined templates in order to 'enforce corporate standards'.
   a Explain why most large organisations use templates.
   b In what way does making everyone use a template 'enforce corporate standards'?

3 There are a number of health problems associated with using computer equipment.
   a Name three such health problems and briefly describe how they are caused.
   b Describe what can be done to eliminate or prevent the health problems you identified in (a).

4 Multi-page documents should always have a consistent style from one page to another.
   a Explain four things that would constitute the 'style' of a document.
   b Explain the terms widow and orphan.
   c Explain the terms header and footer.

5 Information is a corporate resource and must be protected. The method of protection used must ensure the confidentiality and the security of the material.
   a What precautions can an organisation take to protect against:
      i viruses

     ii theft

     iii accidental loss?

  b What is the difference between confidentiality and security?

6 Often a document that you are working on using a computer will go through several revisions before the final version is produced. Each version is best stored using a different file name. What is the main reason for doing this?

7 Backup copies of data should always be dated. Why?

8 The following questions refer to documents and their construction.

  a What is the name for a document used to inform people of a meeting and outline the topics that are to be discussed?

  b What is a detailed account of what was discussed at a meeting and the actions to be taken called?

  c To shorten the time taken to prepare commercial documents, a file can be set up using word-processing software in which the user simply fills in the variable information. What is this type of file called?

  d What name is given to a piece of text placed at the top of every page of a document?

  e A photograph is to be scanned into a document using a scanner. Which is the most appropriate graphic type for storing the picture – vector or bitmap?

9 Here is a list of popular business documents used in a business. For each one, state its purpose and give a brief example of its use.

  a Memo

  b Agenda

  c Invoice

  d Purchase order

  e Newsletter

  f Questionnaire

10 In a sales order processing system there are three important documents – the customer order, the invoice and the dispatch/delivery note. Explain the purpose of each of these documents and explain the order that they are sent in.

Assignment 1.1

A holiday company needs to write letters answering complaints that the holiday makers have made following their return from their holiday. Generally, the complaints they make relate to:

- the quality of the food
- noisy bedrooms (e.g. room is over the disco, an all-night bar across the road, etc.)
- the flight delay
- the surroundings (e.g. building work being carried out, facilities advertised not being available, etc.)
- the bedroom (e.g. dirty, basic, problems with plumbing).

Letter 1

```
120 Coppice Drive
Emmerdale
N Yorks
BD3 5RT

Dear Sir/Madam

Re Holiday Hotel Apollo, Corfu, Greece: Departure
Date 17/5/01

When we arrived at the above hotel we found the room
very dirty with the floor looking as though it had
never been cleaned. Also, the shower did not work and
we had to use the shower near the pool for three
days until our shower was fixed.

We would like you to compensate us for the
inconvenience.

Yours faithfully
```

[signature]

```
Mr F A Jones, Mrs R Jones and family
```

Letter 2

```
                    Mrs A Smith
                    34 Elmdale Ave
                    Crosby
                    L23 6TT
```

Dear Sir/Madam

Re our holiday departure date 9/9/01 at the Hotel Ibis, Budapest.

I wish to draw your attention to the following problems that I experienced whilst on this holiday.

1. The flight BT675J was delayed for 5 hours, during which time we were not told about the problems.
2. The food in the hotel was far below an acceptable standard and I had to eat out after the first three days because the food was so bad.
3. The swimming pool advertised in your brochure was little more than a children's pool, being only about 15 feet long and four foot deep in the deepest part.

I would like to hear what you have to say about the above comments and I await your reply with interest.

Yours faithfully

[signature: Anne Smith]

Anne Smith

Tasks

1 Here are two letters that have been sent in by holiday makers. You have been given the job by the travel company of replying to them. In your letters you need to apologise for the inconvenience and to offer them a goodwill gesture of £50 per person.

   Produce these two letters using word-processing software. As your evidence you will need to produce:

   - copies of your drafts for each letter (handwritten or word-processed)
   - a copy of your draft after it has been typed into the computer and checked for grammatical and spelling mistakes by the word processing software
   - copies of each letter after they have been proofread and are in a form in which they can be sent to the customer.

2 The company often need to send very similar letters out. For example, a letter of complaint by a customer may only differ in their details, the name of the hotel and resort and the date on which they travelled. To save time in answering such letters, produce a template for any one of the letters you have already produced and print it out as evidence.

3 The company have thought that templates could be used in other areas of the business. They would like you to explain, using visual aids, what a template is and how it may be useful in order to produce standard company documents. Produce a 5-minute presentation on templates. You should use slides to show the main points of the presentation and you should also include some instructions (no more than four pages) to show how templates may be constructed.

Assignment 1.2

For this assignment you will need to create the following.

1   Six original documents for different purposes to show a range of writing and presentational styles. The documents may be in printed form or shown on-screen. They must include one designed to gather information from individuals and one major document of at least three A4 pages.

2   A report describing, comparing and evaluating two different standard documents used by each of three different organisations (i.e. a total of six documents).

Get the grade

To achieve a grade E you must:

- Create new information that is clear, easy to understand, uses a suitable style and is at a level that suits the intended readers.
- Use text styles, page layout, paragraph formatting and, where appropriate, common standards for layout that suit the purpose of each document.
- Use and combine text, graphics, tables, borders and shading effectively.
- Locate use and adapt existing information to suit a presentation and list your information sources in an appropriate form.
- Describe each of the six collected documents clearly and accurately, identifying the common elements of similar documents.
- Show that you have carefully checked the accuracy of the layout and content of your six original documents and your report, and proofread them to ensure that few obvious errors remain.

To achieve a grade C you must also:

- Show, by presenting original draft copies with proofreading corrections and annotations, how you achieved a coherent and consistent style, made good use of standard formats, placed information in appropriate positions and ensured correct and meaningful content.
- Describe in detail the content, layout and purpose of the six collected documents, accurately evaluating good and bad points about the writing and presentation styles of similar items, commenting on their suitability of purpose and suggesting how they could be improved.
- Show you can work independently to produce your work to agreed deadlines.

To achieve a grade A you must also:

- Show a good understanding of writing style, presentation techniques, standards for special documents and attention to detail by organising a variety of different types of information into a single coherent, imaginative, easy to read presentation of several pages.
- Show effective skills in the appropriate use of software facilities to automate aspects of your document production, such as bullets and numbering, paragraph and heading styles, standardised layout, contents lists and indexes.
- Make appropriate use of lines, borders, shading, tables, graphics and writing style to create a form that is easy to understand and easy to use to enter data and retrieve the information collected.
- Show effective skills in the use of graphics to improve a presentation by making appropriate use of pictures, drawings, clip art, lines and borders, graphs or charts.

Key Skills Opportunity

You can use this opportunity to provide evidence for the Key Skills listed opposite.

Communication C2.3	Write two different types of documents about straightforward subjects. One piece of writing should be an extended document and include at least one image.
C3.1(b)	Make a presentation about a complex subject, using at least one image to illustrate complex points.
C3.3	Write two different types of document about complex subjects. One piece of writing should be an extended document and include at least one image.
Information Technology IT2.1	Search for and select information for two different purposes.
IT2.3	Present combined information for two different purposes. Your work must include at least one example of text, one example of images and one example of numbers.
IT3.1	Plan, and use different sources to search for, and select, information required for two different purposes.
IT3.3	Present information from different sources for two different purposes and audiences. Your work must include at least one example of text, one example of images and one example of numbers.

ICT serving organisations

What is covered in this unit:

2.1 *Types of organisation*
2.2 *Functions within organisations*
2.3 *Information and its uses*
2.4 *Management information systems*
2.5 *Standard ways of working*

This unit is about the way in which ICT serves organisations. To understand the unit you need to understand how organisations operate and how information flows internally and externally to the organisation. You will also need to understand how organisations are structured and how they use and exchange information.

You must be able to evaluate how well ICT can help organisations and consider how it supports many different activities within an organisation. In studying this unit you will see how ICT offers new opportunities, such as just-in-time stock management.

You will produce a case study on how an organisation collects, disseminates and uses information, how it manages the flow of information between sections or departments and the way it uses ICT to access and exchange information. This unit links with the advanced units:

- Unit 1 Presenting information
- Unit 3 Spreadsheet design
- Unit 6 Database design.

This unit is assessed through an external assessment. The grade on that assessment will be your grade for the unit.

Materials you will need to complete this unit:

- computer hardware, including a printer
- word-processing software such as Microsoft Word
- a drawing package – if you do not have a specialist package then you can use the drawing features in Word
- access to the Internet
- access to the school/college management information system.

2.1 Types of organisation

What is an organisation?

We often come across the word organisation when talking about IT, so it is useful to have a clear idea of what it means. In general, an organisation consists of:

- a group of people
- working together with a common purpose
- to satisfy a series of objectives and to carry out all the activities that go towards those objectives.

We use the word organisation collectively to mean all types, such as banks, production companies, charities, government departments or professional practices. Organisations can vary in size, from one person working from home to a large multinational company with offices throughout the world and employing tens of thousands of people.

Organisations can be grouped under three headings: commercial, industrial and public.

Commercial organisations

The prime **function** of a commercial organisation is the purchase and sale of goods and **services**. A shop, for instance, buys goods from its suppliers, adds its profit and sells the goods on to the public. Some shops do not deal with a manufacturer direct but buy through a **wholesaler**. The wholesaler, like the shop, adds its profit to the goods it buys before selling them on to the **retailer**, so it is also a commercial organisation.

Commercial organisations are not just those organisations that sell goods for profit. Many commercial organisations sell a service. For instance, a private nursing home sells care facilities, food and accommodation; a bank sells banking facilities. The bank's profits are obtained from the charges, loan and overdraft interest that they levy on their customers. Insurance companies, car rental firms, video hire shops and telephone companies are all commercial organisations that supply a service.

Industrial organisations

The prime function of an industrial organisation is to manufacture products, process raw materials or construction. The main feature of such an organisation is that it makes something from something else.

For example, Imperial Chemical Industries (ICI) mines rock salt in Cheshire. The raw material is processed by removing all the rock and sand from it before it is sold to customers for a variety of uses. ICI is therefore an industrial organisation. ICI also uses salt in the manufacture of sodium hydroxide, another important chemical. Unilever buys sodium hydroxide from ICI and uses it (along with oil or fat) to produce soap. Unilever is another industrial organisation.

Many industrial organisations make things up from components supplied from other industrial organisations. Car manufacturers such as Ford, household appliance manufacturers such as Hotpoint and computer manufacturers such as IBM are all industrial organisations.

Public service organisations

The prime aim of a public service organisation is to provide services and goods from public funds. Making a profit is not a primary aim, although many public service organisations do aim to be profitable. Their main aim is to obtain the maximum benefits for the public they serve. This means that they are under constant pressure to keep their costs down while providing a good service. Like other kinds of organisation, they have turned to ICT to help them perform their work in the most efficient manner.

Define the term 'organisation'.

Activity 1

a Each of the following organisations can be classed as commercial, industrial or public service. Construct a table like the one below and put each organisation into the correct column.

- oil company
- building society
- stockbroker
- local area health authority
- college
- dating agency
- computer manufacturer
- supermarket
- garage

- private nursing home
- quarry
- coal mine
- bank
- Department of Social Security
- car manufacturer
- council tax department
- insurance company
- insurance broker.

Commercial	Industrial	Public service

b Draw a table similar to that in (a), and place the following organisations into the most appropriate columns:

- Sainsburys
- HP Bulmer (cider-makers)
- Liverpool Area Health Authority
- Driver and Vehicle Licensing Authority (DVLA)
- Esso (an oil company)
- The Ford Motor Company
- Manchester Airport
- Halifax PLC
- British Airways
- Metropolitan Borough of Sefton Education Department
- IBM (a computer manufacturer)
- Glaxo (a pharmaceuticals company)
- Liverpool Community College (a college of further education)
- Direct Line Insurance
- Imperial Chemical Industries (ICI)

Ways in which organisations are structured

Organisations can also be classified according to how they are structured. There are two basic groups: **hierarchical organisations** and **flat organisations** (Figure 2.1).

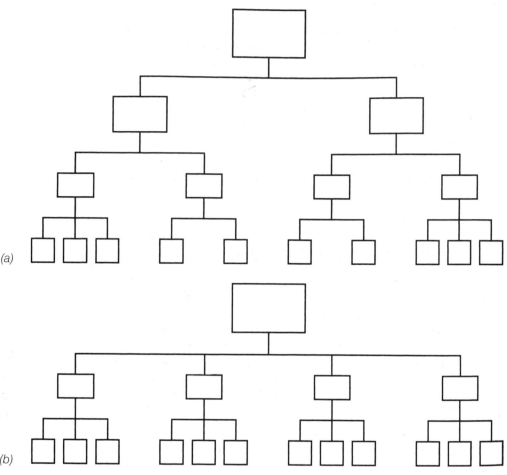

(a)

(b)

Figure 2.1 The two types of organisational structure. (a) Hierarchical; (b) Flat

Hierarchical organisations

Some organisations have many levels and grades of staff, with a tree-like management structure and strong patterns of vertical communication. This means that there are many different grades of staff between people lower down the organisation and the person at the top. This kind of structure can be represented as a triangle, with a greater number of staff in the lower tiers and fewer towards the top, more senior, posts. Large traditional types of company tend to have this structure, as do many government departments.

Hierarchical organisations suffer from problems with bureaucracy, as information needs to be directed through the correct channels before appropriate action is taken.

- At each level several staff are responsible to the person at the next level up. The process is repeated until the top of the organisation is reached.
- In a limited company the person at the top is the Managing Director, who is ultimately responsible for the whole organisation.
- As you ascend the levels within the organisation, the number of people at each level decreases. This gives the organisation a pyramidal structure.

Flat organisations

In an organisation with a flat structure there are fewer levels or grades of staff and much more emphasis on communication *across* the organisation. This is more likely to be the structure of a small business where everyone knows each other and works together more as a team.

Examples of flat organisations include some shops and small family-run businesses. The main advantage with the flat organisation structure is that staff are able to work more on their own initiative, which brings flexibility and more creativity to their jobs.

Activity 2

a Explain the difference between a flat and a hierarchical organisation structure.

b Which type of organisation structure would you prefer to work within? Why?

Drawing organisation charts

An organisation chart is a diagrammatic illustration of the structure of an organisation (Figure 2.2). Such charts show the organisation, functions, activities, posts, lines of responsibility, levels of authority, accountability, lines of communication and span of control. Some organisation charts refer to particular people and others refer to particular functions or activities within an organisation. Organisation charts can also do both, showing a name alongside a particular post. By looking at such a chart it is easy to determine a particular person's position within the organisation, who they are directly accountable to (i.e. their immediate boss) and which staff are at the same level. You can also see whether they too are responsible for a group of staff.

Organisation charts are useful aids for systems analysis: like most diagrams they present the information more concisely than is possible in words.

Why is it necessary for organisations to have an organisational structure?

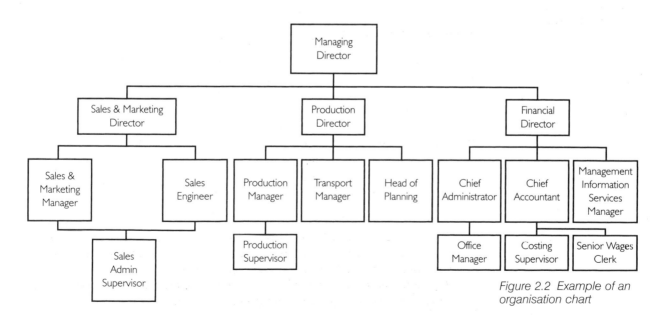

Figure 2.2 Example of an organisation chart

A number of steps must be taken when drawing an organisation chart:

1   Find out the name of each job and to whom the job holder reports (i.e. who their boss is). In some organisation charts you can attach a person's name to each of the jobs shown, but staff frequently change, so in large organisations the chart will require frequent updating.
2   From the information gained in step 1, complete a table similar to the one below.

Name of post	Responsible to
Costing Supervisor	Chief Accountant
Managing Director	
Chief Accountant	Financial Director
Management Information Services Manager	Financial Director
Production Supervisor	Production Manager
Sales and Marketing Director	Managing Director
Transport Manager	Production Director
Office Manager	Chief Administrator
Senior Wages Clerk	Chief Accountant
Head of Planning	Production Director
Sales Engineer	Head of Marketing
Sales and marketing Manager	Head of Marketing
Chief Administrator	Financial Director
Production Director	Managing Director
Production Manager	Production Director
Head of Marketing	Sales and Marketing Director
Financial Director	Managing Director
Sales Admin Supervisor	Sales and Marketing Manager and Sales Engineer

3   The person who reports to no one is obviously the boss or the person in charge of the organisation. Place a box with the name of this post (and the name of the person if required) at the top of the page.
4   Find out from the table the posts that report to the person in step 3 and add them one level lower down.
5   Continue to do this until all the posts have been placed.

Activity 3

The table lists the various members of the senior management team in the computing department at Tesco. Produce an organisation chart to show the relationships.

Position	Reports to
Retail New Technology Manager	Divisional Director, Computer Systems Development
Divisional Director Computer Systems Development	Director of Computing
Retail Systems Manager	Divisional Director, Computer Systems Manager
Project support Manager	Divisional Director, Computer Services
Technical support Manager	Divisional Director, Computer Services
Commercial and Estates Manager	Divisional Director, Computer Systems Development
Service Initiatives Manager	Divisional Director, Computer Services
Operations Manager	Divisional Director, Computer Services
Contracts Manager	Divisional Director, Computer Services
Finance and Personal Systems Manager	Divisional Director, Computer Systems Development
Project Support Manager	Divisional Director, Computer Systems Development
Stock Control and Distribution Manager	Divisional Director, Computer Systems Development
Divisional Director, Computer Services	Director of Computing

▶▶

Choose suitable software and explain reasons for your choice. You may find that a graphics package such as PowerPoint (in the Microsoft Office suite of programs) has a special template for the production of organisation charts.

The information requirements of organisations

All organisations need to communicate information both within the company and between it and external agencies such as customers, suppliers or government departments. In order to design and implement successful ICT systems, it is necessary to understand the different types of information that need to be communicated. The types of information used within a company depend ultimately on the nature of the business, but most businesses use at least some of the following types of information.

Sales information

Most companies have a sales department where goods or services are sold and records are kept of these sales. In many cases, customers are given credit so they have a certain period between receiving goods or services and having to pay for them. The value of goods a customer is allowed to receive without advance payment is called that customer's credit limit.

When customers place orders they can do so in a variety of ways. Sometimes they use their own order forms; at other times they use one provided by the company. Many orders are made over the telephone or using the Internet. Some companies use a service called **electronic data exchange** (**EDI**), through which one computer places orders directly to another computer with no corresponding paperwork.

A sale usually takes place as follows:

1 The customer places an order for goods or services.
2 The goods or services are supplied.
3 The customer is invoiced (an invoice is a request for payment).
4 The customer makes payment.

In a perfect world, once the customer receives an invoice, he or she will make payment for the correct amount without further requests. However, it is often necessary to generate further requests for payment. The letters generated by the system become more severe and threatening as time elapses. Take, for example, an electricity bill: the first bill is usually printed in blue, the second one in red and a couple more reminders are sent after this before the supply is cut off.

In addition to non-payment, all sorts of other problems can occur. For instance, customers might send the wrong amount or they could send goods back and require refunds.

Sometimes, if an order is made and no invoice produced, a receipt is given to record the transaction. This is often the situation where goods are ordered over the telephone using a credit card. Another example is when a purchase is made from a catalogue shop such as Index or Argos. With this system the customer looks through a catalogue and writes down the catalogue number and price of the goods they want. They take this to the point-of-sale terminal where the details are entered, payment is made and a receipt issued. Finally the receipt is presented at a storage area where the goods are collected.

Activity 4

Investigate two different systems for dealing with sales of goods or services with which you come into contact in your everyday life. You might like to base your investigations on:

- the system used by a retailer who uses a catalogue to advertise goods, such as Index or Argos
- the system used for payment of petrol at a petrol station
- the system used for booking a holiday at a travel agent
- the system used when buying a ticket at a railway station
- the system used for buying a meal at a self-service 'burger bar' restaurant.

In your investigation you need to consider:

- the forms of information used in the system (verbal, documents and electronic)
- how information is passed from one place to another. (You could draw a simple diagram to show this along with a brief explanation of your drawing.)

Purchase information

Some organisations, such as retailers, purchase goods from suppliers to which they add their profit before re-selling to someone else. All companies must purchase goods or services from other organisations, if only basic services such as water, electricity and gas. Nearly all companies also buy stationery, office furniture, cleaning and toilet supplies and so on. The purchasing department of a company can be huge. Take, for example, all the equipment and services needed by the armed services – a whole civil service department deals with this. Manufacturing companies usually have their own purchasing departments which order raw materials.

As well as making sure that orders go out, it is also necessary to keep track of goods received and to check the invoices for them against the original orders before payments are made. It is quite common in industry for an order to be only partly fulfilled, and this makes keeping track of the purchases more complicated.

Market information

It is important for companies to match the supply of goods with the demand for them. For instance, a car manufacturer needs to know how many cars have been sold at certain times of the year in previous years so that they can match the production to probable demand.

Marketing information is used:

- to plan marketing initiatives that increase consumer awareness
- to increase the company's market share
- to devise new pricing strategies and
- to direct development of new products.

Market information also includes details about competitors and their products. It is useful to know whether rivals are offering special promotions (such as 'buy one get one free') to encourage customers to buy *their* products. Such offers are often used as a way of increasing brand awareness and may increase a company's share of the market. A company may react to market information by reducing the prices of their products to match their competitors, or offer a special deal, to ensure that they maintain their market share.

It is also important to make sure that the goods or services provided are what customers actually want. Market information includes details of customer opinions, obtained by talking to customers or getting them to complete questionnaires. Such information can be processed by computer to give valuable market knowledge.

Sources of market information include such things as the volume of sales for a certain item over a period of time or why consumers behave in certain ways.

How is marketing different from sales?

Design specification

A **design specification** is a document that specifies the company rules and the statutory regulations that govern the manufacture of the goods being produced. It also ensures that the goods are manufactured in accordance with the customer's wishes. It is hard to generalise about what a design specification should contain – the details will depend on the nature of the goods being made. It would usually involve:

- materials from which the goods are to be made
- tolerances
- special finishes
- sizes
- colours.

Operational information

Operational information relates to the internal workings of an organisation. Examples of operational information include instructions, decisions, responsibilities, holiday dates and times of opening.

Customer information

Most organisations hold information about their customers, including:

- name of company
- invoice address
- delivery address
- contact name (often this is someone in the purchasing department)
- credit limit
- average value of order
- main products ordered.

Supplier information

Organisations also need information so that they can place orders with their suppliers:

- name of supplier
- invoice address
- delivery address
- contact name (usually this is someone in the sales department)
- credit limit.

Activity 5

An organisation uses many different documents to cope with the day-to-day running of the business. Identify the different situations which could give rise to each of the following documents:

- quotations
- requisitions
- invoices
- orders
- statements
- delivery notes
- credit notes
- expense claim forms.

Information exchange

Different organisations exchange different types of information about different things. In this unit you will need to learn about the types of information that may be exchanged between or about:

- customers and clients
- wholesalers and retailers
- distributors
- suppliers (of services or goods)
- manufacturers
- managers and employees
- products
- briefs
- services
- goods.

Customers and clients

Customers are the organisations or people who buy a product or service from a company. The word client is used instead of customer when there is a professional relationship between the organisation and a customer. For example, an accountant has a professional relationship with the owner of a small business whose end of year accounts they are drawing up – the small business owner is a client of the accountant. Accountants, solicitors, insurance brokers, financial advisers, agents, etc. usually refer to their customers as clients.

Let us look at the exchange of information that takes place when you buy a new car. If your mind is set on the make and model then you need to find suitable garages near to your home. Adverts in the local newspapers or *Yellow Pages* could tell you the location. You could also use the Internet to search the manufacturer's **Web site** for a list of dealers. If you had a vehicle to part-exchange you would need to visit each garage to see how much they will give you for it as well as the price they would charge for the new car. On visiting the garage they will want some details about you and the car you are part-exchanging along with the details of the car you are looking to buy. The details normally required are shown in the table on p. 79.

Once the salesperson has collected this information, there will be some information flows.

<table>
<tr><td>Your personal details</td><td>Your car</td><td>The car you are looking to buy</td></tr>
<tr><td>Name
Address
Telephone numbers
  (Work, homo, mobile, etc.)</td><td>Make
Model
Registration number
Year of registration
Engine size
Mieage
Extras (CD player, leather
  upholstery, etc.)
Number of owners</td><td>Model
Engine size
Extras</td></tr>
</table>

- The detail about your part-exchange vehicle may be passed to a mechanic who will check the car out.
- The salesperson will check with the computer system if a car that meets your description is available and what price they can get it for (this information will come direct from the manufacturer). They will then add on a suitable amount of profit to arrive at a price they are happy to sell the car for.
- The mechanic reports back about the condition of the part-exchange vehicle and the salesperson consults the computer system for a guide as to the market price of the car based on that information.
- He or she tells you how much they'll give you for your car and how much they want for a new one, and there will probably be some negotiation on price. You may want to drive a hard bargain but the salesperson might need to contact their manager before agreeing to the deal you want.
- In most cases the customer will borrow some or all of the money to pay for the car. If the dealer is arranging the finance, then the customer will be asked to supply information that will given to the bank or finance company. The information required is quite detailed in order to ensure that the person is who they say they are and that they can afford the payments. It will take some time for the company to do their checks and they will let the dealer know if there are any problems.
- The car dealer will need to perform some checks on the car put in for part-exchange to make sure that it is not stolen or that there are no outstanding loans on the car (i.e. to check that the person trading the car in actually owns it).
- The dealer will order the new car from the manufacturer and obtain a firm delivery date. Any special accessories will be fitted by the manufacturer or the dealer.

Wholesalers and retailers

Wholesalers are companies who deal in the bulk purchase of a limited type of product for re-sale to their customers who are traders or small retailers. The wholesaler uses their bulk purchasing power to get the goods cheaply – this allows them to add their profit and still sell to traders and retailers, who sell the goods on to their customers (often members of the public).

Fruit and vegetable merchants are wholesalers who buy products direct from the growers and sell them on in smaller quantities to the retailers (the people with the fruit and vegetable shops.

If an organisation is large, it will have the bulk purchasing power on its own and will be able to deal directly with the manufacturers, thus cutting out the need to buy from a wholesaler. Small retailers such as corner shops cannot buy in bulk and will need to buy from a cash and carry warehouse (i.e. a wholesaler). They will usually pay for the goods as they collect them.

Retailers all have some mechanism for determining what goods they need to buy (usually checking the stock on the shelves and in the stock room to see what is running low). More sophisticated systems will use details taken directly from the goods that are sold (e.g. using the bar code) and automatically deduct the numbers sold from stock. Although this system may seem ideal, it might not show the actual stock position because goods could have been returned or removed from stock because they are damaged. Stock checks are still needed, although not as frequently as with a manual system.

Distributors

The manufacturers of larger items (cars, motorcycles, boats, etc.) do not often sell direct to the customer. Instead they use **distributors**, who act as intermediaries between themselves and the customers. They do not allow just anyone to become a distributor and usually control the way in which their distributors operate – for example, the staff employed by a distributor are usually trained by the manufacturer and are given the support to sell and maintain their products. Because of this close relationship, the distributor and the manufacturer will often use the same ICT systems so that they can communicate easily with each other and reduce the administrative burden. This free and efficient flow of information allows the manufacturer to find out about sales of their products and the reasons the customer chose the product in preference to that of a competitor. It also allows the distributor access to all the technical details about the product and helps them track an individual order from manufacture to delivery.

Ford, the car manufacturer, have distributors (or dealerships as they are sometimes called) who sell, repair, service and supply accessories for Ford cars. Because distributors are given a specific location in which to operate, they can often help each other so a customer can walk into any showroom and pick a car of the make, colour and with extras they require from another showroom, knowing that the car will be delivered to their local showroom.

Briefs

Briefs are summaries as to what is required. Briefs allow architects, engineers, designers, etc. free range to develop a design or product using their skills to the full. Detailed instructions are not required because it is assumed that the person contracted to do the job will perform it to the best of their ability.

Services

Many organisations provide a service rather than a product – generally a task that needs to be done. Services used include catering, security, maintenance and payrolling facilities. Service organisations employ large numbers of people and, although they do not hold products or stock and have no manufacturing facility to worry about, they still need to hold information about their contracts and their staff. Some service organisations are set up to deal with the ICT requirements of another company. This means that a company can own the computing facilities in another business.

2.2 Functions within organisations

The functions within an organisation relate to broad areas of activity that must be carried out to achieve the company's objectives. A study of any organisation will reveal that it is divided into certain functions such as finance, marketing, personnel, operations, purchasing, design and sales.

Functions can be classed as internal or external.

Internal functions

Internal functions cover the distinct areas into which an organisation can be divided. These functions obviously vary depending on the type of organisation, and some functions appear only in certain organisations.

Accounts or finance

The finance department (sometimes called the accounts department) is responsible for all the company's financial records. This involves logging all money coming into and going out of a business, either manually or by computer. The finance function also covers payment of wages and pension contributions, collection of tax and National Insurance contributions and making sure that legal requirements for collection are adhered to. Finance departments also set up departmental budgets and make sure that managers do not overspend. The accounts or finance department would typically be responsible for:

- dealing with payments made by customers and payments made to suppliers
- dealing with overdue customer accounts
- sorting out account queries
- preparing end of year accounts as required by law
- preparing budgets and forecasts
- producing sales contribution analyses
- evaluating investment opportunities
- arranging loans, overdrafts, etc.
- appraising capital projects (e.g. investigating the purchase of new buildings or equipment)
- dealing with takeovers/acquisitions
- dealing with the returns that need to be made to government departments (the Inland Revenue for tax, Department of Social Security for National Insurance payments and Customs and Excise for VAT)
- dealing with the payroll.

Activity 6

Many ICT systems are set up to keep track of money coming into and going out of organisations. If you eventually become involved in ICT in a business context, you will need to understand some of the accounting terms and procedures for keeping track of money.

Using suitable books on accountancy/business studies from your library, explain what is meant by the following accountancy terms:

- general ledger
- sales ledger
- purchase ledger
- profit and loss account
- balance sheet
- end-of-year-accounts.

Marketing

The marketing function is the process of identifying, anticipating and satisfying customer requirements profitably. Marketing is as relevant to services as it is to goods – for instance, your school or college will need to do a certain amount of marketing to attract students. Large companies have a separate marketing function but in smaller ones its role tends to merge with the sales function.

The marketing department would typically be involved in:

- conducting market surveys and evaluating the results
- producing reports based on sales information (e.g. reports by product or geographical region)
- preparation of market share reports
- evaluation of brand performance against targets
- conducting advertising campaigns
- organising special promotions
- organising the production of special display or promotional material
- evaluating a competitor's products.

Human resources/personnel

Only large organisations have a separate personnel department; smaller ones tend to leave this function to individual managers rather than employ specially qualified personnel staff. The main tasks undertaken within the personnel function are the recruitment of staff, dealing with industrial relations and staff training.

Staff in the human resources department would typically be involved in:

- liaising with department managers regarding staffing requirements
- wording advertisements for jobs and placing them in relevant newspapers
- sending out application forms following requests
- processing applications for employment
- dealing with reference requests
- interviewing applicants and recording the results
- keeping accurate and up-to-date records of personal details (surname, address, telephone numbers, e-mail address, etc.)
- dealing with staff lateness, absence and sickness
- dealing with disciplinary matters
- preparing induction material for new employees
- producing the staff newsletter
- keeping track of holiday rotas
- dealing with trades union matters.

Operations

Operations may be thought of as the production function in a manufacturing company. Some organisations do not have an end product, but provide a service. In this case the day-to-day task of providing the service is akin to the 'production' part of the business. Either way, this part is sometimes referred to as the operations function. Businesses that would have large operations departments include building societies and travel agents.

Purchasing

Manufacturing companies buy raw materials and components, which are then processed in some way to produce finished goods. The purchasing function is a crucial one in many manufacturing companies, because lack of one type of component could hold up a whole production line. To remain competitive, it is important for all organisations to ensure that they are buying the highest possible quality components at the cheapest possible price and that the timescale for delivery is suitable.

Whether or not they are involved in manufacturing, all organisations have a purchasing function – even if it is just essential services (gas, electricity, telephone), car rental, office equipment, stationery or office space that they are buying. A system must be in place for dealing with the purchasing function.

Basically, the purchasing function can be broken down into the following tasks:

- finding a suitable supplier
- establishing payment and discounts and negotiating delivery schedules/dates
- placing an order
- taking delivery of the order or chasing up delivery
- checking that the goods received are the same as those ordered
- paying the invoice.

Design

Not all organisations have a design function. Manufacturers and some service organisations (architects, engineers, etc.) do use an element of design in their work. IT is used in design because of the ease with which plans and drawings can be changed. **Computer-aided design** (**CAD**) software is used to produce three-dimensional views from two-dimensional plans.

Sales

The sales function of a business is very important in an organisation, because it secures income to drive the rest of the business. The sales function aims to persuade customers to buy products or services, and this involves coordinating travelling representatives, telephone sales, preparing and sending mailshots. It also involves interaction with the marketing function.

The sales department would typically be involved in:

- taking customer orders
- dealing with customer complaints
- monitoring sales cost information and maintaining a budget
- working out the profitability of each product
- producing sales targets for a product
- dealing with the expenses of sales staff
- setting sales targets for individual members of staff
- checking the availability of credit for customers
- visiting customers
- preparation of catalogues and price lists
- dealing with customer queries
- setting up customer accounts
- chasing up orders required urgently
- dealing with special requests.

Distribution

Many companies supply products and need a mechanism for getting the goods to the customer – the goods arranged on shelves in a warehouse will need to be picked according to the orders, packaged and dispatched. In most cases the picking will be done manually but there are robot systems that pick the customer's orders automatically.

The distribution department would typically be involved in:

- picking out items to make up customers' orders
- dealing with returns from customers (e.g. wrong goods ordered, goods damaged, etc.)
- making orders up and packaging them
- arranging for delivery of orders
- dealing with delivery queries
- chasing up lost packages
- controlling stock levels
- arranging stock on shelves/bays for easy location
- accepting deliveries from suppliers.

External functions

There are also several external functions/entities that influence the administration of an organisation.

Supplier/customer functions

Sometimes the customers or suppliers determine the administration procedures adopted by an organisation. For example, the staff of suppliers of parts for military aircraft may have to sign the Official Secrets Act as a condition of their contract.

Legal and statutory bodies

Various legal and statutory bodies influence the way in which administration procedures are performed.

- Health and safety at work regulations govern the working environment and some working patterns.
- There are income tax, VAT and National Insurance requirements for ensuring that correct deductions are made and passed on to the Inland Revenue and Customs and Excise.
- In addition, there are certain obligations that apply to limited companies under company law.

Activity 7

Find out what the following organisations do, and say how they can influence the running of a business:

- The Health and Safety Executive
- HM Customs and Excise
- Trading Standards Department
- the environmental control department of a local council.

Information flows

A manufacturing company typically has the following functional areas:

- sales and marketing
- purchasing
- production, stores and distribution
- accounts and finance
- research and development
- personnel.

Information needs to flow within and between these functional areas and the efficiency of these flows is a measure of the efficiency of the whole organisation.

The main concern of the sales and marketing department is to sell the products to the customers. To do this the company has to make sure that the goods they are making satisfy a market need and are priced competitively.

In most cases the customer will place an order and, before it is accepted, the person taking the order will check with the stores that the order can be satisfied. Once the order is accepted a confirmation is sent to the customer and the accounts department is sent a copy of the order so that an invoice can be raised and sent to the customer. They may also at the same time check on the credit-worthiness of the customer (if they require credit) or determine whether an existing customer has satisfactorily settled their previous account. At the same time, the systems in place for dealing with sales will generate information as the transactions are made and that can be used by management for comparing actual sales compared to targeted sales, etc. Any problems that crop can also be dealt with on an *ad-hoc* basis. For example, a large customer may place a large order before an outstanding invoice is paid, and management will need to decide whether to extend their credit.

To ensure that the components are in stock ready to assemble to make the finished article, the production department needs to send information about the orders and current stock situation to the purchasing area so that goods and materials can be replenished. When stocks fall below the minimum stock level, an order is sent to the suppliers and a copy of the order is sent to the accounts department so that they can expect to pay their suppliers when they receive an invoice. Again, management must be kept informed of problems.

The production/stores/distribution function has the task of making sure that the components are always available to keep the production department going. Once the goods are made they need to be delivered to the customer.

Constructing information flow diagrams

Information flow diagrams show how information flows between the functions within an organisation and between the internal and external functions.

Consider a mail-order company, which buys goods from a series of suppliers and then advertises and sells them through a catalogue. For the sake of simplicity we will assume the following list of functions:

- sales
- supplier/customer requirements
- purchasing
- accounts
- dispatch.

When constructing information flow diagrams it is useful to adopt the following steps.

1 Consider which functional areas you are going to look at. If you are investigating a large organisation you may at find it easier at first to limit the diagram to a few functional areas, although it can be quite difficult to isolate areas in this way. Look at the internal functions first and then see if any of them communicate with external functions in any way. In our example, both suppliers and customers are external functions, the rest being internal functions.

2 Add lines showing the information flow. Identify and add the type of information that passes between functions and the form that this information takes (documents, verbal or electronic).

3 Don't worry if the design is not right first time. One problem you may encounter after drawing several of these diagrams is that the flow lines start to cross each other and the diagram ends up looking like a plate of spaghetti! If so, redraw the diagram, moving the boxes to prevent overlaps.

4 Try not to include too much detail at first. Draw a simple diagram and then redraw it, adding in more detail. Remember that the whole point of an information flow diagram is to be able to identify all the functions and the way information flows between them.

Information flows to and from the purchasing department

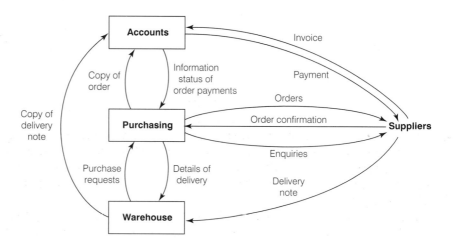

Figure 2.3 The information flows to and from the purchasing department

Information flow diagrams are used to illustrate the flows of information between internal and external functions. Using the diagram in Figure 2.3, we can describe the information flows.

- The warehouse will inform the purchasing department when any items are low in stock so that an order can be raised and sent to the suppliers.
- The purchasing department will send a copy of the order to the accounts department to let them know that an invoice requesting payment will arrive shortly.
- On receiving the order, the suppliers will send back confirmation of the order to inform the purchasing department that the goods are in stock and give a date for the delivery of the goods.
- The goods that arrive at the warehouse will be accompanied by a delivery note listing what has been delivered. This *should* match the order but it might not – some of the items might be sent in a different delivery, for instance.
- A copy of the delivery note is sent to the accounts office. The accounts staff will compare the delivery note with the original order to see if all the goods have been sent. If they have, as soon as the invoice arrives from the supplier, they will pay for the order.

Why might goods ordered not be sent all together?

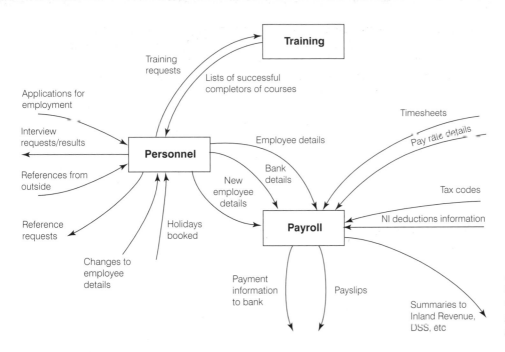

Figure 2.4 Information flow diagram showing the flows of information from the personnel, training and payroll functions

Activity 8

a Figure 2.4 shows the information flows between the personnel, training and payroll functions in an organisation. Explain the diagram in terms of the functions and the information flows between them. Ask yourself whether someone could draw the diagram just using your description, without any further information.

b Draw an information flow diagram for the sales, accounts and dispatch functions. Here is some information to help you construct the diagram:

Customer orders are processed by the sales department. As soon as the order is received the current stock and delivery position are determined from the dispatch/warehouse department. If the order can be satisfied, confirmation of the order is sent to the customer, and a copy of the order is sent to the accounts department, who raise an invoice and send it to the customer's accounts department. Another copy of the sales order is sent to the dispatch department, where it is used to make up the customer's order. The address on the order tells the dispatch department where to send the goods. The goods are picked, packed and sent out, and the delivery note accompanying the package informs the customer what the package contains. If the customer does not pay the amount on the invoice within 14 days, the accounts department sends the customer an overdue letter. Information is exchanged between the dispatch and sales departments to answer customer queries such as when an order is likely to arrive or whether a certain item is in stock.

c Investigate the functions and information flows for two of the types of organisation described below (choose the two with which you are most familiar). Some questions are included to get you to think about the system. You don't have to answer them, but will need to bear the answers in mind when explaining and illustrating the system.

▶▶

- *Your college/school library.* When a person joins the library, what documentation is produced? How is it stored? When the library buys a new book, what information about the book is needed? How is it stored? When a member of the library borrows a book, how are the details recorded? What needs to be done when a book is returned? What happens if the book is overdue by several weeks?
- *A college enrolment system.* You should probably attempt this one only if you are at a college. Think about what happened when you first came to college. What forms did you fill in? Did you have an interview? Did you receive an acceptance letter? Did you have to turn up at a certain time on a certain day to enrol or was it done by your course tutor? What information did you need to supply in order to enrol? What forms were filled in during the enrolment process? What does the college do with them?
- *A video library.* What information needs to be recorded when a new video is bought? Where is it held? What information (e.g. proof of identity, name, etc.) does a person need to join the library? In what form are these details stored and where? What information needs to be recorded when a video is hired out? In what form are these details kept? What information is needed when letters are sent out to people who have not returned their videos?
- *A mail-order company.* What forms of information (order forms, phone orders, Internet orders, etc.) are used when placing an order? What payment methods are used? What information needs to be included in each order? What methods can a customer use to pay for goods? What information is needed so that the goods may be picked from stock? What documents are sent with the goods (if applicable)?

2.3 Information and its uses

Suppose an organisation keeps its customer details in three different files, in the sales, accounts and distribution functions. If a customer changes their address the staff in each of the functional areas will need to be told so that their files can be changed. This is very wasteful and there is the danger that some of the files will not be changed so a customer could have different addresses depending on which file you look at. Although it sounds obvious to use just one central store of data about the customers that can be used by all the applications, many companies still keep separate databases in their different departments. To keep a centralised store of data, it is necessary to make use of a network and hold the data on a file server.

The main reasons why companies do not use a central store of data, called a data warehouse, is because of the development time and cost. To use a central store of data may require alteration of all the systems in the organisation and the company might not have the time to do this.

Methods used for communicating information

Information comes into an organisation and flows around it in a variety of forms. In this section we will be looking at the different forms that information can take.

Verbal information

Verbal information includes **face-to-face** and telephone conversations either in real time or recorded using an answering machine; it also includes using the voice mail facility that is available on newer PCs.

Although verbal information is usually obtained quickly it often needs to be backed up by some form of documentation. For example, if you want to book a last-minute holiday, you could look at Teletext on the television, select one of the many holiday companies offering last-minute deals and ring them up to check availability. You can then book, quoting your credit card details over the phone. The company you choose will still send you a receipt and, if there is time, tickets. The strength of this system is the fact that it is fast, but misunderstandings can occur and, unless the telephone conversations are recorded (which they frequently are), there is little evidence to back up either side in a dispute. This type of system is very popular and is used by many catalogue, direct insurance, direct holiday and direct mortgage/banking companies.

Face-to-face meetings

Face-to-face meetings are an important way of communicating information. If you wanted to tell your sales staff about a new product that was being developed, the easiest way would be to conduct a meeting. Face-to-face meetings are useful for developing new ideas because each person is able to think things up. Many face-to-face meetings are conducted on the premises but where the organisation is geographically dispersed, such as in a large multiple retailer, it is expensive to bring all the employees to a central point.

Off-site face-to-face meetings are very popular with business people because they allow them to conduct meetings in more convivial surroundings – and are a way of getting out of the office and doing some travelling in the process. A lot of business is conducted over lunch, dinner, in hotel rooms or on the golf course.

Videoconferencing

Pressures of time and money mean that people need to be able to see and talk to each other as well as to pass documents around without leaving the office. Videoconferencing provides an ICT answer – it allows two or more people on different sites to hold a conference over a computer network, transmitting sound and video data. Each person has a microphone and a camera attached to their computer, which allows everyone involved to see, hear and talk to the other participants, no matter where in the world they are.

e-mail

Internet e-mail

Using the Internet you can send e-mail and file attachments to anyone who has an e-mail address.

LAN e-mail

LAN e-mail is used to provide an e-mail facility within the organisation. This allows anyone who is logged onto the local area network to send and receive e-mail. It is an ideal method for sending letters, memos and brief messages within the company.

Documentary information

Under this heading we will be talking *only* about paper documents; documents passed in electronic form will be dealt with later. Most companies try to reduce the amount of paperwork to a minimum but it is very difficult to eliminate it

Explain the difference between an intranet and the Internet.

completely, as customers, suppliers and other outside agencies might have no facilities for dealing with **electronic mail** (e-mail), EDI (electronic data interchange), etc.

Paper documents (letters, brochures, price lists, memoranda, contracts, invoices, etc.) are used throughout society as the main medium for the transmission of business information among organisations and between organisations and the public (Figure 2.5).

Figure 2.5 Just some of the many paper documents a traditional organisation could use

Fax

A conventional **fax** (an abbreviation of 'facsimile copy') transmission is sent from a fax machine at one end of a line to another at the other end. The sending machine scans a paper document placed in it (containing text, graphics or both) into digital form. The modem in the fax converts the data to analogue form, and sends it along the telephone line to the receiver. The receiving fax uses its modem to convert the analogue data back into digital form, recreates the image and prints it out.

It is clear that all these functions could be carried out perfectly well by a typical small computer and this is the purpose of the fax modem. A fax modem enables one computer to send material generated on the computer directly to another, eliminating the need to scan a printed copy into a fax machine. If a fax is sent to the computer on which you are working, the fax modem automatically takes charge, answering the phone and storing on your hard drive the data from the fax transmission. You can view the fax using special software supplied with the fax modem; a hard copy will be made only if you need one. This system saves both time and paper.

All fax files are sent in graphic format, so you cannot simply transfer them to your word-processing software – you need optical character recognition (OCR) software to recognise the shape of each character and convert it into the ASCII code that can be imported into any word-processing package.

Electronic information

Information can be passed from one place to another in electronic form. For instance, a stock list can be sent on disk to all the branches of a shop. A more streamlined way would be to transfer the data over the telephone lines electronically, using computers and modems. Many organisations now network their computers so that there is no need to transfer data via floppy disks. Nearly all networks have an e-mail facility, and this can be used to send documents in electronic form around a company. Widespread use of e-mail for external mail is restricted because not everyone yet has e-mail. **Electronic data interchange (EDI)** allows companies to place orders and make payments, without the need for any paperwork.

Many organisations make use of both intranets and the Internet, which gives them the opportunity to complete most of their day-to-day transactions electronically. Suppose a large retailer wanted to streamline the system for ordering goods from their suppliers. One way would be to send orders automatically and in electronic format straight to their suppliers but the ICT system they use will need to be compatible with the ones used by all their suppliers. The use of **e-commerce** in this way will eliminate the need for paperwork, allow automatic ordering of stock and pass the ordering data along communication lines directly into the supplier's computer, reducing the costs at the supplier's end in processing these orders. There needs to be a close relationship between the retailer and the supplier for the system to work but the lower costs of dealing electronically rather than in traditional ways brings clear advantages for both parties. One advantage to the retailer is that the administration of orders is reduced, leaving more time to make strategic buying decisions. From the supplier's point of view there are also advantages – for example, they can see how well their products are selling and can adjust their production in order to meet demand.

All the savings made by the system can be used to lower the prices of goods on the shelves. This means increased sales – for both the retailer and their supplier – and increased profits for them both.

Telephone

The telephone is a good method of communicating with people, especially as now many people have both a land line (ordinary telephone) and a mobile telephone. If, however, the person you are trying to contact is continually on their phone, it can be quite difficult to get in touch. You might have to leave messages on an answerphone – many phone companies have special answerphone services where you can leave a message. Mobile phones often allow you to send text messages.

Telephones are a good way for an organisation to keep in touch with its customers. For example if you buy a new car, the garage may ring you a few weeks after you have received it to see how happy you are with it.

Many companies use call centres where customers can telephone to order goods, take out insurance, deal with problems relating to gas, electricity, water or the telephone supply. Although these can be quite frustrating for the customer to use, he or she uses the keys on their own phone to route the call to the most appropriate section of the organisation and by doing this is actually acting as the telephone operator because they are effectively directing the call.

The main advantage in using the telephone is that it is quick and people can actually talk to someone on the other end and seek advice if necessary. This 'personal touch' is lost when you place orders for goods using the Internet.

Internal/external post

It would be a mistake to think that all organisations make use of all the latest ICT developments. There are still many organisations that do not have e-mail facilities, instead using paper documents such as memos or letters to communicate internally and externally. Unless the organisation is very small, these paper-based documents need to be passed around the different departments so there needs to be a way of

Sending the same message to everyone in an organisation, whether they need the information or not, is not a good thing. Explain briefly why not.

doing this internally. Instead of the originator of a document having to get up and deliver it by hand most organisations have an internal mail system. However, because there is always a danger that a document could be lost, people frequently photocopy them, which not only adds to the expense but also increases the bulk of paper that needs to be stored. In an internal mail system, a person collects the mail from trays in each department and delivers it to the other departments. There is a delay (just like with the ordinary postal system) so this system is not as immediate as e-mail, but in organisations that do not use networks it usually works well.

The external post system you are all familiar with. It generally works well – everyone understands it and has faith in it, and we are all able to receive post through our letter boxes. Until every home has e-mail the traditional postal system, called 'snail mail' by users of e-mail, will still be around. Eventually, more and more organisations will make use of e-mail and the use of external and internal postal systems for sending routine correspondence will decline.

Centralised database systems

A **centralised database** is a computer system where all a company's data is held in one place and the computers in the company are networked. This central database can be used by any of the separate applications run on the computer and the data can be obtained from any terminal in any functional area of the business, provided that the particular terminal has permission to access the data. As the data is held centrally, it needs to be entered only once, which avoids the problem of data duplication. Centralised database systems are used by most large companies and are a very efficient way of storing company data.

Activity 9

a Many companies choose to communicate using a variety of different methods, including post, e-mail, telephone and face-to-face meetings. A power supply company wishes to contact all the people in a certain area to discuss a special deal for the gas and electricity supply. They want to be able to convince the existing customers of other supply companies to buy their power off them instead.

Compare and contrast the relative advantages and disadvantages of using these communication methods for making contact with all these people.

b Which form of information (verbal, documentary, electronic) best describes each of the following?

- A telephone conversation
- A pick list to be given to warehouse staff
- A memo sent using e-mail
- Payment made for goods using a debit card
- An answerphone message
- A CV sent to a potential employer using e-mail
- Invitations to a party sent by post
- A voice mail message
- A discussion at a meeting
- Minutes of a meeting written in shorthand
- A requisition slip
- The winning lottery numbers obtained from the TV using Teletext.

Commercial transactions (e-commerce)

Most companies have their own Web site and use it to advertise their products and services. Some Web sites are simply interesting public relations exercises; others are used to conduct e-commerce and make on-line orders. The Internet allows companies to do business on-line, which is a lot cheaper than doing the business in a more traditional manner. When you type in an order for goods or services using the Internet you are actually performing the data input into a company's system for them. In the past they would had to pay someone within their organisation to do this, so they can save money on wages. It does not matter where in the world the organisation is situated, so they do not need to be near the customer. As long as the customer is happy with the price and the delivery of the goods he or she will shop again and recommend the service to a friend. Sometimes the savings made by companies engaged in e-commerce are passed on to the customers in the form of cheaper goods and services. If you know what you want, the Internet can be the best place to look. Many companies have been started with the aim of conducting e-commerce over the Internet and some have gone on to become the largest companies in their field. For example, Amazon is the largest bookseller in the world yet it has no bookshops – it does all its business over the Internet. Amazon does not own any shops: all it has are distribution centres in different parts of the world. As there are fewer staff to employ and no bookshops to pay for, Amazon is able to offer books at large discounts and deliver them straight to the customer's door.

Even small businesses can have a world-wide presence 24 hours a day, 7 days a week using the Internet. To have this kind of presence in the 1980s would have cost around £70 000 per month; it is now about £3500 per month.

The quality of service that Internet customers expect is rising all the time so ICT systems need to support these expectations. Systems need to provide immediate, accurate and secure information exchange, automatic delivery of goods and accurate tracking of goods and parcels until they arrive at the customer's premises.

E-commerce is similar to the mail-order business because the customer is not normally in front of the supplier. When an order is placed over the telephone, the signature on the back of the credit card cannot be confirmed, but with e-commerce you can track where the customer is 'calling' from to help reduce fraud. Another advantage of e-commerce over mail order is that suppliers are able to give the customer more information about their purchase, further information about the specifications of the product and provide a link to the delivery company to find out exactly when the goods will be delivered.

Most e-commerce sites have four sections involving information being passed to and from the customer:

1 Entry and search
2 Results
3 Invoice
4 Thank You.

The customer enters the site and selects what they want to see or performs a search. The 'Results' are then shown. These will usually be information about a product or products. The customer is given the option to collect the products they want to buy and add them to their electronic shopping basket or trolley. Once they have selected all the goods they want they can proceed to the checkout, where payment details are recorded and an invoice is produced. At this stage the customer will be asked to supply certain payment information (e.g. name and address, telephone numbers, e-mail number, credit card number, type of card, expiry date). Once the customer has completed all the information for the invoice, he or she is thanked for their custom.

Web-based marketing or advertising

Many companies do not yet do business over the Internet, preferring to use it for advertising and marketing purposes only. These organisations are happy to use the Internet to make the public aware of their products and services. For example, many manufacturers, such as car manufacturers, do not sell directly to the public. It is obviously important that the public know about the quality of their products.

In order to make their Web sites more interesting and to encourage potential customers to leave their details (names, addresses, telephone numbers, etc.) many companies have free competitions where you have to answer questions about information contained on their site.

When surfing the Web, have you ever noticed those adverts for products that look like thin strips at the top or bottom of the screen? These are called banner adverts and clicking on them takes you to the advertising company's Web site.

e-commerce and Tesco
In some areas of the country, Tesco (the large retailer) is using e-commerce, allowing customers to shop from home (Figure 2.6). Tesco packs and dispatches a customer's on-line orders from the nearest supermarket rather than fulfilling the orders from large warehouses like other e-commerce companies.

Figure 2.6 The opening screen of Tesco direct, Tesco's Internet shopping service

Toyota: reaping the benefits of banner advertising on the Internet
Toyota, the car manufacturer, put lots of banner adverts about their cars all over the Internet. Over a 12-month period around 152 000 Web users typed in their names and addresses to request a brochure about a car. Toyota then used these names and addresses to see how many of these brochure requests actually resulted in a purchase of one of their cars by comparing the names and addresses collected on the Web site with the details obtained when customers bought a car at one of their dealers. The management at Toyota were surprised at how many of these brochure enquiries actually resulted in sales: from the 152 000 brochure requests they sold 7329 cars. This meant that about 5% resulted in sales. Because of the success of this advertising campaign, the Internet has now become the main source of sales leads for Toyota.

Activity 10

a E-commerce is very popular and it is now possible to buy almost anything over the Internet. All you need to order goods or services using the Internet is a computer/TV/phone connected to the Internet and a credit card for making the payment. Here are some e-commerce sites for you to look at.

- www.cd-wow.com (a company that sells chart CDs)
- www.easyjet.com (the low-cost airline).

Investigate each one of the sites listed above and for each site find out what information you are asked to supply in order to complete a transaction. You can see the forms that you have to fill in to complete a transaction (e.g. book a flight, order products, etc.).

For each site, produce a list of the information you are asked to enter.

b Managers in organisations will often need to send the same information to certain groups of staff. As these groups of people will frequently need to be sent different e-mails, explain in general terms how the manager can accomplish this.

How organisations use key information in their systems

Most large organisations use key information similar to that in other organisations. For example, information requirements for the systems outlined below are generally similar.

Personnel and training

The personnel department will hold the personal details of an organisation's employees – employee number, name, address, telephone numbers, date of birth, next of kin, educational qualifications, previous employment, position within company, etc. The personnel records will also contain details about sickness, other absences, holiday records and any disciplinary actions.

Training records will also need to be kept to record the training courses and instructions the employee has received from the organisation. This is particularly important for assessing the training requirements for each employee and making sure that the skills of each employee are suitably matched to their job.

The personnel system will usually link to the training and payroll records using the employee number, which is a unique number that is given to each employee who joins the organisation. Keeping the personal details of employees in only one place ensures that data can be kept secure. If the personal details of the employee change, then only the details in the personnel file need to be altered, rather than in the training and payroll files as well. The staff involved in running the payroll would be able to know a person's salary only if they can put a name to the employee number so the data is kept confidential.

Accounts and finance

Accounts and finance will monitor and control the money coming into and going out of the organisation, thus ensuring its financial success. All the other departments will have links back to this important department.

Payroll processing

All employees expect to be paid the correct amount and on time, so payroll processing is crucial to any organisation. Employees can be paid in different ways: salaried employees have their annual salary divided by twelve to be paid monthly, other staff have an hourly rate of pay and are paid weekly according to the number of hours they have worked (these hours may be at a basic rate or at an enhanced overtime rate for working weekends, bank holidays, etc.) In addition to their basic pay, some employees receive bonuses, commission and so on.

Not all employees receive the same pay, even if they are doing identical jobs. Some may be given more pay because they have worked in the organisation longer. People's pay may differ because the amounts deducted vary. For example the amount of tax a person pays depends on their tax code, which in turn is determined by their individual circumstances. Each employee is supplied with a detailed breakdown of the amount they are paid, including deductions, in the form of a payslip.

In addition to the money it pays to its employees, an organisation has to send money to many other organisations such as the Inland Revenue for income tax and National Insurance contributions, pension agencies for the payment of pension contributions or trades unions for the payment of subscriptions.

Research, design and development

Research

Many organisations perform research as part of their operations. Pharmaceuticals companies spend huge amounts of time, effort and money researching new drugs. Computer models are built to understand how these drugs can fight disease and large amounts of data concerning this research are stored on computers. The results of clinical trials of drugs are also recorded on computer systems.

- Manufacturing companies are continually looking for new products or ways of improving their existing products.
- Computer chip manufacturers are continually researching new ways of making chips faster and more powerful.
- Manufacturers of software are always developing new versions of their software that are easier to use and have more features.

Design

Because changes in design are so easy to make using a computer, more and more aspects of designing are being performed on computer. CAD packages are used to manipulate the designs of, for example, buildings, cars, engineering components, structures (bridges, tunnels, etc.) and computer chips. Designs can be made up from previously stored shapes and drawings in a fraction of the time it would take to do a manual drawing. In addition, alterations can be made rapidly without the need for redrawing.

Development

Once a product has been researched and designed it needs to be developed. Sometimes a prototype is tested to show any problems with the design (it may be necessary to go back and alter the design). The problems involved in the mass manufacture of the product can then be looked at and ironed out.

Sales and purchase orders

Purchase orders are the orders placed by an organisation with their suppliers for goods or services. These orders frequently involve the purchase of components, raw materials, goods or services. To ensure the constant supply of essential components, a manufacturing company must make sure that it promptly pays the invoices it receives from its suppliers. It must also make sure that the goods or services supplied were actually ordered and it must check that the invoice matches the goods or services received.

Sales orders are orders coming into the organisation for the supply of goods or services from customers, who may have dealt with the organisation before or after placing their first order. Some organisations will allow the customer some credit so they do not need to send payment with the order. These customers will have a certain period to settle their bill or be given an account where interest accrues on the outstanding balance.

Stock control

Most organisations require some stock to be kept, even if only supplies of office stationery, and it is important to keep track of the amount of stock so that items can be re-ordered when stocks are getting low. Many companies are involved in the manufacture of goods from raw materials or components and the lack of a single item can stop a whole production line and cause a company to lose large amounts of money. Accurate stock-taking must therefore be an integral part of a business. One way to avoid this situation is for a company to buy so much stock that it is unlikely ever to run out. The problem with this approach is that stock costs money, so money is tied up in stock rather than used more profitably. Also, the more stock that is held the more difficult it is to find a particular item, and the greater the amount of resources needed to deal with it (storage space and staff).

A computerised stock control system maintains a balance between keeping the amount of stock held at any one time to a minimum and making sure that the demand for stock can be satisfied. Food retailers such as Tesco need very efficient stock control systems because many of the items they sell have a shelf-life of only a couple of days, and if they buy too much they may have to sell it at a reduced price or throw it away. The case study later in this unit looks at how Tesco deals with stock control.

Supermarkets are masters of stock control and use a system called '**just in time**', which requires the goods to be delivered just in time to satisfy customer demand. This means that large stocks of products do not need to be kept because fresh stock is delivered frequently.

Stock control systems should provide accurate and up-to-date information on the following:

- number of items in stock
- prices
- number re-ordered, date ordered, agreed delivery date from supplier
- minimum stock level
- re-order quantity.

In some cases the system will give a warning if there is too much or too little stock and will generate management information – such as outstanding orders from the suppliers that need to be chased up, particularly fast moving or slow moving items or how closely the system managed to meet customer demands.

What is meant by 'just in time' stock control?

Order processing

Any organisation involved in the supply of goods or services must have an administrative procedure to deal with customer orders. **Order processing** systems

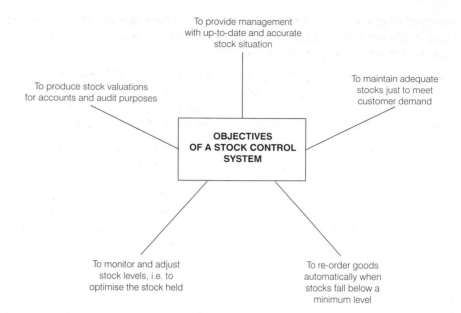

Figure 2.7 The main objectives of a stock control system

record the details of each order as it comes in, check to see if the goods are in stock, check to see if the customer owes the organisation any money or if the value of the order will bring the customer's account above their credit limit, arrange for the order to be made up and delivered and for the customer to be invoiced for the price of the goods. Order processing is dealt with in more detail in the Tesco and HP Bulmer case studies later in this unit.

Because of the security and cost problems of dealing with the transfer of large amounts of money, many organisations use electronic funds transfer (EFT) using **BACS** to transfer funds between the organisation's bank account and their employees' accounts. They can also use the system to pay income tax and National Insurance contributions that have been deducted from employees' pay.

Activity 11

Research, using suitable material, the BACS system for transferring money from one bank account to another. Explain briefly how it works.

Case study

WH Smith Business Supplies

WH Smith is a familiar name, with shops on nearly every high street in Britain. There is, however, another important part of the business, called W H Smith Business Supplies, which is the UK's largest direct supplier of business stationery. The whole business is centred around giving customers a next-day delivery service and to do this the company has invested large amounts of money in very sophisticated computers and computer-controlled equipment. The business operates from a huge warehouse using fully automated put-away systems which automatically place stock in designated places as it arrives from suppliers. There are 12 000 pallets and 1000 m of computer-controlled carton and pallet conveying, and the whole system is the most automated in the UK.

The system at work looks like something out of a science fiction film, all the tasks being performed automatically. Thousands of boxes and pallets move along conveyor belts while automatic cranes deliver or retrieve the pallets. Robotic arms are used to remove individual boxes from the line.

The warehouse control system (WCS), linked to the ordering and stock control systems, is able to identify at any time the exact location of any item whether in store, somewhere along the picking process or waiting to be dispatched. The WCS knows the size, shape and weight of every article in stock, how many are held in stock and where each is located in the warehouse. The size, shape and weight of each item is important because, with this information, the computer can decide what size carton is needed to pack the various items making up a mixed order. It even knows how much space to allocate in the delivery vehicle and how to load the vehicle so that goods delivered first are nearest the door.

Bulk stock is stored on shelves right up to the ceiling in about 4000 storage locations. A combination of manned and unmanned cranes put away stock when it arrives and retrieve goods when needed. Some of the unmanned cranes travel at speeds of around 50 km/h in confined areas. Staff are not allowed in such areas because of the obvious danger.

The main picking operation, where individual orders are made up, is situated in a low bay area where 'full cases' of products are picked. In this area, full cases (usually of rapid turnover products such as photocopying paper, envelopes, etc.) are labelled with the customer's name and address and sent to the dispatch sorter, which unites them with the other parts of the order. Another area deals with 'split cases' where a customer has not ordered a complete case of an item. At this site, one of three different-sized cartons is automatically erected according to WCS instructions and is conveyed to one of a number of picking stations, where individual items are added to it. This part of the order then joins the 'full case' part of the order at the dispatch sorter. A picking list of goods is printed and added to the carton, along with an advice note.

The whole warehouse system is geared to next-day delivery. At present the system deals with 3000 orders per day, requiring 16 000 picks. The range of items picked is huge, from office desks to printers, pens and pencils.

Activity 12

a Why do you think WH Smith feels that it needs to have such a heavily automated system in place rather than perform all the tasks manually?

b All the goods are measured and weighed. Explain why this is so important.

c Frozen-food warehouses are often very similar to the WH Smith system, with few or no staff working in them. What are the main advantages of such a system in a frozen-food warehouse?

Case study

Britannia Airways

If you have ever taken a holiday abroad with Thomson Holidays, then you will have probably flown on a plane owned by Britannia Airways, which is a subsidiary of Thomson Holidays. It is a charter airline (as opposed to a scheduled airline), which means that it books whole aeroplanes full of passengers through a holiday company rather than individual seats on planes.

Because the cost of travelling to and from the destination is only part of the total price of the holiday, costs need to be kept low so that the company will be profitable while being able to provide a high level of customer service. Information technology is therefore used throughout Britannia to reduce costs, maximise efficiency and customer satisfaction and increase profits.

Central to Britannia's operations is the planning and management of the flight programme. This involves the air staff (pilots and cabin crew), the aircraft planning department, engineering, finance and some other areas. Before IT systems were introduced, the only way that information could be passed from one of these areas to another was by paper (memos, letters, etc.). An integrated approach was needed to enable a system in one area to communicate with the systems in other areas.

The first step in trying to develop such a system was to build a business model that charted the main business areas. Britannia's business cycle starts with the production of an outline flying schedule, which is discussed at an annual International Air Transport Association (IATA) conference at which airport take-off and landing slots are agreed. A revised schedule is then produced, from which the operations, crew, engineering, procurement (the buying of supplies and services) and finance departments of the business make their plans. The objectives of the new system were to automate all the elements of this process.

Britannia conducted a world-wide search for a software package around which it could base the core system. The integrated flight plan software was chosen out of a possible 45, and was used as the basis for choosing the hardware. The advantage of the software chosen was that it was based on open systems so it was not necessary to purchase hardware from any particular company. Britannia's computer hardware and software cost about £6.5 million, but the system soon delivered benefits in the areas of scheduling and operating the flying programme, which previously had been very labour intensive.

The best way to look at how the new system improved upon the old is by looking at the savings made. In the year the new system was introduced it contributed an extra 1–2% to the profit of the airline compared with the previous year.

The on-the-day decision-making tools are a very important part of the software. These are used for deciding what to do if aircraft have to be diverted, are delayed on take-off or affected by an air traffic control problem. The task performed by the software is to make the best decision, taking all the factors into consideration, to maximise profit and minimise customer inconvenience.

▶▶

Finance

The financial system was developed by Britannia in house. The key element in this system is a 'direct operating cost' system which is based on a large database with information about the tariffs charged at all the airports Britannia uses. This tariff information includes everything from parking an aircraft overnight to loading it with breakfasts for passengers. The system checks the invoices that arrive from the various airports against the main database and makes sure that the costs are correct. Invoice errors that might previously have gone undetected are picked up. This saves Britannia a lot of money and adds an extra 1% to the profit. EDI is also used to exchange information with the suppliers.

The new system supplies important management information and it is possible to segment different areas of the business and look at their contributions to costs and overall profit. It can also be used to determine the holiday routes that are the most profitable.

Engineering

Using the engineering system, aircraft maintenance work is carefully scheduled so that minimal time is wasted between the completion of one task and the start of the next. It is anticipated that in the future this new system could save about 1.5% on maintenance costs.

Activity 13

Read the Britannia Airways case study carefully.

a There was a reference in the case study to a system called EDI. Find out what this system is, how it works and why so many large companies now find it necessary to use it. Produce a document outlining your findings.
b Many of the benefits of the new system have been identified in the case study. Produce a list of the main benefits.
c It is very important for Britannia to know exactly how much everything costs. Why do you think this is?
d Classify the following systems as either commercial or industrial:

- cabin temperature control
- route information displayed for the benefit of passengers using the plane's navigation computer
- the aircraft's automatic pilot computer
- the small computers used by the cabin crew for recording drink and duty-free purchases
- crew rostering
- procurement (i.e. the purchasing of supplies and services)
- cabin air-conditioning system
- engineering maintenance scheduling
- air traffic control.

Case study

Tesco

Everyone is familiar with Tesco, one of Britain's top food retailers. Tesco used to have many small, high street shops but now concentrates on developing the huge superstores we see today. Each one stocks over 14 000 food lines and has a sales area of over 2400 m².

The laser scanning system (barcode reading system)

Tesco was one of the first high street stores to use a barcode reader, which is now called a laser scanner. The objectives of the scanning system were to improve the service to customers and to increase company productivity and profits.

The scanning system uses a laser beam to read the barcode on the goods. As the barcode is passed across the scanner the price and description of the goods are obtained from the computer, the sale is registered and an itemised receipt is produced (Figure 2.8).

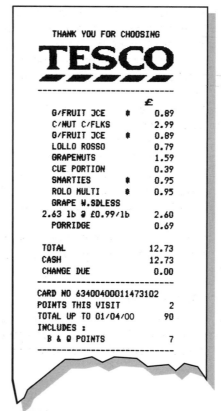

Figure 2.8 An itemised receipt

Benefits of the system to the customers

There are numerous benefits to customers.

- With the old system, prices were entered into the cash register manually. With the scanning system this is done automatically, which eliminates typing errors.

▶▶

- The scanning system is reckoned to be 15% more efficient than the manual system, so customers spend less time waiting to be served.
- Produce such as loose tomatoes are weighed at the checkout so the customers no longer have to queue twice – once at the pricing point and again at the checkout.
- Customers can have their cheques and credit card vouchers printed automatically.
- Customers using a debit card such as Switch can withdraw up to £50 in cash from any checkout.
- More promotions can be offered, such as 'buy two and get one free' ('multisaver').
- An itemised receipt like that in Figure 2.8 is produced. Notice the detail it contains.

Benefits of the system to the company

- Improved checkout accuracy: it is no longer possible for the till operator to key in the wrong price, so there are fewer errors and less fraud.
- Faster and more efficient throughput: there is, on average, a 15% saving in time to register the goods in a shopping trolley compared with the manual system.
- Improved customer service: new services such as the Clubcard, multisavers, etc. mean that customers enjoy a better service.
- Improved productivity: there is no longer the need to price each item individually. Prices are provided on the front of the shelves on which the items are displayed. Weighing and pricing at the checkouts eliminates the need for separate price points.
- Sales-based ordering: sales information from the checkout is used to create the orders for stock replacement.
- Reduced stock levels: more efficient stock control means lower stock levels are needed, so less money is tied up in stock and there is less likelihood of running out of certain items on the sales floor.
- Reduced wastage: perishable goods such as fresh meat and salads can be ordered accurately using the sales information obtained from the checkout.
- Promotional and sales analysis: scanned data can be used to assess the effectiveness of special promotions and can provide information about the sales of certain goods.

Disadvantages of the system to the company

- The stores become totally reliant on their computerised systems and if the system goes down, even for a short time, chaos can result.
- Shelf prices need to be checked carefully to ensure that they match those on the computer.

The barcoding system

Figure 2.9 shows a barcode from a tin of baked beans. At the bottom is the European Article Number (EAN), a number that is allocated to all product manufacturers by the Article Number Association (ANA).

- The first two digits represent the country where the goods are produced.
- The next five digits identify the supplier of the goods.
- The following five digits identify the product.
- The final number is a check digit and is used to check that the other 12 have been entered correctly. ▶▶

Figure 2.9 A barcode from a tin of baked beans

EFTPOS and the use of debit cards

EFTPOS (electronic funds transfer at point of sale) is the method used by Tesco to transfer money from customers' credit or debit cards directly to the Tesco bank account. A debit card is rather like a cheque in that the money comes straight out of a person's bank account, except there is no limit to the amount a person can spend using one of these cards – provided the money is available in the account. With cheques there is a limit, usually £50 or £100, to the amount a customer can write.

Using checkout information for planning bakery production

Sales information from the checkout is used by the in-store bakeries to plan the production for the same day of the following week. This reduces waste and means stores are less likely to run out of bread.

Sales-based ordering

Sales-based ordering is the automatic re-ordering of goods from the warehouse using the sales information from the checkouts. If, for example, 200 tins of baked beans were sold from a certain store in one day, then 200 tins would be automatically re-ordered and delivered to the store the following day from one of the Tesco distribution centres.

Stock control

All ordering is performed by computer and there are fast electronic communication lines between the shops, the distribution centres and the head office. There are also direct links to the major suppliers, which means that orders can go straight through to the production lines. The stock arrives just when it is needed, so it is always fresh. Another advantage of this system is that money does not need to be tied up in stock and may be used for more productive purposes, such as reducing the firm's borrowing, advertising or improving production techniques.

Electronic shelf labelling

Tesco is developing a system using liquid-crystal shelf labels containing the price, description and ordering information of goods. The label is operated from a computer using radio signals, which means that if a price is changed on the computer database, the price displayed on the shelf is changed automatically at the same time. This avoids human errors such as a price change on the computer not being transferred to the shelf.

▶▶

EDI

EDI is a method of speeding up the transfer of orders to suppliers. Using EDI eliminates the need for paperwork because the ordering is carried out between the supplier's computer and Tesco's computer. This system is less expensive and quicker than sending the orders by phone, post or fax and cuts out errors such as lost or wrongly printed orders. Tesco can send information to suppliers regarding sales forecasts and stock levels so that the suppliers can plan their production appropriately.

Once an electronic order has been placed, the invoice is generated automatically by the supplier's computer. This is sent back and checked by the Tesco computer before payment is made. The HP Bulmer case study later in this unit looks at EDI in more detail.

The hardware

As you can imagine, the computers used to run Tesco's systems need to be extremely sophisticated and powerful. The mainframe computers are situated in two computer centres, each one being capable of running all of the company's systems on its own. They can deliver 216 million instructions per second (mips) and are among the fastest commercial computers in the world.

As computers are so vital to Tesco's operations, backup procedures are in place so that, if one of the computer centres were completely destroyed, the other would be able to re-establish the vital systems within 48 hours. The back-up procedures are tested each year so that staff know exactly what to do if a disaster were to occur.

Designing store layout using CAD

Tesco no longer uses drawing boards for planning new stores and redesigning existing ones: instead, they use CAD. This has reduced the time taken to plan new stores: a data bank holds designs and plans from existing stores which can be adapted, rather than new ones having to be generated each time. CAD is also able to show three-dimensional views of the stores, and colours, lighting and different finishes on materials can be altered simply by moving the mouse.

When a new store is planned, photographs of the proposed site can be used in conjunction with CAD to see what the area will look like with the Tesco store in place.

CAD is also used to design the warehouse layouts, and the roads and areas surrounding the distribution centres. This is important because the access roads need to be suitable for large articulated vehicles, and there must be ample room around the distribution centres for them to turn round.

Warehouse systems

Computers are used in the warehouse to monitor complex stock control procedures and make the best use of space, time and labour. Like all areas of retailing, better operating methods need to be found to ensure Tesco's continued success. As with all the other systems Tesco has in place, paperwork has been eliminated wherever possible, and the thick binders containing lists of stock items replaced by computer terminals. These terminals are mounted on fork-lift trucks so that the operators can receive information regarding the movement of the pallets to enable them to move stock items quickly and efficiently. If some stock goes out of the warehouse, for example, then a slot is available for the new stock arriving and the operators are notified of this by the terminal. Efficient use of the available space means the trucks have to travel shorter distances and the whole process is therefore faster. The computer system also monitors where each fork-lift truck is situated in the warehouse so that a particular job can be given to the ones best able to complete it in the least amount of time.

Electronic mail

Tesco, like most other large companies, has realised the benefits of using e-mail. Conventional methods of communication can have a variety of problems (lost post, unanswered telephones, engaged fax machines, people not at their desks, etc.). Use of e-mail eliminates many of these problems.

Advantages to Tesco of using e-mail:

*   Unlike with a telephone call, the recipient does not need to be there when the message is sent. People can receive their mail at any terminal connected to the system at any time.
*   The sender can be sure that the messages have been received.
*   It is possible to send mail to a whole department or a group of people without knowing anyone by name.
*   The e-mail system is used as a vehicle for company information and a noticeboard. You can, for example, find out about the latest job vacancies and appointments and look at the latest share price.
*   It is possible to send e-mail to the major suppliers, thus speeding up orders, etc.

Tesco and the Internet

Tesco was the first UK supermarket to offer a shopping and delivery service on the Internet for all the items it sells. At present, the service is available at selected stores only; customers register for the service and can browse through and select items from a range of 20 000 product lines. Each customer selects the method of payment and a suitable time for the goods to be delivered to their home. There is a fixed charge of £5 for the delivery, which is added to the customer's bill. The delivery service is available over quite a wide area in the vicinity of the store. To pay by credit card, the customer keys in his or her details, which are then sent over the Internet. The customer then signs an authorisation slip when the goods are delivered.

The Internet business consultant for Tesco hopes that the Internet shopping service will attract working people who dislike spending the little spare time they have in the supermarket. Customers can also shop 'off-line' using a product list available on CD-ROM, and send in the order on-line. You can take a look at this service yourself at www.Tesco.co.uk

Activity 14

a Sales-based ordering systems are used to keep track of the current stock position as customers buy goods at a computerised till (point-of-sale terminal). As soon as the number of stock items falls below a certain level (called the re-order level) an order is automatically issued by the computer which is then checked and sent to the supplier.

Use the following sources for information: local shops; textbooks; the Internet (most of the large retail chains have a Web site); articles in *Computer Weekly*, *Computing* or similar publications (e.g. the *Guardian* computer section) to investigate the following:

- How the data is captured at the till (in most cases this will be with a barcode reader but there are other methods).
- The advantages to the management of the shop.
- The advantages to the shoppers.
- The advantages to the till operators.
- The disadvantages to the till operators.
- How the point-of-sale system links up with the stock control system.

Produce your findings in the form of a report.

b Many of the larger retail chains have their own Web sites. As we have seen, Tesco has now extended this so that goods can be ordered over the Internet. Investigate other retailers' Web sites and produce a brief report to be submitted to a magazine for publication comparing the different sites and facilities offered.

Case study

HP Bulmer

HP Bulmer is a medium-sized British company based in Hereford and has been involved in cider making for over 100 years. Cider is an alcoholic drink made from apples; its consumption in the UK is growing at an annual rate of 8%. Currently, around 450 million litres per year are consumed. The brands produced by the company include Woodpecker, Max and Woodpecker Red (Figure 2.10). Roughly 900 staff are employed, with about 500 using PCs and 50 using terminals. There are around 28 permanent IT staff, and 7 employed on a contract basis, with most of these involved in systems development.

The main activities of the company are:

- manufacturing (making the cider by the fermentation of apples)
- trading and marketing (dealing with customers who buy the cider and suppliers who supply bottles, labels, equipment, etc.)
- finance (payment of wages, accounts, etc.)
- distribution (stock control, dispatch, delivery, etc.).

One of the problems that Bulmer had with its old systems was that a large number of applications had evolved separately in different areas of the business, which meant that the data from the different areas was not consistent. For instance, instead of a single customer file, there were three: one for order entry, one for invoicing and another for sales outlets (pubs, off-licences, etc.). The information obtained about customers thus depended on which of the customer files was being used and when the company wanted to produce sales analyses they obtained conflicting results. It is for this reason that the company decided to rationalise and centralise its customer and product information. ▶▶

Figure 2.10 Three of the ciders produced by HP Bulmer

The IT director had to develop a new system that was controlled centrally but with clear ownership of each application by the department concerned. Steering groups and working parties were set up, consisting of members of staff from all the departments involved. Agreement was reached on all the definitions and concepts used so that everyone knew they were talking about the same thing. All the data for the system is kept in an Oracle data dictionary and the systems development and support teams all use computer-aided system engineering (CASE) design tools. CASE involves combining a variety of software tools to assist computer systems development staff produce and maintain high-quality systems. Oracle, which should not be confused with the Teletext service, is a relational database management system suitable for large, multi-user systems.

Within the IT department there are three systems development teams, with each team developing a group of associated applications. Between them these teams cover the four core business areas mentioned above. Each team reports to the IT Director as well as to the head of the user department. The budgeting and delivery of the systems is the responsibility of the project sponsor, who is a senior director of the company.

The main objectives for the new systems include:

- efficient, auditable sales order processing
- accurate monitoring of revenue generation
- efficient distribution
- exploitation of opportunities in the marketplace.

The distribution management system

The distribution management system covers sales order processing, warehouse management and distribution. Although the initial development costs were £2 million, the system is already making savings of £400 000 per year.

The distribution operation is large, with over 400 product lines being delivered to the main brewers, the 'big six' supermarkets and outlets such as the smaller off-licences. During heavy trading periods, the company is involved in the delivery of over 3000 pallets a day on 130 trucks.

Warehouse control

Facilities for controlling the goods in the warehouses are available at the Hereford factory and the depots throughout the country. The system handles stock location and enables staff to order replacement stock and to obtain accurate, up-to-date information on any items of stock in any of the warehouses. The system also helps with planning routes for the delivery lorries and their stops along the way.

Customer service

All the telesales staff are located at the Hereford head office. Their job is to take orders, confirm the availability of the goods and confirm delivery dates over the telephone (see Figure 2.11). At the busiest times telesales staff can take more than 1000 orders per day, ranging from a single line to several pages. On receipt of an order the order details are automatically passed to the warehouse where a picking list is produced, which lists the positions in the warehouse where the goods are stored. A place in the warehouse is also assigned to the pallet on which the goods are placed. Once the goods have been placed ready for dispatch a dispatch note is produced.

The new system has resulted in many benefits besides the obvious financial ones. It is now possible to track all stock movements, and produce accurate dispatch notes. Under the old system, the dispatch note was printed before the stock was picked, and sometimes had to be amended by hand because some of the order was unavailable. Management information about orders is now vastly improved, and customers are more satisfied with the service they receive.

Figure 2.11 Inputting telesales orders

Profitability management system

An account manager is responsible for the sales and servicing of a group of major customers. In order to assess its sales success under the old system, Bulmer looked only at the volume of cider it managed to sell. However, this was misleading because it did not take into account the profitability of the sales: after all, sales could easily be increased by negotiating a price at which the company is making very little profit. Now the system assesses the account managers on their sales by taking into account how much they have spent on marketing, advertising, special promotions and discounts.

Weekly sales report

A sales report lands on the desk of each director and senior manager every week, and is the most important piece of management information they receive. Basically it presents the sales of each product and expresses them as a percentage of the budget figure.

EDI

There is now a move towards transferring information between Bulmer and its main customers electronically. This is how Bulmer and one of the large supermarket chains use EDI to cut out the use of paper (see Figure 2.12):

* prices for the goods are agreed between Bulmer and the supermarket chain
* Bulmer enters the agreed price on the system
* this price file is sent electronically to the supermarket's head office

- there is confirmation that the retailer has received this file
- the retailer can then place an EDI order
- Bulmer loads the vehicles with the pallets containing the order, and as each pallet is loaded a barcode on its side is scanned and the information used to build up a delivery note
- the delivery note is sent electronically to the retailer
- the goods are delivered
- the retailer confirms delivery using EDI
- when this confirmation is received, Bulmer automatically issues an invoice
- the invoice is automatically paid and money is transferred from the retailer's bank account to Bulmer's account.

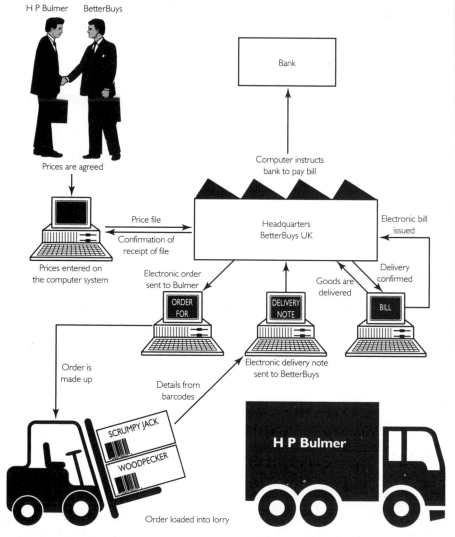

Figure 2.12 How EDI works

Bulmer and the Internet

Bulmer has its own Web page on the Internet (http//www.bulmer.com) and it plans to make more use of it in the future. Bulmer uses the Internet mainly for marketing, and for e-mail contact with its customers.

▶▶

The keg information system

Bulmer sells draught cider through tens of thousands of pubs, wine bars and restaurants throughout the UK. The keg technicians are responsible for installing and repairing the equipment needed to dispense the draught cider at the bar – the pump you see on the top of the bar, the cooler unit and all the pipework which goes to the aluminium barrel, called a keg.

When something goes wrong with the equipment or new equipment is to be installed, a telephone call is made to head office where the scheduling is organised. The schedules for each of the 19 technicians scattered throughout the UK are transferred overnight to the technicians' hand-held terminals. During the day the technicians record what they have done that day using their hand-held terminals, and this information is relayed back over the telephone lines to Bulmer's head office where the central computer files are updated.

Using this system, the technicians are able to deal with more calls each day than under earlier systems, and the staff involved in coordinating this activity can be engaged in other tasks. A future development might be to give technicians portable modems so that they can update files in real time. If any problems arise in the course of a day, then the technicians can be alerted and the one closest to the location of the problem can deal with it as soon as their schedule permits.

Process control

The cider-making process at Bulmer's Hereford cider mill was previously carried out by the traditional methods of the nineteenth century, which had changed little since Percy Bulmer founded the business in 1887. Today, the old equipment has been replaced by the latest computer-controlled technology.

Activity 15

a Bulmer has listed the four main areas of its business as:
  - manufacturing
  - trading and marketing
  - finance
  - distribution.

  Using both business studies textbooks and the information given in the case study, write a series of short paragraphs outlining what tasks would be performed in each of the above areas.

b Bulmer identified certain problems with the old systems. Identify which systems they were and how the new system is an improvement.

c Certain terms used in the case study are not explained in the text. Give a brief explanation of the following terms:
  - Oracle
  - data dictionary
  - CASE
  - Unix
  - server.

d Bulmer hopes to use EDI with all its major customers. Produce a diagram on computer, using suitable software, to explain how this works.

2.4 Management information systems

The difference between a data processing system and an information system

Data processing systems are frequently called transaction processing systems because they are the systems that record, process and report on the day-to-day business activities of an organisation. Data processing concerns all the routine transactions (another name for bits of business) that need to be performed on a day-to-day basis – such things as processing customer orders, dealing with returns, arranging deliveries, processing payments made, dealing with queries, processing the payroll, dealing with stock and placing orders with suppliers. The main body of data processing concerns operational data. Operational data is essential for the running of the organisation or business. Data processing systems are mainly used by the operational level of staff in an organisation.

Information systems and management information systems

Information systems are used to obtain information that enables users to make effective and timely decisions. An example of this is an order processing system – when a customer rings up to place an order the member of staff taking the order will check that the item is in stock and if not, how soon it can be obtained. The member of staff will then decide whether to order extra stock or not. As managers are the people who have the responsibility in organisations for making decisions, information systems used to supply managers with information on which they can base their decisions are called **management information systems** (**MIS**). MIS go hand in hand with data processing systems but are used for different purposes.

What do managers do?

Before looking at MIS we need to consider the role of a manager and the tasks involved in management. Managers are present in many layers in an organisation – from junior managers through middle managers to the senior managers and directors. What they do depends on their level within the organisation: junior managers tend to deal at an operational level and therefore management of the day-to-day issues in an organisation; senior managers deal with strategic matters and make major decisions.

All managers have to make decisions using information obtained from internal and external sources. The types of decisions can be classed under the following headings:

- planning
- directing
- controlling
- forecasting.

The lower layers of management are responsible for:

- the day-to-day management of the operations staff
- allocating work to subordinates
- arrangement of staff rotas, dealing with staff sickness/absence
- motivating staff
- handling a departmental budget.

The higher levels of management are responsible for:

- strategic planning (this involves the setting of overall objectives and policies)
- market share
- cash flow
- profits
- growth in profits.

Management information systems help managers to plan, organise and make decisions. They can provide a variety of different types of information or support, such as easy to understand tables, responses to direct queries, graphical output, answers to a 'what if' scenario or a signal to warn that data is exceeding a set limit.

It is important to note that MIS are really just a specialist part of information systems.

Activity 16

Figure 2.13 shows the location of the shops for a large supermarket chain. Lorries deliver the goods from several large distribution centres situated strategically for easy access to all the stores. At present all the shops shown on the map are fed from a distribution centre in the Midlands but more shops have now opened in northern England and a new distribution centre covering this area is to be built. The company now has to decide where to site the new centre.

a Explain the difference between internal and external information.
b What are the main factors for choosing the location of the new distribution centre?
c What information would be needed in order to make an informed decision?
d Where would the information come from on which to base the decision?

Typical MIS

A comprehensive database is used to hold all the different types of information processed by the organisation, and provides managers with ready-made reports such as comparisons of actual and targeted sales. This allows the analysis and comparison of data in the database over a period of time to provide information in graphical form about sales, purchases, wages, stock levels, etc. Warning signals are used to indicate that decisions are required – when stock levels are low, expenditure exceeds income or the number of faulty manufactured items exceeds a set expectation, for example. Managers can use the MIS to look at the main identifiers of company performance on a daily basis if needed. This could include:

- the daily calculation of productivity levels by analysis of costs and output
- monthly graphs of price comparisons with competitors' goods or services, resulting from regular market research
- audio and visual warnings are made when the quantity of incoming orders exceeds the production capacity
- a model of the organisation that enables 'what if' queries to be input to forecast the effects of policy decisions, changes in market conditions, production rates, VAT or tax rates.

Figure 2.13 The locations of supermarkets in a chain

Activity 17

a Investigate a small business from the following list. It is best to choose one that you are familiar with or one it is fairly easy to find out about.

- A video library
- A car rental company
- A dental practice
- A mail order clothes catalogue business
- A coach travel company
- A clothes/hat hire company
- A mail order computer supplier

Produce a brief report outlining the following.

- What information the organisation relies on
- How it processes that information
- Whether a management information system would be useful
- What specific purpose or purposes it could serve
- What information is needed for the management information system to operate
- How the ICT system could best provide the management information required

b You need to be able to distinguish between a data processing system and a management information system. Basically, if the system processes raw data to produce routine information, then it is likely to be a data processing system. However, if the data produced is communicated to managers at different levels, in an appropriate form, to enable them to make effective decisions for planning, directing and controlling the activities for which they are responsible, then it is a management information system.

Categorise the following as either data processing systems or management information systems.

- A payroll system for processing time sheets and printing pay slips
- A system that compares the sales made of the same make of car over the same month for the last 5 years.
- Production of a list of all the main dealers in the country with a certain colour and model of car in stock to be given to a customer.
- A list of the items to be ordered from a supplier, produced from an EPOS system.
- A sales analysis system to investigate trends in sales over a certain time period.
- Production of a list of debtors to send to the manager for a decision on further action to take.
- The sales figures for previous similar periods for the planning of production levels.
- A system that analyses a competitor's prices for similar products.
- A system producing a report outlining the frequency of the calls to a help-desk and the average time each problem took.
- A system for producing a list of students who have paid their course fees, to be given to a course tutor.

MIS in schools and colleges

Your school/college will almost certainly make use of a management information system. If you are in a school then a system called SIMS (Schools Information Management System) will probably be used. In a college a variety of different systems could be used.

A school or college is an example of a service organisation that needs to be carefully managed to make the most of the money allocated to them.

Activity 18

Your teacher/lecturer will arrange for someone who uses your college's management information system to come and explain to you how the system works. Take notes as they talk and produce a written summary of the system.

2.5 Standard ways of working

As part of Unit 2 you will have to show that you have used standard ways of working. The whole point of having a standard way of working is to help you to manage your work more effectively. You can save a lot of time and effort if you think logically about how you intend working before you start. All of the standard ways of working are looked at in detail in Unit 1. Here is a list of the ways of working you should apply to this unit.

* Plan your work to produce what is required to given deadlines.
* Edit and save work regularly, using appropriate names that remind you of their contents.
* Store your work where others can find it easily in the directory/folder structure.
* Keep dated backup copies of your work on another disk and in another location.
* Keep a log of the ICT problems and how they were resolved.
* Make sure that confidentiality (i.e. Data Protection Act 1998) and copyright laws are obeyed.
* Ensure that you avoid bad posture, physical stress, eye strain and hazards from workplace layout.

Key terms

After reading this unit you should be able to understand the following words and phrases. If you do not, go back through the unit and find out, or look them up in the glossary.

BACS
Brief
CAD (computer-aided design)
Centralised database system
Customer information
Data processing system
Design specification
Distribution
Distributor
e-commerce
EDI (electronic data interchange)
EFTPOS (electronic funds transfer at point of sale)
Electronic information
Electronic mail
Face-to-face meeting
Fax
Flat organisation
Function
Hierarchical organisation
Human resources

Information flow diagrams
Information flows
Information system
Internet e-mail
Just-in-time stock control
LAN e-mail
MIS (management information system)
Marketing information
Operational information
Order processing
Organisation
Purchase information
Retailer
Sales-based ordering
Sales information
Service
Stock control
Supplier information
Verbal information
Web site
Wholesaler

Review questions

1 **a** Most organisations make use of a management information system (MIS). Explain the purpose of a management information system.

   b Explain why an MIS is particularly useful to managers.

   c Give an example of one MIS that you have come across and explain briefly how it was used.

2 Figure 2.14 shows two different systems, A and B, along with their subsystems. Explain which of the two systems is likely to be better, and give your reasons.

System A: The other subsystems overlap with the order processing system

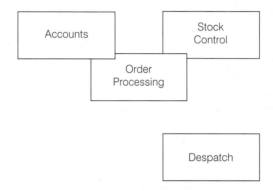

System B: The other subsystems do not overlap with the order processing system

Figure 2.14

3 A manufacturing organisation is made up of the following functional areas:
   * accounts
   * sales
   * marketing
   * production
   * research and development
   * purchasing
   * distribution
   * human resources/personnel
   * ICT services.

Decide which one of the functional areas of the organisation would be most likely to undertake the following tasks.

a Carrying out a survey to find out why their customers bought their goods rather than a competitor's.
b Visiting customers on their premises to take orders.
c Making sure that the organisation is selling the right product for the right price.
d Planning the route the delivery vehicles should take to minimise the mileage the vans need to travel.
e Processing sales staff expenses.
f Processing the payroll for all staff.
g Keeping the organisation's network running on a day-to-day basis.
h Making up and packing customers' orders.
i Sending out reminders to customers that their account is overdue.
j Sending a stationery order to a supplier.
k Picking out goods in the warehouse to make up customers' orders.
l Checking on references for a potential employee.
m Planning the weekly production targets.
n Placing advertisements in the trade magazines.
o Finding better ways of producing the goods.
p Paying the suppliers' invoices.
q Preparing delivery notes.
r Issuing credit notes.
s Dealing with returned goods.
t Dealing with holiday entitlements.

4 Companies that supply goods or services need to know as much about their customers as they can.
a Explain what details, other than name and address, a telephone company would be likely to store about their customers.
b Explain what details, other than name and address, an Internet book company such as Amazon would be likely to store about their customers.

5 a Explain what is meant by an organisation.
b There are three different types of organisation. Name them.
c Give five functional areas that would be found in each of the organisations that you named in (b).
d Not all functional areas are found in every organisation. Give the names of three functional areas in a manufacturing business (such as a company involved in the manufacture of washing machines) and a service business (such as a company that provides an office cleaning service).

6 Information flows take place from one functional area to another. Describe one information flow that takes place between each of the following departments.
a Accounts and sales
b Sales and production
c Purchasing and accounts
d Marketing and sales
e Human resources and accounts

7 Give three advantages and three disadvantages of communicating using the following methods in a business.
a Face-to-face meetings
b Fax
c e-mail
d External post

8 Company data is frequently stored in a single computerised database and is used by all the departments that need it. Give three reasons why a centralised database system is a very efficient way of dealing with the company's data.

9 More and more companies are taking advantage of the e-commerce revolution.
   a What is meant by the term 'e-commerce'?
   b What are the main advantages to an organisation in conducting business using e-commerce?
   c Explain briefly how an e-commerce business that you have studied works.

10 Explain what each of the following are and how they differ from each other.
   • Data processing systems
   • Information management systems
   • Management information systems

11 Many organisations are now choosing to trade electronically, thus eliminating much of the paperwork.
   a A supermarket uses EDI to trade with its suppliers.
      i Explain what is meant by EDI.
      ii What information passes between the supermarket and the supplier using EDI?
   b A college uses e-mail internally on their local area network.
      i Explain what is meant by a local area network.
      ii In what ways will the use of e-mail improve the efficiency of the college?

Assessment

There is no assignment for Unit 2, as this unit is externally assessed. This means that the awarding body for your GNVQ will send materials to your school/college that tell you what to do. These materials should be used to assist you in preparing for the external assessment.

Assessment criteria

1 Describe clearly, with the aid of diagrams, the main function(s) of the organisation, its associated customers (or clients) and suppliers, the function of each department, the structure of the organisation and the relationships between the main departments and outsiders.

2 Describe the ICT provision for each of the organisation's departments (or functions) and identify possible extensions or improvements to the use of ICT that would benefit the organisation.

3 Show clearly, using diagrams, how information essential to successful operation moves within the organisation and to and from outsiders.

4 Describe in detail the purpose and operation of an important ICT application used within the organisation, including examples of input and output data and the job functions and personnel involved.

5 Ensure that your case study is presented clearly as a coherent report and is checked for meaning and accuracy.

6 Produce a well-structured case study that shows fluent use of technical language, contains appropriate conclusions and makes suitable references to the information sources used.

7 Explain in detail how information used in the organisation is processed, including details of the data capture techniques, any processing or calculations involved and the specification and style of data output.

8 Show that you can work independently to produce your work to agreed deadlines.

9 Explain in detail, with the aid of diagrams and definitions of the data, how information moves from a customer or client through the organisation to result in the delivery of a product or service.

10 Use examples to recommend improvements to the organisation's internal ICT system (this may cover items such as integration of existing systems, specialised equipment or software, database development, LAN or WAN systems).

11 Describe in detail how the organisation might benefit from more extensive use of new communication technologies, such as the Internet, mobile communications, e-mail, e-commerce or EDI.

12 Describe how the organisation might use a management information system to monitor or control activities, improve decision making and improve efficiency.

Get the grade

To achieve a grade E you must fulfil criteria 1–5.

To achieve a grade A you must also fulfil criteria 6–12.

Key Skills | Opportunity

	You can use this opportunity to provide evidence for the Key Skills listed opposite.
Communication C2.3	Write two different types of documents about straightforward subjects. One piece of writing should be an extended document and include at least one image.
C3.3	Write two different types of documents about complex subjects. One piece of writing should be an extended document and include at least one image.

Spreadsheet design

What is covered in this unit:

This unit is about designing spreadsheets and then using spreadsheet facilities to make a spreadsheet for others to use with their own data. You will learn how to design spreadsheets that meet users' needs and which are thoroughly tested before handing over to the user.

You will be assessed in this unit through your portfolio of evidence – the grade you get for this will be your grade for the unit.

Materials you will need to complete this unit:

- access to computer hardware, including a printer (a colour printer will be needed on an occasional basis)
- access to Microsoft Excel (preferably Excel 2000 or Office 2000, although there is very little difference between Excel 2000 and Excel 97 or Excel 95)
- access to the Internet, for detailed information on more advanced features of Excel.

3.1 Developing a working specification

If you are working in IT, then as well as developing systems to use yourself you will also develop systems for other people to use. In fact, most of your time will be spent developing systems for others. Users have their own requirements and these need to be detailed in a working specification. He or she will want to enter the data into the spreadsheet and process this data in some way to produce the required output. There are various considerations when developing systems for others.

What problems might
there be if the
spreadsheet developer
produces a spreadsheet
they think the user
wants?

What output information does the user want?

We always work from the end result, the output from the system, because this determines what input is needed and the type of processing that must be done. The person who develops a system, such as a spreadsheet, should be clear about what the user requires. For example, the output from one system may be a **chart** for a presentation whereas the output from another might be a list of numerical values. It is no good developing a system that does not give the users what they really want.

Constant consultation with the user at all stages is necessary so that everyone agrees on what is required from the system. The output must be chosen to present the output in the way most useful to the user.

How is the information currently obtained (if at all)?

In some cases the information may already have been produced in some form. For example, a college may need information about course results, percentage attendance, or the percentage of students who completed the course. This information will come from a variety of sources – registers, mark books, student registration details and so on. It may have been prepared manually and the output written down on a form.

The college might need to produce a summary for all the courses in different subject areas. They would need to produce a blank spreadsheet that can be sent to all the course tutors, who will enter the data about their own students. This makes it easy to merge the data to produce the large spreadsheet with the data for all the courses.

Where does the data to be input come from?

Data may need to come from different departments in order to complete the spreadsheet. There must be ways for the information to be routed from the various departments to the person who inputs the data into the spreadsheet. The data may come from another package such as a database or via file transfer using the Internet.

What data capture methods can be used?

Data capture is the term for the various methods of entering data into a computer ready for processing. The most common method of data capture is to use the keyboard, but this can be slow and is prone to errors.

Other methods of data capture include voice recognition, **optical mark reading (OMR)**, **optical character recognition (OCR)** and barcoding.

Results of questionnaires are often analysed using either OCR or OMR, and the data obtained is passed to spreadsheet software for statistical analysis and for the production of graphs and charts.

The choice of a method of data capture is mainly influenced by cost and expertise. Keyboard entry, although cheap as far as hardware is concerned, is expensive in terms of the time needed to type data in. For example, if you wanted to analyse the results of a questionnaire that contains 20 questions and has been filled in by 20 000 students data entry would be an awesome task. OMR (or possibly OCR) could read the answers and provide an analysis in much less time, although it would take some time to set up the system and read the questionnaires, and the questionnaires would have to be carefully designed.

The data used as input to a spreadsheet package could come from another package such as a database – for example, a database of all the details about customer orders. In order to perform statistical manipulation of the values of these orders, the data can be imported into a spreadsheet (it is much easier to perform statistical analysis using a spreadsheet, and you can also perform 'What if?' analyses).

Data can also be directly input from instruments. For example, sensors used to record the conditions during a laboratory experiment can relay the data straight to the spreadsheet package.

What data processing is needed to get the required output?

Data processing is not just about doing calculations on data. It includes merging data from several sources, sorting data into numerical or alphabetical order, grouping and classifying data. Processing can mean presenting the data in different ways – rather than just as numbers in a table for instance, graphs and charts can be produced. Sometimes a spreadsheet will be used to make logical comparisons, such as comparing the number in one **cell** with that in another, and printing an appropriate message.

Spreadsheet software is most powerful in its calculating ability and most spreadsheets consist of a large number of formulas for calculating results.

Users should not simply assume that these formulas are correct – they should always be thoroughly tested. The spreadsheet should be designed so that the formulas can be checked easily. Values that are likely to change – for example the VAT rate (currently 17.5%) – should *never* be placed in formulas because if they change the user would have to change the formulas in the spreadsheet. The idea is that the user should be able to alter figures on the **worksheet** without having to alter formulas, which they may not understand. A good spreadsheet would ensure that the user never has to worry about formulas or the design of the spreadsheet. The should be able to just concentrate on the task in hand – that of entering the correct data to produce the output they require.

Once all the formulas have been entered into a worksheet it is a good idea to protect them so that they cannot be altered without first entering a password. This prevents people tampering with the formulas.

Write a list of the things that constitute processing.

How does the output information need to be presented?

Output information can be presented in many ways – as tabular data, graphs and charts, data to be placed on a Web site, attachments to e-mails, etc. Again the user will need to be asked to what use the output is going to be put – some users may take a spreadsheet and then import it into a presentation package such as Microsoft PowerPoint or into a word-processing package such as Microsoft Word, for example. You might be able to suggest a better way of presenting the output using your IT knowledge. At the end of the day, though, it is the user who should decide how they want the output presented. If the aim of the spreadsheet is to produce a printout then it might be necessary to create an application that no longer looks like a spreadsheet.

What aids can be provided to assist with the data input or processing?

Rather than enter data directly into the spreadsheet matrix, you can set up forms for the user to enter the data. It is quicker for an experienced typist to enter the details into a form rather than straight into the individual cells.

You could also add comments to cells so that when the user moves the cursor to the cell, a comment automatically appears telling them what they have to do.

Macros can be used to automate some of the keystrokes or events, thus saving the user a lot of effort.

Buttons can be added to initiate certain procedures, such as running macros or drawing charts.

It is also possible to restrict the parts of a spreadsheet that the user can change to only those cells where data is entered. Locking all the other cells protects the **integrity** of the spreadsheet and stops the user accidentally (or deliberately) altering the formulas used. By building **validation** checks into the cells where the user has to enter data you can protect the integrity of the data and make sure that invalid data is not processed by the spreadsheet.

Producing the specification

When you are developing a spreadsheet you need to plan your design and consider each of the headings above when considering the user's requirements. It is a good idea to prepare some notes under each of the headings at your meeting with the user to discuss their requirements. Once you have understood what is needed, you can discuss each item and get agreement on each point. It is a good idea to get the user to sign an agreement so that you both know exactly what is being developed. This avoids conflicts about what was agreed at the meeting.

If you produce a good design specification, there should be no doubt about the scope of the project (i.e. how far the project goes) or the work that has to be done.

The integrity of a spreadsheet must be maintained. What is meant by the word 'integrity'?

Why is it important to the user as well as the spreadsheet developer to reach agreement over a system specification?

You can use Activity 1 to provide evidence for key skills for Communication C2.1a or C3.1a.

Activity 1

The developer of a spreadsheet system is meeting with the user to discuss the user's requirements. Here is what the user said when describing what they want from the new system.

'As the manager of the Chester branch of the Chester Building Society, I have to be able to give details about mortgages. There are many special deals at the moment to entice people to take the plunge and buy a house, as well as many to attract customers from other banks and building societies to re-mortgage. The system we use gives details of mortgages on offer at the moment. These are usually fixed rate deals where the interest rate stays at a low value for a period (anything from one to seven years) and then goes back to the normal rate for the remaining term. Many customers ask about what the payments will be if the rate goes up. They may also want to see the effect of altering the term of the loan on the monthly periods. Some like to know how much money they will be likely to pay back over the whole period of the loan.

'For a quotation for a mortgage, I will ask the customer their name. The date of the quote also needs to be printed along with a suitable heading.

'I then ask them how much they wish to borrow, and over what period they want the mortgage (this is given in years).

'I then type in these figures, together with the interest rate (e.g. 7.5%). The spreadsheet would need to calculate from this the monthly repayment, the annual repayment and the total repaid over the period of the loan.

'All of the details need to be displayed on the screen and the customer may want a printed copy of the quotation to take away.'

For this task you are required to work in teams. The idea of working in teams is to allow the team members to 'bounce' ideas off each other. Each part of the team needs to make a contribution to the activity.

Produce a design specification for this system.

Once you and the user have established what needs to be done, you can get down to the work of planning and developing the spreadsheet.

Design the output

Designing the output from the spreadsheet is best done on paper. Don't just do one design. Play around with the design and bear in mind that the most important information should be in a prominent place. Choose colours and fonts carefully. Remember you can put data in boxes to make it stand out.

Output in the mortgage spreadsheet in Activity 1 will be either the display on the screen or a printout on paper.

The design of the output from any spreadsheet is important as it determines the layout and structure of the original spreadsheet. If possible, the user should be consulted regarding the design because they will understand the purpose of the spreadsheets better than you and will be using the spreadsheet on a day-to-day basis.

Construct the spreadsheet

From your final sketches, you can start using the spreadsheet software to construct the spreadsheet. You will need to think about what the user needs to input, the processes (formulas, logic expressions, etc.) and the output.

The idea is to make the spreadsheet easy to use. Explanations should be given when users are expected to enter their own data. They should never be left in any doubt about what they have to do.

How could you provide evidence that agreement has been reached with the user?

Why is it much better to plan a spreadsheet design on paper before using the computer to create it?

Spreadsheets: the basics

You will probably have used spreadsheets before as part of your GCSE IT or ICT course, so much of what follows will simply be a reminder. In the next section you will be looking in more detail at spreadsheet features so it is important to understand the basics.

Areas of the spreadsheet

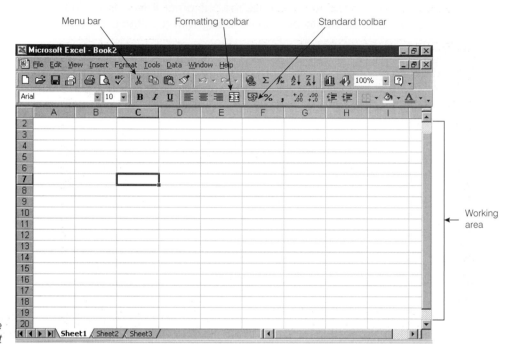

Figure 3.1 Areas of the spreadsheet

The top part of the spreadsheet is the command area, where commands to do certain things are issued.

The large area in the middle is the working area of the spreadsheet and it is in this area that you enter your data. This area is commonly called the **worksheet**.

You can move the cursor (sometimes also called the active cell) by using the mouse or the cursor keys.

The working area of the grid, into which you put your own data, is often referred to as a worksheet (the words spreadsheet and worksheet are often used interchangeably). The worksheet is the working area of the spreadsheet and is arranged as shown in Figure 3.2.

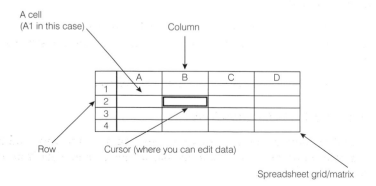

Figure 3.2 Arrangement of the working area of a spreadsheet

Cells

The small rectangles produced by the crossed lines of columns and rows are called **cells**. To refer to a particular cell, we use the column letter followed by the row number. For example, cell C3 is the cell where column C intersects row 3. S4 is a proper **cell reference** but 3E is not. Case does not matter, so you can use b3 or B3.

The menu bar and toolbars

When you load Microsoft Excel the screen shown in Figure 3.3 appears.

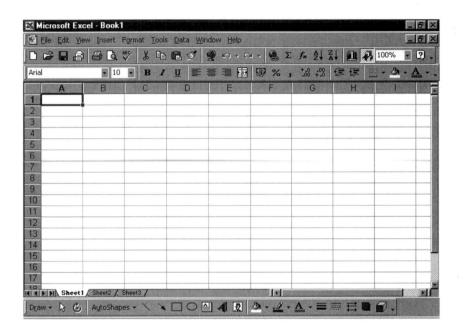

Figure 3.3 The opening screen in Microsoft Excel

There are many different **toolbars**, and you can choose which will be displayed by clicking on View and then Toolbars. A tick is shown beside the toolbars to be shown.

Using Undo

It is easy to issue the wrong command, which makes the spreadsheet do something you did not expect it to do. You can also press the wrong buttons by accident (some of the keys on the keyboard, when used together, issue commands and you might hit these by mistake). Suddenly something happens or the screen goes blank. What can you do?

Don't panic. You can undo the last command:

1 Go to the menu bar, click on Edit and select Undo from the list.
2 If more than one command has been issued, then you can use step 1 several times.

Undo is also very useful if you delete something by mistake.

Inserting rows and columns

When you are designing spreadsheets of your own, you may find you need to insert an important column between two columns that already contain data. You might also need to insert an extra row.

To insert a column:

1 Use the mouse to position the cursor on any cell in the column to the right of where you want the column to be inserted.
2 Click on Insert in the main menu bar and select Columns from the list.
3 Check that the column has been inserted in the correct place. Notice that insertion of the column pushes all the data on the left of the cell the cursor is on to the left. Any formulas in these cells will be adjusted automatically.

To insert a row:

1 Place the cursor on any of the cells in the row below where the new row needs to be placed.
2 Click on Insert in the main menu bar and select Rows from the list.
3 Check that the row has been inserted in the correct place.

Deleting a row or a column

1 Move the cursor to the row or column that you want to delete.
2 Select Edit from the menu bar and then choose Delete.
3 The box shown in Figure 3.4 will appear (this box is called a dialogue box).
4 Select either Entire row or Entire column and click OK.

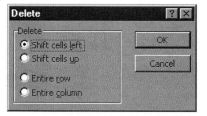

Figure 3.4

If you are deleting a column, the cells in columns to the left will move and fill the gap.

If you are deleting a row, the cells in the rows below the one being deleted will move up to fill the gap.

Deleting groups of cells

First, select the group of cells you want to delete. Choose Edit and then Delete. A box will appear to let you select the way in which the remaining cells will fill in the gap.

A note about deleting cells

You have to be careful about deleting cells because the remaining cells will be shifted to fill the gap. This can cause unexpected results, so remember you can use Undo to undo an unwanted action.

If you have just a couple of cells to delete, it is best to use the backspace key (the key at the top of the keyboard with the arrow pointing to the left). Another way is to use the delete key when the cursor in positioned on the cell.

/ **Activity 2** /

Explain how you would:

a Delete the contents of a single cell
b Delete the contents of an entire column
c Delete the contents of a block of cells.

Using the mouse

Single and double mouse click

As with all Windows software, you make selections using the mouse.

- Single click – press the mouse button once. This will be the left button if there is more than one.
- Double click – the mouse button is pressed twice in quick succession.

Dragging

Click and then keep the mouse button pressed down. Dragging is used for sizing and moving items.

Toolbars

Toolbars appear at the top of your worksheet. Their purpose is to make it easy for you to find and issue commands to tell the spreadsheet to do something for you.

The menu bar

The menu bar (Figure 3.5) is a special toolbar where you can select menus for printing, editing and viewing. Clicking on any item in the menu bar pulls down a list from which further selections can be made.

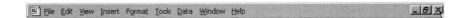

Figure 3.5 The menu bar

The standard toolbar

The standard toolbar (Figure 3.6) contains buttons for opening, saving, printing, cutting, pasting, etc. There are also buttons to sort data in ascending and descending order, to add up cells, produce graphs and even produce a map from your data. The button at the end with the question mark on it is the Help button.

Figure 3.6 The standard toolbar

Formatting toolbar

The formatting toolbar (Figure 3.7) enables you to format cells. This means that you can alter the appearance of the data. You can also format numbers to a fixed number of decimal places. In addition, you can use borders and colours to give the worksheet greater impact.

Figure 3.7 Formatting toolbar

Correcting mistakes

What happens if you enter the wrong data? There are three ways to correct mistakes:

1. Delete the contents of a cell (text, number or formula) by moving to the cell and using the backspace key.
2. Simply type over what is already there.

3 Highlight those cells whose contents need deleting. Then go to Edit then Clear and finally All.

Saving your work

If you want to come back to your work at a later date or if you want to print it out, you will need to save it. Usually you will save your work onto floppy disk or the hard disk. It is important to save your work before printing in case something goes wrong with the printer – sometimes the computer freezes if it can't print the work out and your only option will be to switch it off and then on again. If you do this without saving your work, you will lose it.

There are several ways of saving work. These include:

1 Go to the save button (it has a picture of a floppy disk on it) on the standard toolbar ▣ .

2 Go to the File menu and chose one of the save options from the list.

Saving for the first time

When you save a worksheet for the first time you will be asked to give it a name – give it one that explains what it does. Remember that you might have to come back to the spreadsheet months or years later, so titles such as Spreadsheet 1 are not very helpful. Names such as Favourite_Crisps.xls and Weekly_Budget.xls are much more informative.

To save a worksheet for the first time:

1 Click on File and then click on Save As. This will open up a dialogue box where you can do several things (Figure 3.8):
 • change where the worksheet will be saved. In many cases you will want to store your worksheets on the floppy disk
 • type in the file name you are giving to the worksheet
 • check everything in this box is correct and then click Save.
2 When you have saved your work once it is easy to save it again because the computer will remember what you called it and where you last stored it. You just go to Save in the File menu. It will choose the same again unless you choose otherwise. If you intend to change the name or location where you want to store it, then you will need to go to the Save As option to make your changes.

Figure 3.8

Closing Excel without saving

Sometimes you might just want to do a quick calculation using the spreadsheet and then exit the package without saving your work. To do this, go to the File menu on the menu bar and select Exit from the list of options. You will then be asked if you want to save your work. Click No to exit the package.

Printing formulas

When a worksheet is printed, it will always show the results of any formulas along with any text or numbers. In order to check the spreadsheet it is useful to be able to see these formulas so that they can be checked. The following worked example shows how to display the formulas on the screen and then print them out.

Worked example: displaying formulas and printing

Using this spreadsheet you will learn how to:

- uco tho AutoSum button
- show the formulas in the worksheet.

1 Set up the worksheet shown in Figure 3.9.

	A	B	C	D	E	F
1	Item	June	July	August	September	Total
2	Swimwear	600	150	142	112	
3	Ski-Wear	50	35	60	98	
4	Shorts	85	87	82	50	
5	Golf Clubs	104	69	75	82	
6	Football	20	32	146	292	
7	T-Racquets	85	90	50	80	
8	Totals					
9						

Figure 3.9

2 Total column B and put the result in cell B8. The quick way to do this is to press the AutoSum button ( ) on the toolbar with the cursor on cell B8. Be careful how you use AutoSum. Check that the cell range indicated by the dotted rectangle is correct. If it does not work as expected, just type in the formula instead.
3 Total row 2 and put the result in cell F2 by positioning the cursor on cell F2 and then pressing AutoSum.
4 Copy the formula in cell F2 down the column to cell F7.
5 Copy the formula in cell B8 across the row to cell F8. Your worksheet will now look like Figure 3.10.

	A	B	C	D	E	F
1	Item	June	July	August	September	Total
2	Swimwear	600	150	142	112	1004
3	Ski-Wear	50	35	60	98	243
4	Shorts	85	87	82	50	304
5	Golf Clubs	104	69	75	82	330
6	Football	20	32	146	292	490
7	T-Racquets	85	90	50	80	305
8	Totals	944	463	555	714	2676

Figure 3.10

▶▶

6 Now we will insert the formulas wherever there are calculations. Select the Tools menu and then click on Options. The screen in Figure 3.11 will appear.

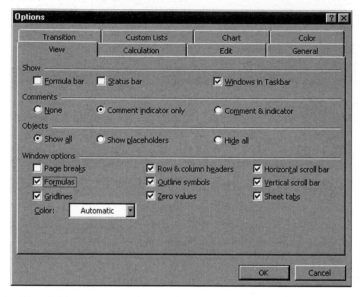

Figure 3.11 The dialogue box used to show formulas

7 In the bottom part of the dialogue box you will see Formulas. Click on this box to produce a tick in the box because we want to show the formulas, then OK. The worksheet screen will now show the formulas (Figure 3.12).

	A	B	C	D	E	F
1	Item	June	July	August	September	Total
2	Swimwear	600	150	142	112	=SUM(B2:E2)
3	Ski-Wear	50	35	60	98	=SUM(B3:E3)
4	Shorts	85	87	82	50	=SUM(B4:E4)
5	Golf Clubs	104	69	75	82	=SUM(B5:E5)
6	Football	20	32	146	292	=SUM(B6:E6)
7	T-Racquets	85	90	50	80	=SUM(B7:E7)
8	Totals	=SUM(B2:B7)	=SUM(C2:C7)	=SUM(D2:D7)	=SUM(E2:E7)	=SUM(F2:F7)

Figure 3.12

8 The formulas often require more space than the current column width so you might need to alter the column width. Do this and save, then print a copy of the worksheet containing the formulas on a single sheet. To do this click on File, and then Page Setup . . . and make sure that the settings are the same as those shown in Figure 3.13. Click OK.

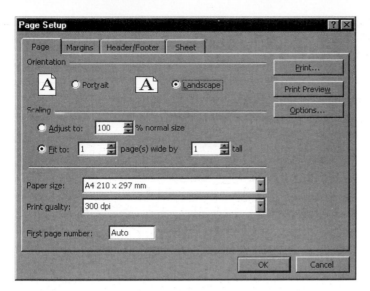

Figure 3.13 Setting up the page for printing

Finding your way around the package

Excel is a very big package and contains many features – and in this book we cover only some of the ones used most. About 80% of users use only around 20% of the available capabilities of any software, often because they do not know certain features exist. After you have finished the exercises in each section of the unit, you should spend some time experimenting so you can learn as much as possible about the package.

There is not enough room in this unit to go into every detail about Excel. You will have to find out things for yourself. There are a number of ways you can do this. For example:

- use a reference manual
- use the on-line help
- ask an experienced user to go through the program with you for a couple of hours
- use some of the tutorials on the Internet.

Using the Help facility

The Help facility is part of the software package and is therefore free (once you have bought the software). The idea of the Help menu is to give you help to perform your task, and it saves having to wade through thick manuals to find something out. When you call up the on-line help and are working something through the screen will split into two parts – one part contains your work and the other contains the help on the topic you have requested. This means you can apply the help to your own work because you can see both screens at the same time (in the past you would have had to flip from the Help screen to the screen containing your work or have had to remember or print the help).

In Excel 2000 Help is called Office Assistant. Using Office Assistant you can:

- obtain a new tip every time the software is loaded
- get it to answer your questions
- learn how to do things more quickly and easily.

Activity 3

Explain the purpose of a Wizard. Give a brief description of a task in Excel that a Wizard can be used for. If you do not know the answer straight away, use the Help facility to find out.

Using Help on the menu bar

If you look at the menu bar, you will see that there is a menu item Help. Clicking on this brings down the menu.

One very useful feature of Help is the What's This? option. If you click on this option the cursor becomes an arrow with a thick black question mark attached. This can be moved to any item on the screen on which you require help. For example, if you move it to the Bold icon on the formatting toolbar and then click the mouse, a screen appears telling you about this icon (Figure 3.14).

Figure 3.14

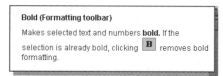

Bold (Formatting toolbar)

Makes selected text and numbers **bold.** If the selection is already bold, clicking **B** removes bold formatting.

Using on-line help to help with your assignment

It is interesting to look at the on-line help now and again to find out what can be done using the spreadsheet facilities. Unless you know about the facilities you won't be able to use them so the on-line help could give you some ideas when it comes to project work.

For the end of unit assignment you will need to do your own project, and it will be better to use the on-line help rather than ask your tutor for assistance – you can only gain the higher grades for the assignment if you have worked independently and unaided.

Activity 4 will give you practice in using this help.

You can use Activity 4 to provide evidence for key skills for Communication C2.2 or C3.2.

Activity 4

Here is a list of some of the things you might want to do using Excel or some of the facilities you will find it useful to know about:

* How to use the Cut and Paste feature.
* How to fill a column with a series, starting with 2, and going up in steps of 2 (2, 4, 6, 8 etc.) up to 100. NB: You must use the fill command.
* Move a group of cells using drag and drop.
* Use the ROUND function.
* Use the UPPER, LOWER and PROPER functions.
* Use the LEFT, RIGHT and MID functions.

Use the on-line help to find out about each item and write a short explanation in language that a novice user could understand. You can use screen shots (and screen capture programs if you have them) to help with your description.

Backing up a worksheet

For a variety of different reasons, sooner or later most computer users will experience some loss of data. Usually this happens at the most inconvenient time – such as the day before your GCSE project work has to be handed in. The many reasons why people lose data include:

- virus attack
- hardware fault
- a bug (error) in the software
- operator error
- loss of an important disk.

It is therefore important to keep a separate, backup copy of your work. The process of copying your work onto another disk is called backing up your data. Back up regularly and keep your backups in a safe place away from the computer, and also away from where you normally keep your disks. I always save my work on my hard disk and then onto a floppy at regular intervals (usually every half an hour).

Activity 5

a Why should you never keep your backup copy in the same place as the disk containing the original?

b Microsoft Office has a facility called AutoSave. Use the Help menu to find out what it is and how it can be useful. Write a short paragraph about AutoSave.

Taking a backup

When you are working on a spreadsheet you should periodically take a copy of your work. If you simply save then the latest version of the worksheet will replace the previous version, and if you mess up a spreadsheet and then save it you cannot easily go back to the previous version. Instead of using save you can use Save As (click on File to see this) and use different version numbers for each save – for example you could have 'Car Hire 1', 'Car Hire 2' etc. Once you are happy with the final version you can delete the earlier versions.

Activity 6

a Produce a screenshot of the screen in Microsoft Excel by pressing down the Alt key and, keeping it down, press the Print Screen key. Nothing will happen, because the screenshot is stored in the clipboard. Now load your word-processing software, create a new document, then go to Edit and Paste. The picture of the screen will appear in your document.

b Mark on your diagram (you can do this by hand or by using the drawing tools in the word processor):
- the active cell (sometimes called the cell pointer)
- the worksheet tabs (these tell you which worksheet in a workbook you are using)
- the title bar
- the close button

- the minimise button
- the restore/maximise button
- the menu bar
- a scroll bar
- the formatting toolbar
- the standard toolbar
- the column headings
- the row headings.

3.2 Using spreadsheet facilities

The aim of this section is to build up your skills in using Microsoft Excel. In order to build your portfolio of evidence in this area you will need to be a proficient user of the software and should understand the basic and more advanced features. This section takes you through a series of carefully graded exercises that will demonstrate many of the features you will use as evidence in your portfolio.

A spreadsheet package such as Microsoft Excel offers a huge number of facilities and you will have to be familiar with many of them. This section details some of the main facilities you need to know about.

Setting cell formats

Cells have to be capable of holding the data you want to put into them. Excel will interpret the data you put into a cell. What is displayed in a cell depends on the **cell format**.

Cells are given the general cell format by default and with these cells you simply enter the data. You have to change the cell formats to make a cell capable of holding the data you put into it.

/ Activity 7 /

Complete the table below, then open a worksheet and enter the data to check your answers.

Type of data	Data	Tick if you think you can enter it as shown	Tick if you were correct
Integer number	300		
Decimal number	32.9		
Decimal number	0.0008		
Date	12/12/99		
Percentage	17.5%		
Fraction	1/2		
Fraction	2 2/3		
Text	Correct		
Scientific	5.00E+7		
Scientific	5.56E-5		
Decimal number	0.0000000000004		

The general number format

Unless you tell Excel otherwise, it will set all cells to the general number format by default. A cell containing a formula will show the results of the formula, not the formula itself. The results of a formula are shown using eleven digits, which includes all the numbers before and after the decimal point.

Although each cell is set to the general number format, the program can change it, depending on the data you type in.

- If you type in a pound sign followed by a number, the spreadsheet will realise you are dealing with currency and will format the cell to currency automatically. It will only show the currency to two decimal places so if you typed in £1.349 only £1.35 would be shown.
- Large numbers are often written with a comma to make them easier to read (e.g. 3,000,000). As soon as such a number is entered, the spreadsheet will apply the number format with the thousands separator and use a maximum of two decimal places.
- If you enter a number ending in a % sign (e.g. 4%), then the spreadsheet will set the cell automatically to the percent format with two decimal places.
- If you enter any data that matches any of the date or time formats then the cell will be automatically set it to the nearest matching date format. Some of the date formats are shown in Figure 3.15.

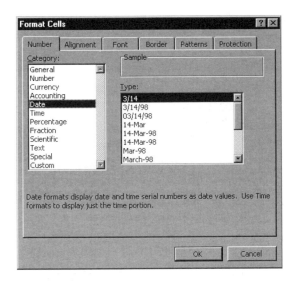

Figure 3.15 Date formats

When does 3 + 3 = 7?

Three plus three *can* equal seven if you don't consider the cell formats. Suppose you have 3.51 in cell A1 and 2.86 in cell B1 and both cells have been set to integers. If you place the formula =A1+B1 (to add these two cells) in cell C1 and cell C1 is set to integer, the result of the calculation will be 7 because Excel has rounded 3.51 to 4 and 2.86 to 3.

You can see that you have to be *very* careful with cell formats. Make sure that you match the cell format to the number of decimal places of the numbers being entered. In the example above all the cells should have been set to two decimal places.

Worked example: changing the cell formats of cells

1 Move the cursor to the cell.
2 Click on Format and then on Cells . . .
3 The Format Cells screen appears (Figure 3.15). Notice the list of categories you can have for a cell format. You can move to any of the formats in the list and click on OK to format the cell.

Activity 8

Open a worksheet and set the following formats.

- Set cell A1 to Number
- Set cell B1 to Date in the format 30/04/98
- Set cell C1 to Percentage (and to two decimal places).
- Set cell D1 to Fraction
- Set cell E1 to Currency
- Set cell F1 to Text

The fraction format

To format a cell to contain a fraction, open the Format Cells menu, then select Fraction (Figure 3.16).

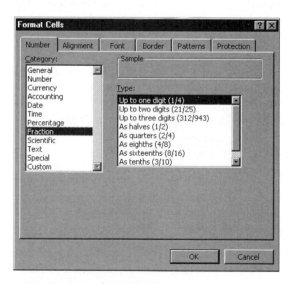

Figure 3.16 The Format Cells menu

You can select the type of fraction to use – for example, Up to three digits (312/943). Once a cell has been set to fraction format, you can enter data.

You enter the fraction ³⁄₁₆ into cell A1 by first entering 0 then leaving a space and entering 3/16. Double-click on this cell to see the decimal number that is equivalent to this fraction.

Try entering the fraction ⁴⁵⁄₉₈ into cell B1. Double-click on this cell to see the decimal equivalent of this fraction.

Enter the mixed fraction (a mixed fraction is one that also contains whole numbers) 3 4/5 into cell C1. Did you remember to set the cell format to fraction? What happens if you forget to change the cell format to fraction and enter 3/5 into cell D1?

Custom or special cell formats

To select a custom/special cell format, click on Custom in the Format Cell menu. If you scroll down the list of types, you will see the date format (dd/mm/yy) we are familiar with (Figure 3.17).

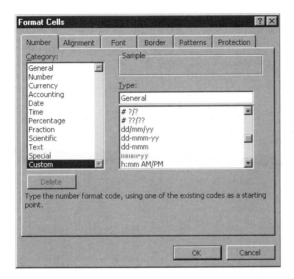

Figure 3.17 There are many custom formats to use

Worked example: using fractions in worksheets

1 Load Excel and create a new worksheet.
2 Format cells A1 to C4 to accommodate fractions with two numbers as the numerator (the top number) and two numbers as the denominator (the bottom number).
3 Enter the fractions in the positions shown in Figure 3.18.

	A	B	C
1	3/4	1 1/2	
2	5/8	2/3	
3	1 2/3	4 1/3	
4	10 1/13	12 3/25	

Figure 3.18

4 Enter a simple formula in cell C1 to add the fractions in cells A1 and B1.
5 Copy this formula down the column to show answers for the other fractions. Your final worksheet will now look like Figure 3.19.

	A	B	C
1	3/4	1 1/2	2 1/4
2	5/8	2/3	1 7/24
3	1 2/3	4 1/3	6
4	10 1/13	12 3/25	22 13/66

Figure 3.19

Activity 9

The answers in column C of the spreadsheet in Figure 3.19 are assumed to be correct. After all, the additions were all calculated by the computer. Surely nothing could have gone wrong?

a Describe how you could use a test strategy to make sure that the results displayed on the worksheet are correct.

b Use the test strategy you have described to test the answers the worksheet is producing.

c One of the answers in the worksheet is not quite right. Which one is it and why is it not producing an exact answer? Alter the worksheet so that the correct fraction is shown.

Cell presentation

Data can be presented in cells in a variety of different ways.

Aligning cells

When you enter data into a cell the package automatically aligns (i.e. positions) the cells.

- Numbers are aligned to the right.
- Text is aligned to the left.

Do *not* put any spaces in front of numbers in order to align them as this will make it impossible for the spreadsheet to use the numbers in calculations.

If you want to align the data differently, you can use the buttons for alignment in the formatting toolbar to align them to the left, right or centre (Figure 3.20).

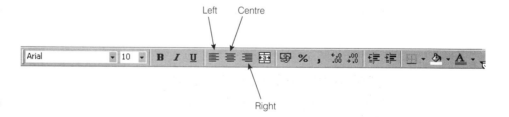

Figure 3.20

You can pre-set the arrangement of cells using the Format Cells screen (Figure 3.15). First highlight the cell or cells to be set, and then open Format Cells. Click on the Alignment tab and the screen changes (Figure 3.21).

There are some very useful things in this menu. If you click on the drop-down arrow for Horizontal text alignment the screen shown in Figure 3.22 appears. All the settings in the drop-down list can be applied to data in a cell to control the horizontal position of the data within that cell.

You will see a similar screen for controlling the vertical alignment of data in a cell in Figure 3.23.

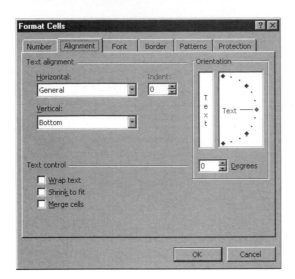

Figure 3.21 The alignment screen in Format Cells

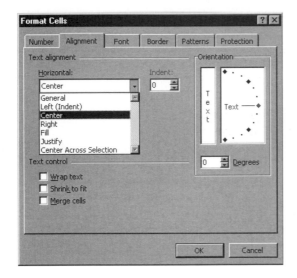

Figure 3.22

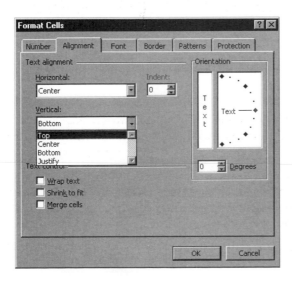

Figure 3.23

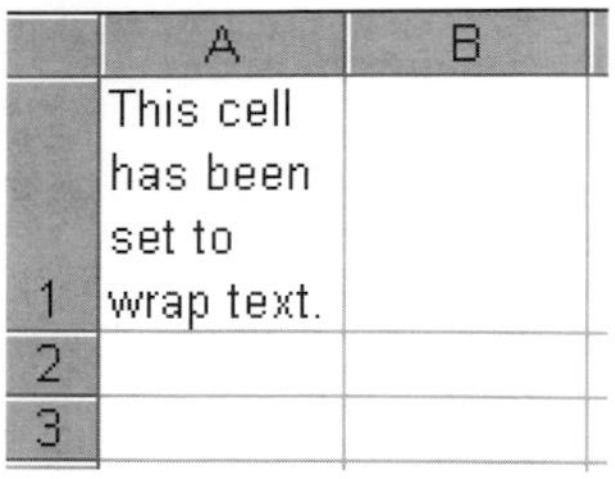

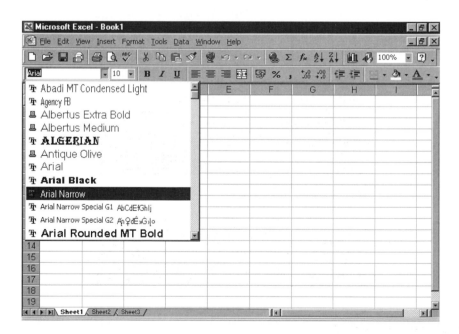

Here are some other features you may find useful.

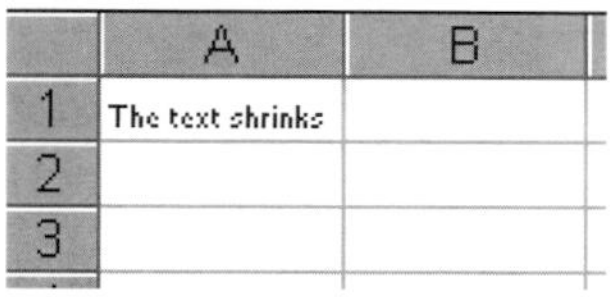

Wrap text

Allows the text to flow down, enlarging the height of the cell in the process (Figure 3.24).

Shrink to fit

Makes the data smaller so that it will fit the existing cell size (Figure 3.25).

Merge cells

Allows the contents of one cell to move into another. The lines separating the cells are also removed in the process (Figure 3.26).

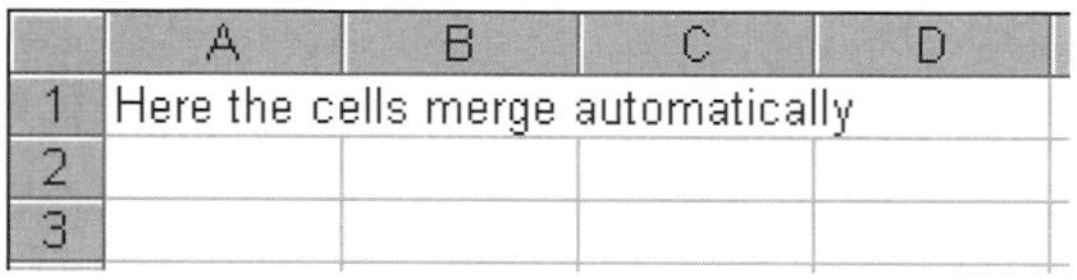

Figure 3.24

Figure 3.25

Figure 3.26

Some of the features outlined can be used together. Which ones?

Fonts

A **font** is the typeface used to display a character. The selection of fonts available is obtained by dropping down the font list (Figure 3.27).

Figure 3.27 A large number of fonts are available for use in Excel

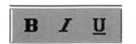

Font style buttons

Font style

The font style includes **bold**, underline, *italics* etc. To select the font style, highlight the cell or cells and then press one or more of the buttons opposite.

Font size

A font's size is measured in points. One point is 1/72 of an inch. The default font size in Excel is 10 point but you can change it (Figure 3.28).

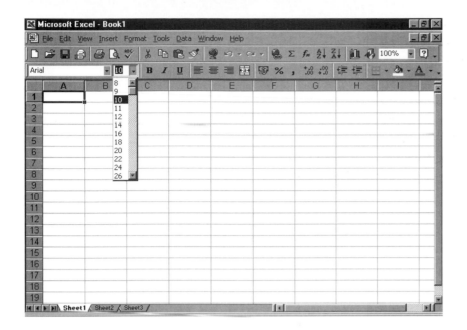

Figure 3.28 Changing the font size in Excel

Font colour

The font colour can be changed by selecting a colour in the drop-down palette in the Format Cells screen in Figure 3.29 or by using the font colour button on the toolbar.

Font colour button

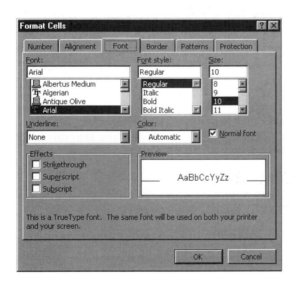

Figure 3.29

Adding borders

To add borders around a cell or a group of cells you highlight the cell or cells and then use the Format Cells function. Click on Border and a new menu appears (Figure 3.30).

You can now decide on the border around the cells.

* First, go to the Line part of this menu and select the type of lines to make up the border. The thick black lines and the double line are quite useful to make part of a spreadsheet stand out. In this part of the menu, you can select the colour of the border.

Explain the differences between font style and font size.

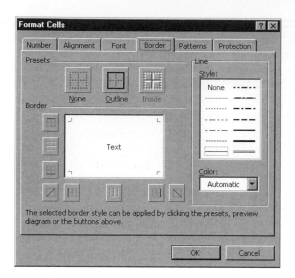

Figure 3.30 The Borders tab in Format Cells

- Choose whether you want a border around just the cells you have selected or outside as well as inside the spreadsheet.
- You can also select just part of a border and you can have diagonal lines across borders.

Fill colours and patterns

Cells can be filled with colour, as a background to the data in the cell or simply to mark out areas where there is no text. Either way, colours can considerably improve the appearance of a spreadsheet. However, choose the combination of text colour and cell fill colour carefully.

As well as the fill colour you can use patterns. Patterns are best used to divide up the worksheets. If you try to use a pattern as a background for those cells that contain data, the data becomes almost impossible to see. It is therefore very important to make sure that you do not destroy the legibility by 'brightening up' the spreadsheet too much.

Also bear in mind the purpose to which the worksheet will be put. Things that work well in colour may not work so well if the worksheet is printed out in black and white.

Worked example: improving the appearance of a spreadsheet

1 Set up the following worksheet shown in Figure 3.31. Proofread the data.

	A	B	C	D	E	F
1	Forrest View Garden Centres					
2	Annual Sales By Department					
3		Spring	Summer	Autumn	Winter	Total
4						
5	Seasonal	9800	14500	15600	21900	
6	Plants	21800	36900	14500	2800	
7	Tools	8900	12700	6400	3900	
8	Compost/Fertiliser	10900	12300	6100	1900	
9	Totals					

Figure 3.31

2 Insert a formula in cell F5 to work out the total of the amounts in cells B5 to E5. Copy this formula for rows 6–8. Insert a formula to work out the totals of the columns in columns B to F.

3 Highlight the text Forrest View Garden Centres and change the font to Arial Black, 18 point.

4 Select the text Annual Sales By Department. Increase the font size to 12 and apply the bold and italics font styles.

5 Embolden the headings in cells B3 to F3. Do the same with the text in cells A5 to A9.

6 Check that your worksheet is identical to the one shown in Figure 3.32.

	A	B	C	D	E	F
1	**Forrest View Garden Centres**					
2	*Annual Sales By Department*					
3		Spring	Summer	Autumn	Winter	Total
4						
5	Seasonal	9800	14500	15600	21900	61800
6	Plants	21800	36900	14500	2800	76000
7	Tools	8900	12700	6400	3900	31900
8	Compost/Fertilise	10900	12300	6100	1900	31200
9	Totals	51400	76400	42600	30500	200900

Figure 3.32

7 Did you notice that some of the text disappeared? Emboldening text makes it slightly bigger and some of it can not be displayed in the current column width. To correct this, widen column A. Move the cursor to the line between the A and B headings of the column, click and then drag to adjust the column width so that all the text is visible.

8 Format all the numbers in cells B5 to F9 to currency without any decimal places.

9 Select cells A1 to F1. Format the font colour for the text in these cells to red.

10 Select cells A1 to F8 and set the fill colour to the palest yellow on the palette.

11 Select cells A2 to F2, then select Format Cells. Now use the Alignment tab to alter the horizontal alignment to centre the text across the selection (Figure 3.33). This will put the heading in the middle of the selected cells.

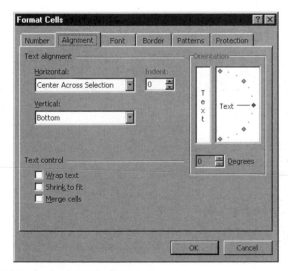

Figure 3.33 Altering the horizontal alignment

147

12 Select cells B3 to F3 and centre these cells in each of their columns.
13 Select cells from A9 to F9 and open the Format Cells menu. Now select the Borders tab (Figure 3.30). Select the double line in the Style box, choose the Outline border button and then click OK. This puts a double line around the edge of the selected cells.
14 Select cells A9 to F9 and set the fill colour to a darker yellow.
15 Check that your worksheet looks like Figure 3.34.
16 Save the worksheet using the filename 'Forrest View' and print a copy.

	A	B	C	D	E	F
1	**Forrest View Garden Centres**					
2	*Annual Sales By Department*					
3		Spring	Summer	Autumn	Winter	Total
4						
5	Seasonal	£9,800	£14,500	£15,600	£21,900	£61,800
6	Plants	£21,800	£36,900	£14,500	£2,800	£76,000
7	Tools	£8,900	£12,700	£6,400	£3,900	£31,900
8	Compost/Fertiliser	£10,900	£12,300	£6,100	£1,900	£31,200
9	Totals	£51,400	£76,400	£42,600	£30,500	£200,900

Figure 3.34

Using and manipulating spreadsheets

Going to a specified cell

On a large spreadsheet it can take some time to scroll up and down, left and right to find a particular cell. There is a quick way of moving to a cell if you know the cell reference. Click on Edit and then Go To. The screen in Figure 3.35 is displayed.

Figure 3.35

Type in the cell reference (L44) and then click OK. The cursor moves straight to the required cell.

You can also type in cell ranges (e.g. K3:P3). Try this out. Notice that the whole cell range will be highlighted.

Find data

If you do not know a cell reference, it is useful to find a cell containing certain data in a worksheet.

Consider a worksheet containing the names of countries and their capitals. Suppose we wanted to find the country whose capital was Copenhagen.

Click on Edit and then Find. The screen shown in Figure 3.36 is displayed.

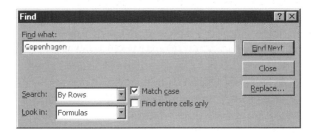

Figure 3.36
Using the Find screen

Type in 'Copenhagen'. If you click on Match case the search will find an exact match (if there is one). Click on Find Next.

The cursor will now appear on Copenhagen in the worksheet.

Search and replace

If you look at the Find screen in Figure 3.36 you can see that it is possible to replace an item with another one automatically each time it occurs. This is similar to the search and replace facility in word-processing packages.

Cut, copy and paste

Cut and paste
To cut cells select them (the selected cells will be highlighted) then click on Edit and Cut. The cells will be stored in a temporary storage (the clipboard) by the computer. You can now move to another part of your worksheet where you want the cell or cells to be moved to, click Edit and then Paste. You can move cells to a different position in the same worksheet – or a different one – using this method.

If the worksheet contains other cells with formulas that refer to the cells you are moving, do not worry – the formulas will be adjusted to take into account the moved cells positions.

Copy and paste
When we cut cells, they are taken out of the worksheet. To use the same cells in a different worksheet without having to type them in we can use Copy. Copy is similar to cut except only a *copy* of the cells is put into the clipboard. The spreadsheet from which the cells are copied remains unaltered. Once copied, the cells can be pasted anywhere in the same or a different worksheet.

Using Paste Special
One way of moving and manipulating data on a worksheet is by making use of Paste Special. This feature copies the contents of one or more cells to the clipboard.

To use Paste Special copy the contents of the cells, then click Edit and Paste Special. The screen shown in Figure 3.37 appears.

Suppose a computer supplier has a price list to accompany a catalogue they send out to customers. The price list is held as a spreadsheet. If all the prices in a sale are reduced by 20% then you could create a new column and use formulas to put the new prices in this column. If you did not want to show the old prices as well you could use Paste Special.

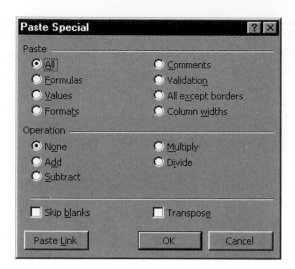

Figure 3.37 The Paste Special screen

Worked example: using Paste Special

1 Create a new worksheet and enter the data exactly as it is shown in Figure 3.38.

	A	B	C
1	**Computer Supplies Price List**		
2	**Product**	**Price**	
3	Kodak DC-240 Digital Camera	£345.90	
4	Olympus C-2000 Digital Camera	£546.00	
5	Agfa CL50 Digital Camera	£350.00	
6	Iomega 250MB Zip Drive	£121.99	
7	Iomega CDRW	£110.00	
8	Trisys Mouse Tablet	£34.99	
9	HP 3300C Scanner	£89.99	
10	Freecom External CD-RW	£165.99	
11	Hayes Accura V90 Modem	£59.99	
12	USB Hub	£31.00	
13	Philips 15" LCD 151AX	£650.00	
14	Taxan Ergovision 17" TC099	£245.00	
15			
16			

Figure 3.38

2 Enter the value you want to reduce the cells by (0.8) into the empty cell D3 .
3 Select cell D3 and copy it to the clipboard.
4 Select the prices in cells B3 to B14 by highlighting them (Figure 3.39).
5 Click on Edit and then Copy to copy the cells in the range selected to the clipboard.
6 Because we want these cells to be pasted in the same position as they were but alter them to 0.8 of their original value, keep the same range highlighted. Click on Edit and then Paste Special. The Paste Special menu (Figure 3.37) will appear.

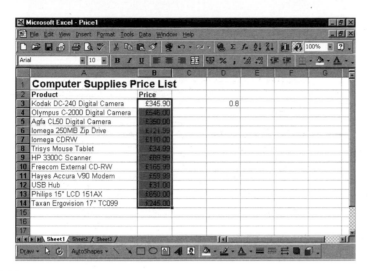

Figure 3.39

7 Click on Multiply – and make sure that you click on Values in the Paste part of the box. This is needed because we want to keep the currency format of the existing numbers. Then click OK.
8 All the values have been multiplied by 0.8 to show the 20% reduction.

Activity 10

Paste Special can be used to flip data in a column into a row and vice versa. The facility to do this in the Paste Special dialogue box is called Transpose. Enter the surnames of ten people in a column and then use Paste Special to put the data into a row.

Cell referencing

There are two ways of referring to another cell in a spreadsheet: **relative referencing** or **absolute referencing**. This is particularly important when you copy or move cells.

- An absolute reference always refers to the same cell in a spreadsheet.
- By contrast a relative reference refers to a cell which is a certain number of columns and rows away from the current cell.

This means that when the current cell is moved or copied to a new position, the cell to which the reference is made will also change position. This is best seen by referring to Figures 3.40 and 3.41.

You need do nothing to specify relative cell references because they are the default. Relative cell references are useful because the spreadsheet will adjust the formulas automatically when they are copied to other cells.

If you need to refer to a particular value in a certain cell (such as an interest rate, VAT rate, etc.) you must make any reference to it absolute. This is usually done by putting a dollar sign in front of the column and row numbers. For example, the relative cell reference C4 could be converted to an absolute cell reference by inserting the dollar sign thus: C4.

Explain the difference between an absolute and a relative cell reference.

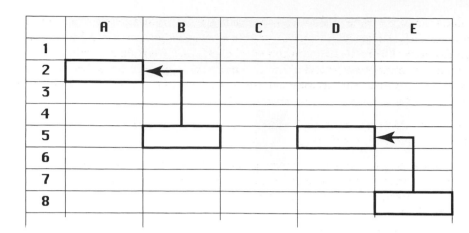

Figure 3.40 Cell B5 contains a relative reference to cell A2. If the contents of cell B5 are copied to E8 then cell E8 will refer to cell D5

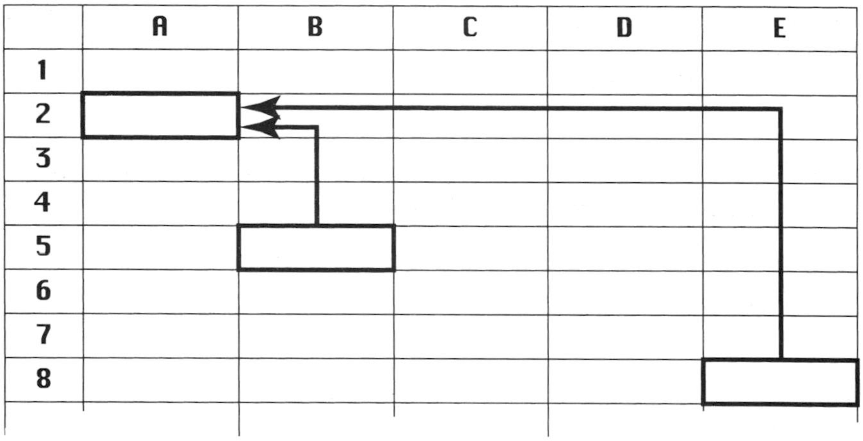

Figure 3.41 If cell B5 had an absolute reference to cell A2, then if the contents of cell B5 are copied to E8, cell E8 will still refer to cell A2

Using operators

Once data has been input to a model, the data may be subjected to a range of different types of **operators**:

- arithmetic (+, -, ×, /, √ etc.)
- relational (=, <, >, IF)
- logical (AND, OR, NOT).

Formulas

If you are constructing a worksheet using spreadsheet software it is important to establish the mathematical relationships between the variables and constants located in the cells in the worksheet. You do this using formulas. A formula in one cell could reference a value in another, thus making it easy to find the sum of a list of numbers, to calculate interest rates or to make financial predictions. All spreadsheets are able to perform a variety of arithmetical operations, from simple addition, subtraction, multiplication and division to more complex operations that calculate statistical derivatives such as mean and standard deviation, or business calculations such as working out monthly payments for loans.

When constructing formulas it is important to bear in mind the order in which the operations should be performed.

1   Brackets (parentheses): any operations in brackets should be performed first. When there is more than one set within the same calculation, the inner ones are worked out first then the next set, until the outermost brackets are reached.
2   Powers: squares, square roots, cubes, etc.
3   Multiplication and division: if both are present in the same calculation it does not matter in which order they are performed.
4   Addition and subtraction: if both are present in the same calculation it does not matter in which order they are performed.

Parentheses

You can control the order in which the steps in a calculation are carried out by using parentheses (also called brackets). Any step in parentheses is evaluated first. Consider the following calculation:

=3 + 5*6

According to the rules governing the order in which calculations are performed, the multiplication is performed before the addition:

=3 + 5*6

=3 + 30

=33

If you wanted to do the addition first, you must enclose it in parentheses:

=(3+5)*6

=8 * 6

=48

To help you understand formulas and the use of parentheses, here is a simple worksheet containing some numbers (Figure 3.42).

	A	B	C
1	10		3
2		4	
3			
4	5		
5			25

Figure 3.42

The table below shows formulas that use the data in this worksheet. The table also shows how each formula should be worked out, and the final result to the calculation is also shown.

Formula	The way it is calculated	Result
=A1+B2*C1	=10+4*3	22
=(A1+B2)*C1	=(10+4)*3	42
=(A4+B2)^2	=$(5+4)^2$	81
=(A1-B2)*(A4-C1)	=(10-4)*(5-3)	12
=A1*B2/A4	=10*4/5	8
=A10+10%	=10+0.1	10.1
=10%*C5	=10%*25	2.50
=5+10%*C5	=5+2.50	7.50
=A4^2+C1^2	=5^2+3^2	34

You can use Activity 11 to provide evidence for key skills for Application of Number N2.2 or N3.2.

Activity 11

Copy and complete the table using data from the worksheet in Figure 3.43.

	A	B	C	D
1	30			
2			6	5
3				
4	4		24	
5				-5

Figure 3.43

Formula	The way it is calculated	Result
=A1+C2*D2		
=C4/6+8		
=C4/A4+C2*D2		
=4+A1*A4		
=(A6+D5)*A4		
=15%+A4		
=A4^2		
=(C2+C4)/D2		
=(A1-C4)/C2		
=A1-C4/C2		
=20+10%*A1		
=C2^2+D2^2		

Text concatenation (text joining)

The **text concatenation** operator used by Excel is the ampersand (&). It is used to join (concatenate) one or more pieces of text. For example, to change 'Forrest' and 'View' to the single piece of text 'ForrestView':

1 Create a new worksheet and enter the text Forrest and View into cells A1 and B1 respectively.
2 In cell C1 insert =A1&B1 to join the text in the two cells together. Your worksheet will now show the concatenated text (Figure 3.44).

	A	B	C
1	Forrest	View	ForrestView
2			
3			
4			

Figure 3.44

Using built-in functions

Functions are pre-defined formulas used for making certain types of calculation.

Each function starts with an equals sign, has a name and also an argument, which is a list of the cell references to be used.

An example of a function is:

=SUM(A1:F1)

There are many different functions, some of which are extremely complex and powerful. Here are explanations of a few of the more widely used functions.

Average
=AVERAGE(A1:A10) gives the average of all the numbers in the cells from A1 to A10.

Maximum
=MAX(E4:E11) displays the largest number from all the cells from E4 to E11 inclusive.

Minimum
=MINIMUM(B2:B12) displays the smallest number from all the cells from B2 to B12 inclusive.

Count
Suppose we want to count the number of numeric entries in the range C3 to C30. We can use =COUNT(C3:C30). Any blank lines or text entries in the range will not be counted. To count a number of items or names of people we need to be able to count text entries. To do this we can use =COUNTA(C3:C30). You need to make sure that headings are not included in the range so that they are not counted as well. Again blank lines are not counted.

INT
The INT function rounds a number down to the nearest whole number. If the number 121.242 is in cell A1 and =INT(A1) is typed in cell B1 then 121 will appear.

MODE
The mode is the number that appears most often in a list. For example, =MODE(A1:A20) returns the most frequent value in the cells from A1 to A20.

MEDIAN
The median is the middle value when all the numbers have been arranged in order of size. If there are two middle values (i.e. if there are an even number of numbers) the mean of these two values will be given. For example, =MEDIAN(A1:A20) returns the median value in the cell range A1 to A20.

RAND
The RAND function creates a random number between and including the numbers zero and one. Try this by typing =RAND() into separate cells. Notice that you get

155

numbers like 0.826455. This is not much use, because what you really need is a way of setting the random number between a high and a low value. You can do this using the RAND function llike this:

$$=RAND()\star(high-low)+low$$

Suppose you want to create a random number between 1 and 10. In this case high =10 and low =1. The function would be constructed as follows:

$$=RAND()\star(10-1)\star1$$

The problem with this is that numbers are given to several decimal places, like 4.398839. We would usually prefer to turn these numbers into integers. This can be done by combining the RAND and INT functions:

$$=INT(RAND()\star(10-1)+1)$$

Each time the worksheet is recalculated, the function produces a random number between and including 1 and 10.

SUMSQ

The function =SUMSQ(B4) will square the number in cell B4. SUMSQ(A1:F1) will find the sum of all the squares of the numbers in cells from A1 to F1.

SQRT

The function =SQRT(A4) will find the square root of the number in cell A4.

IF

If, for instance, you wanted to use a spreadsheet to make the following decision a relational operator could be used.

If an invoice total exceeds £1000 then a discount of 5% of the invoice total is given.

If the invoice total is in cell C4, then to calculate the discount we can use:

$$=IF(C4>1000,C4\star0.05,0)$$

which would put the answer (the amount of discount) in the cell where the formula is located. If the invoice total were less than £1000, a figure of zero discount would be shown in the cell.

The IF function makes use of something called relational operators. You may have come across these in your mathematics lessons but it is worth going through what they mean.

Symbol	Meaning	Examples
=	equals	5 + 5 = 10
>	greater than	5*3 > 2*3
<	less than	−6 < −1 or 100 < 200
<>	not equal to	"Red" <> "White" or 20/4 <> 6*4
<=	less than or equal to	"Adam" <= "Eve"
>=	greater than or equal to	400 >= 200

The most popular operator by far is the equals sign, but sometimes a comparison needs to be made between two items of data. For example, we may need to find a list of employees whose salaries are greater than a certain amount.

Operators can also be used with characters or character strings, so one character can be compared with another. Since each character has a binary (ASCII) code associated with it, the computer can work out that A comes before B and so on. You can also test to see if the contents of a certain cell have a certain word in them. For example you could test to see IF B6='Yes'. The IF function is structured like this:

=IF(Condition, value if true, value if false)

The value can either be a number or a message. If the values are messages, then they need to be enclosed inside quote marks (e.g. 'High').

Logical operators

In addition to their use when making up search criteria with databases, **logical operators** (AND, OR and NOT) can be used in spreadsheets for making decisions based on the values of certain cells. For example, if cells C2 and D2 contain totals which for a 5% discount to be given must both exceed 1000, we can apply the discount to the total in another cell (say D8) by using the following construct:

=IF(C2ANDD2>1000,D8*0.95)

Worked example: choosing between giving a discount and not giving one

We can construct a spreadsheet to do this.

1 Create a new worksheet and enter the data in exactly the same places as shown in the Figure 3.45.

	A	B
1	**Invoice Number**	**Invoice value**
2	A1210	£235.89
3	A1092	£1,218.90
4	A1109	£122.67
5	A1091	£3,425.12
6	A1211	£223.80
7	A1309	£1,200.56
8	A1300	£77.45
9	A1288	£320.09
10	A1200	£9.10
11	A1199	£4.00
12	A1189	£3,932.77
13	A1201	£329.90
14	A1209	£24,900.00
15	A1303	£890.68
16	A1301	£1,000.00

Figure 3.45

2 Put the heading Discount in cell C1 and enter the formula =IF(B2>1000,B2*5%) in cell C2. Copy this formula down the rest of the column (Figure 3.46).
3 Put the heading Discounted invoice value in cell D1. In cell D2 enter the formula =B2+C2 and copy the formula down the column. Your worksheet will now look like Figure 3.47.
4 Save the worksheet using the filename 'Invoice Discount'.

	A	B	C
1	Invoice Number	Invoice value	Discount
2	A1210	£235.89	£0.00
3	A1092	£1,218.90	£60.95
4	A1109	£122.67	£0.00
5	A1091	£3,425.12	£171.26
6	A1211	£223.80	£0.00
7	A1309	£1,200.56	£60.03
8	A1300	£77.45	£0.00
9	A1288	£320.09	£0.00
10	A1200	£9.10	£0.00
11	A1199	£4.00	£0.00
12	A1189	£3,932.77	£196.64
13	A1201	£329.90	£0.00
14	A1209	£24,900.00	£1,245.00
15	A1303	£890.68	£0.00
16	A1301	£1,000.00	£0.00

Figure 3.46

	A	B	C	D
1	Invoice Number	Invoice value	Discount	Discounted invoice value
2	A1210	£235.89	£0.00	£235.89
3	A1092	£1,218.90	£60.95	£1,157.96
4	A1109	£122.67	£0.00	£122.67
5	A1091	£3,425.12	£171.26	£3,253.86
6	A1211	£223.80	£0.00	£223.80
7	A1309	£1,200.56	£60.03	£1,140.53
8	A1300	£77.45	£0.00	£77.45
9	A1288	£320.09	£0.00	£320.09
10	A1200	£9.10	£0.00	£9.10
11	A1199	£4.00	£0.00	£4.00
12	A1189	£3,932.77	£196.64	£3,736.13
13	A1201	£329.90	£0.00	£329.90
14	A1209	£24,900.00	£1,245.00	£23,655.00
15	A1303	£890.68	£0.00	£890.68
16	A1301	£1,000.00	£0.00	£1,000.00

Figure 3.47

DATE

When a date is typed into a cell, Excel converts it to a number. This number is the number of days that have elapsed between 1 January 1900 and the date typed in. To see this number type in a date into a cell formatted for a number.

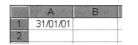

	A	B
1	31/01/01	
2		

Figure 3.48a

	A	B
1	36922	
2		

Figure 3.48b

Create a new worksheet and enter the date shown into cell A1 (Figure 3.48a). Now change the format of this cell to Number without any decimal places (i.e. an integer). The format will appear as shown in Figure 3.48b. So 36922 days have elapsed between 1 January 1900 and 31 Jan 2001.

Because dates are stored as numbers by the computer the computer can perform calculations using dates. For example, two dates can be subtracted to find the number of days between them, which is useful for calculating fines, days overdue, etc.

Be careful with dates – America and many other countries use dates in the format mm/dd/yy, but the UK style is dd/mm/yy.

Now get the form back. To search for an employee called 'Flynn', click on Criteria. This clears the contents of the form (Figure 3.58). Type one of the criteria you're looking for – e.g. the surname Flynn in the Surname field. As soon as you press Enter the record for Flynn will be displayed (Figure 3.59).

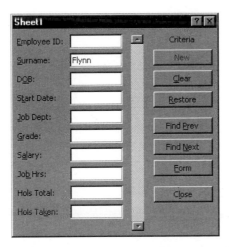

Figure 3.58

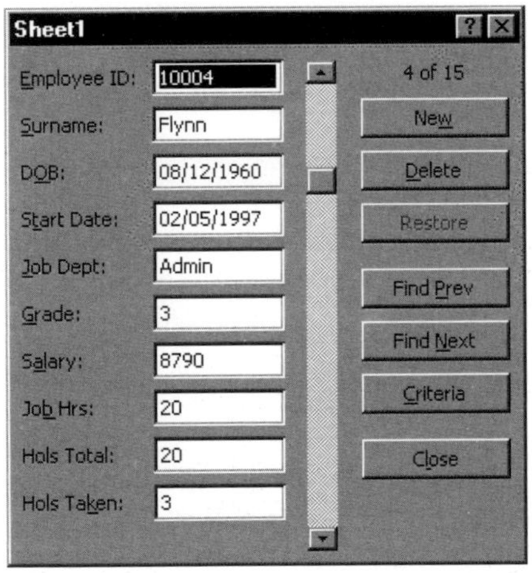

Figure 3.59

You can also add operators to your search criteria (these are listed in the table on page 156).

/ Activity 13 /

Try the following searches.

a   Find the employees whose date of birth is after 1 September 1960. Select the whole worksheet and open the form. Click the Criteria button and enter the criteria as shown in Figure 3.60.

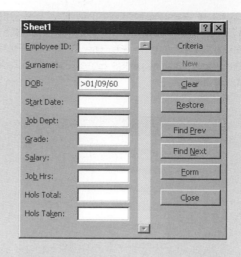

Figure 3.60

Press Enter to find the first matching record. Only one record will be shown at a time so you will have to use the Find Next button to see the others.

b Go back in to Criteria and clear the form by clicking Clear. Look up the details of those employees who are in Production and are Grade 2 by entering the details as shown in Figure 3.61. The spreadsheet will link them automatically with AND, so only those employees satisfying both criteria will be picked.

Figure 3.61

Sorting the data in the database

You can sort on as many as three fields, which is particularly useful if the database is very large and a search on just one field would retrieve lots of records. When the data is sorted all the records are kept together.

You can sort on:

- ascending order (A–Z for letters and 0–9 for numbers)
- descending order (Z–A for letters and 9–0 for numbers).

Worked example: sorting the employee database into different orders

The data in the employee database/list created in the last exercise was entered in order of employee ID, but it is useful to be able to sort the data in different ways.

1 Open the employee database. Click on any surname in the surname column or the field name itself and then click on the ascending order (A–Z) icon in the standard toolbar (Figure 3.62).
   Your database will now be in order of surname. Print out a copy of your worksheet.
2 Click on any item in the date of birth column and sort by descending date of birth. This will put the youngest employee first.
3 You can sort into several orders at once. For example, you might want to sort in order of department and then have the surnames in alphabetical order within each department. Take the following steps to do this.
   • Click on Data in the menu toolbar and then on Sort.
   • Pick Sort by Job_Dept and Then by Surname (Figure 3.63).
   • Click on OK. You should obtain the results shown in Figure 3.64.

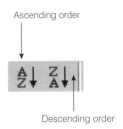

Ascending order

Descending order

Figure 3.62
Sorting by ascending
and descending order

Figure 3.63

	A	B	C	D	E	F	G	H	I	J
1	Employee_ID	Surname	DOB	Start_Date	Job_Dept	Grade	Salary	Job_Hrs	Hols_Total	Hols_Taken
2	10011	Farley	01/01/49	03/09/75	Admin	1	£28,900	40	30	0
3	10004	Flynn	08/12/60	02/05/97	Admin	3	£8,790	20	20	3
4	10010	Jenkins	30/09/70	31/01/90	Admin	2	£17,900	37.5	30	13
5	10008	Smith	01/09/74	03/12/92	Admin	1	£25,900	37.5	30	1
6	10005	Jones	04/03/71	03/04/99	Marketing	2	£20,800	37.5	30	2
7	10015	Sinnot	27/02/50	22/08/80	Marketing	2	£15,600	30	25	0
8	10012	Atkins	03/02/70	09/01/97	Prodcution	1	£34,900	40	35	9
9	10001	Gregson	30/01/43	04/03/67	Production	2	£19,800	37.5	30	12
10	10003	Johnston	01/08/79	09/01/99	Production	2	£15,672	35	30	0
11	10009	Smith	03/08/69	08/12/91	Production	1	£25,800	37.5	30	0
12	10014	South	01/12/49	12/01/69	Production	3	£17,800	40	30	3
13	10006	Doyle	01/12/59	04/01/98	Sales	3	£15,600	37.5	30	20
14	10002	Hughes	12/06/78	07/09/98	Sales	3	£14,780	37.5	28	10
15	10007	Jones	03/02/78	07/11/97	Sales	4	£9,450	37.5	30	23
16	10013	Smith	06/02/67	30/05/90	Sales	2	£12,300	20	22	10

Figure 3.64

Data entry messages

What is the main purpose of a data entry message?

You can attach messages to cells so that when the cursor is moved to the cell your message is displayed. Doing this allows you to add comments and messages without cluttering up the worksheet.

To attach a comment or data entry message to a cell click on the cell, go to the toolbar and choose Insert and then Comment. A pop-up box appears, into which you can put a suitable comment or message. Click on the outside of the box when you have finished the message. As soon as you move the cursor onto the cell, your message will be displayed. An example is shown in Figure 3.65.

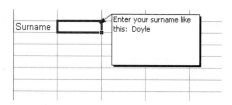

Figure 3.65

Naming cells and cell ranges

Cell references are hard to remember. Rather than use these you can give cells and cell ranges a name, which can be referred to in formulas. Another good thing about using names is that if you insert a row above or to the left of a cell, the cell reference will change but the name stays the same.

To give a name to a cell or a range of cells:

1 Select the cell or cells you want to name.
2 Click on Insert, Name and then Create.
3 In the screen that appears, enter the name (Excel will give it a name but you are free to choose your own).
4 If you have several cells or ranges to name, use the Add button and repeat steps 3 and 4 for each one.
5 Click OK.

Worked example: naming cells and cell ranges

1 Create a new worksheet and enter the data as shown in Figure 3.66.

	A	B	C	D
1				
2		Sales	Costs	Profit
3	1st Quarter	£121,000	£87,000	
4	2nd Quarter	£139,000	£94,000	
5	3rd Quarter	£209,000	£167,000	
6	4th Quarter	£120,800	£109,000	

Figure 3.66

2 Highlight cells B2 to C6.
3 Click Insert, Name and then Create. The box shown in Figure 3.67 appears. Use the same names as in the top row of the worksheet. Make sure that there is a tick in Top row. Click OK. The first column is now named Sales and the second column is named Costs. ▶▶

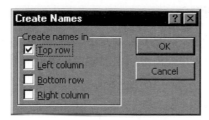

Figure 3.67

4 To see all the names that have been used in a worksheet select the entire worksheet. Click on Insert, Name and then Define. The screen that appears (Figure 3.68) lists the names. Close the window.

Figure 3.68

5 In cell D3 enter the formula =Sales-Costs. The spreadsheet recognises the names and the answer will be displayed.
6 Copy this formula down the column for the other values.

Activity 14

Work out for yourself and write instructions on how to:

a Delete a name
b Select a cell using its name.

Autofill lists

Suppose you want to type the days of the week or months of the year down a column or across a row. Excel is able to anticipate what you probably want to do by the first word alone. So, if you type Monday, then the chances are that you want Tuesday in the next column or row and so on.

To see how this works, go through the next exercise.

Worked example: using AutoFill

1 Create a new worksheet and in cell A2 type Monday.
2 Position the mouse pointer on the fill handle (the small square at the bottom right-hand corner of the cell) and drag it down the column. You will see a box with the day of the week appear as you move from cell to cell. Stop when you reach Sunday.
3 As soon as you remove your finger from the mouse button, the days of the week will be entered. Click on the mouse button to de-select the cells (i.e. remove the highlighting).
4 Type 'Week 1' in cell B1 and use AutoFill to fill in the row up to Week 5. Your worksheet now looks like Figure 3.69.
5 Exit without saving.

	A	B	C	D	E	F
1		Week 1	Week 2	Week 3	Week 4	Week 5
2	Monday					
3	Tuesday					
4	Wednesday					
5	Thursday					
6	Friday					
7	Saturday					
8	Sunday					
9						

Figure 3.69

Activity 15

a There are many ways in which you can fill the data automatically into cells once an example has been placed in one of the cells. Investigate, using the on-line help or by experiment, which of these will work:
- Product 1, Product 2 etc.
- Jan, Feb etc.
- Mon, Tue etc.
- 9:00, 10:00 etc.
- Jan 01, Feb 01 etc.
- Product A, Product B etc.
- 1st Qtr, 2nd Qtr etc.
- Qtr1, Qtr2 etc.
- 1999, 2000 etc.

b A company employs ten sales staff. Their names are:

Hughes, P	Jones, K	Alinson, P	Murphy, J	Monk, J
Freeman, H	Adams, A	Sumner, D	Blair, C	Garrett, A

This list is needed over and over again in spreadsheets. The sales manager has asked if it can be used as a custom list. You have heard that as well as the lists provided by Excel (days of the week, months of the year, etc.) you can also make a custom list. Find out how to do this and add the names to the custom list. Also produce some brief instruction that the manager can follow to add these names quickly to a spreadsheet. **Hint:** To produce a custom fill series click on Tools, Options and then click on the Custom Lists tab. The screen shown in Figure 3.70 will appear.

►►

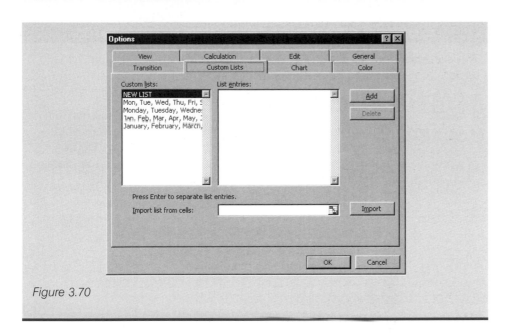

Figure 3.70

Templates

Word-processing packages come with many different **templates**, which allow a variety of different documents to be created (reports, memos, letters, etc.). Templates are really frameworks. They consist of the layout and headings but without the variable content. Excel also provides some spreadsheet templates, although not as many as Word. These include templates to produce invoices, purchase orders and expense reports.

You can also create your own worksheet framework and save it as a template.

Activity 16

Load Excel and look at the templates that are provided by the package.

There are many situations when you need to produce a document with exactly the same format many times with variable data in it. An example might be your yearly accounts. Rather than create the worksheet from scratch each time, you could create a template and save it. The template would ensure consistency between one worksheet and another, which is important if the worksheets need to be linked together using 3D cell references. For example, you could produce a sheet showing how the income, expenditure and profit have varied over several years and show the results using a chart.

What would be included in a template?

A template includes everything except the variable data (i.e. that which changes from sheet to sheet). Headings, labels, comments, cell formats, formulas are all included as part of the template.

Worked example: creating a template

The spreadsheet you created in Section 3.2 (pages 146–148) can be turned into a template for use in future years.

1 Open the Forrest View worksheet (Figure 3.34).
2 Select cells B5 to E8. These cells hold the variable data. Click on Edit, Clear and then Contents Del. This clears the data in the cells without clearing any formulas in the cell or any of the cell formatting. Do not worry about clearing any of the totals because when the variable data on which they depend is removed, they will all go to zero. Your worksheet will now look like Figure 3.71.

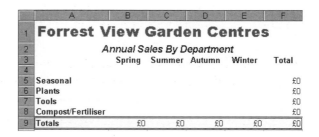

Figure 3.71

3 You can now save the worksheet as a template by selecting File and Save As (Figure 3.72).

Figure 3.72

4 In the Save As type box pull down the list of types and click on Template. When you click on Save the file called Forrest View Nurseries will be saved as a template.
5 Open the template (it may be opened in the same way as an ordinary worksheet) and print a copy.

Activity 17

The accounts of a company are set up in exactly the same way year after year. At present they create each new set of accounts from scratch using the previous year's accounts to guide them on the format. A new member of staff tells them that it would be much better to use a template. Explain what a template is and the main advantages in using them.

Data validation

In some spreadsheets the users have to type in some data themselves. An example of this might be a spreadsheet used to work out loan repayments. The user would need to type in the amount of the loan and the period over which the money is to be borrowed.

It is important that only valid data is entered. For example you might not be able to borrow money over a period greater than 5 years or an amount greater than £10 000.

Validation checks are used to restrict the user in some way when entering data.

What is the main difference between validation and verification?

Worked example: creating validation checks

In this exercise we will look at how to validate a pupil's age in a school.

1 Open a new worksheet and move the cursor to cell A2. This is the cell where a pupil's age is to be entered. Click on Data and then Validation. The data validation menu appears (Figure 3.73).

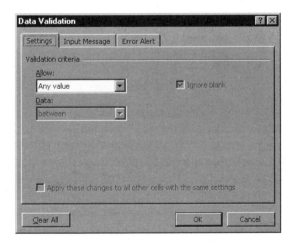

Figure 3.73

2 Click on the pull-down menu for Allow and use the menu to pick the type of data that you will allow to be put into this cell (Figure 3.74).
3 Select 'Whole number' from this menu and then click on the pull-down menu for Data (Figure 3.75).
4 We want to validate the age so that only whole numbers between 11 and 19 can be entered into cell A2. Enter 11 in the Minimum section and 19 in the Maximum section (Figure 3.76). Click on OK. ▶▶

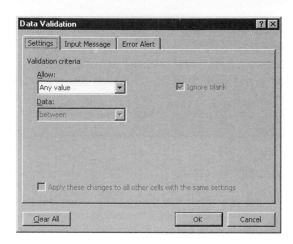

Figure 3.74

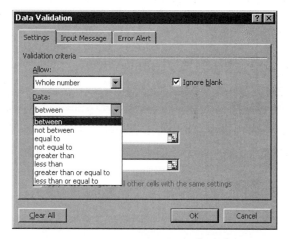

Figure 3.75

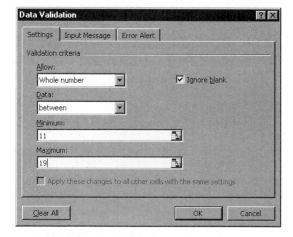

Figure 3.76

5 Click on the Input Message tab (Figure 3.77). In the Title section type in 'Enter a whole number'. In the Input message section type 'Numbers with decimal points are not allowed for this cell'. Click OK to close the box. ▶▶

	A	B	C	D	E	F	G
1	Currency Calculator						
2							
3	Country	France					
4	Currency						
5	Amount in Pounds to convert						
6	Rate						
7	Amount in foreign currency						
8							
9							
10	Country	Austria	Belgium	France	Greece	Germany	USA
11	Rate	21.11	62.06	10.08	506.26	3.02	1.59
12	Currency	Schilling	BFranc	Ffranc	Drachma	DMark	Dollar

Figure 3.87

c Enter the amount to be converted, 500, into cell B5.

d Enter a formula in cell B7 to multiply the amount in pounds by the rate.

e Check that your worksheet is identical to the one shown in Figure 3.88.

	A	B	C	D	E	F	G
1	Currency Calculator						
2							
3	Country	France					
4	Currency	Ffranc					
5	Amount in Pounds to convert	500					
6	Rate	10.08					
7	Amount in foreign currency	5040					
8							
9							
10	Country	Austria	Belgium	France	Greece	Germany	USA
11	Rate	21.11	62.06	10.08	506.26	3.02	1.59
12	Currency	Schilling	BFranc	Ffranc	Drachma	DMark	Dollar

Figure 3.88

f Now test the worksheet by changing the currency in cell B3 and the amount to convert in cell B5.

g Enter Portugal into cell B3. It is important to test the worksheet with incorrect data. You will see that, although Portugal is not in the lookup table, the worksheet is giving you answers. This is much worse than not getting any answers at all. Change the formula in cell B4 to:

=HLOOKUP(B3,B10:G12,3,FALSE)

The FALSE part of the formula makes sure that an exact value is obtained rather than a near match.

h Alter the other formula, and test the new worksheet with names of countries that are not in the table.

i Save the worksheet and print a copy.

Protecting cells

It is important when you create a worksheet for someone else to use that they should not be able to change important formulas or wipe important data. Individual cells in the worksheet can be locked and the whole worksheet protected so that changes to it cannot be made unless the correct password is entered. In some worksheets used for data entry you can lock all the cells other than those where the user has to enter the variable data. In this way you can make sure that the user always enters their data into the correct cells.

179

Worked example: protecting cells for data entry

1 Open a new worksheet and type in the data as shown in Figure 3.89.

	A	B	C
1	**Tour Cost Caculator**		
2	Valid for the following date only	28/11/99	
3	Exchange rate (Pounds to Dollars)	1.58	
4			
5	**Excursions**	**Dollars**	**Pounds**
6	Seven Mile Beach Break	14	£8.86
7	Grand Cayman Flightseeing	64	£40.51
8	Dunn's River Falls	29	£18.35
9	San San Luxury Yacht Cruise	63	£39.87
10	Beach Horseback Riding	79	£50.00
11	Jamaica Snorkelling Tour	31	£19.62
12	Labadee Coastal Cruise	20	£12.66
13	Cancun Tour	31	£19.62
14	Cozumel Island Tour	39	£24.68
15	El Yunique Rain Forest Tour	29	£18.35
16	St John Beach Tour	35	£22.15
17	Sub-Sea Coral Explorer	42	£26.58

Figure 3.89

2 Insert the formula =B6/\$B\$3 into cell C6. Note the absolute reference to cell B3. This is needed so that the formula will refer to the conversion rate when it is copied down the column. Copy this formula down the column. Check that your worksheet looks the same as Figure 3.90.

	A	B	C
1	**Tour Cost Caculator**		
2	Valid for the following date only	28/11/99	
3	Exchange rate (Pounds to Dollars)	1.58	
4			
5	**Excursions**	**Dollars**	**Pounds**
6	Seven Mile Beach Break	14	£8.86
7	Grand Cayman Flightseeing	64	£40.51
8	Dunn's River Falls	29	£18.35
9	San San Luxury Yacht Cruise	63	£39.87
10	Beach Horseback Riding	79	£50.00
11	Jamaica Snorkelling Tour	31	£19.62
12	Labadee Coastal Cruise	20	£12.66
13	Cancun Tour	31	£19.62
14	Cozumel Island Tour	39	£24.68
15	El Yunique Rain Forest Tour	29	£18.35
16	St John Beach Tour	35	£22.15
17	Sub-Sea Coral Explorer	42	£26.58

Figure 3.90

3 Select all the cells in the worksheet. Click on Format then Cells and then click on the Protection tab (Figure 3.91). Make sure that there is a tick in the Locked box. The locking feature will not work yet. We have still to turn the protection on.

4 There is one cell we want to be able to change. This is B3, where the user has to type in the current exchange rate. Click on cell B3, choose Format, Cells and then click on the Protection tab. Turn Locked off for this cell.

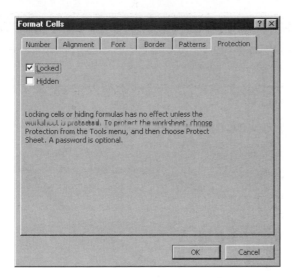

Figure 3.91

5 Now we have to turn the Protection on for the worksheet. Click on Tools and then Protection. A menu will appear, from which you should select Protect Sheet. The screen shown in Figure 3.92 now appears.

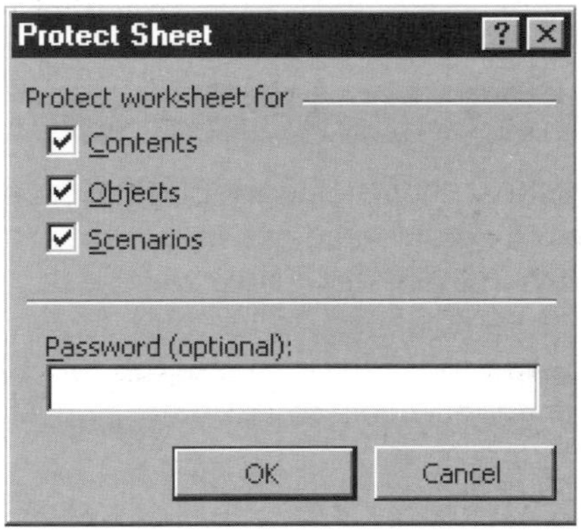

Figure 3.92

There are ticks in all three boxes by default. Keep these settings. At this stage you could protect the worksheet by using a password, but remember that you have to be careful when using a password – if you forget it, you will be unable to gain access to your worksheet. Click OK.
6 Try entering data into any cell except B3. The message shown in Figure 3.93 will appear.

Figure 3.93

/ **Activity 21** /

Using the worksheet you have just created, alter cell B3 to account for a change in the exchange rate from 1.58 to 1.61.

Many spreadsheets contain some cells that are protected. What are the main advantages in protecting some cells?

To remove protection

To remove protection from a worksheet click on Tools, Protection and then on Unprotect Sheet. You can now make changes to the sheet. If you have added password protection, you must enter a correct password before the protection can be removed.

Linking worksheets

There are many occasions when you need to use data from more than one worksheet. To do this, the worksheets need to be linked. You can link cells from one worksheet to another or even from a worksheet in one workbook to another worksheet in a different workbook. A link to more than one worksheet is called a 3D reference.

Suppose there is a total in Sheet1 in cell F6, which needs to be displayed in Sheet2 in cell B2. You would first click cell B2 in Sheet2 (i.e. where you want the results to appear) and enter the formula =Sheet1!F6

Worked example: linking worksheets

1 Create a new worksheet. At the bottom of the page you will see that there are three worksheets. These are here because you can link the worksheets together. In this exercise we will create three worksheets – one for each month from January to March.
2 Double click on the tab for Sheet1. Delete the text Sheet1 and replace it by January. Now change Sheet2 to February and Sheet3 to March. Your tabs will look like Figure 3.94.

Figure 3.94

3 Type in the data in each of the worksheets as shown in Figure 3.95.
4 We are now going to create a fourth worksheet called Qtr 1. Point to any of the sheet tabs and click on the right mouse button. A menu will appear, from which you should select Insert. The screen shown in Figure 3.96 appears. Click on Worksheet and then on OK. The new worksheet tab will now be shown at the bottom of the screen. Change the name of the new sheet to Qtr 1.
5 Enter the data into the worksheet Qtr 1 (Figure 3.97).
6 Enter the formula =SUM(January:March!B4) in cell B4.
   Copy this formula down the column for the other regions.
7 Enter the formula =SUM(January:March!C4) in cell C4.
   Copy this formula down the column for the other regions.

	A	B	C	D
1	Sales by region for January			
2				
3	Region	Used cars	New cars	
4	North	123	56	
5	South	145	209	
6	East	99	178	
7	West	109	103	
8	Midlands	124	56	

	A	B	C	D
1	Sales by region for February			
2				
3	Region	Used cars	New cars	
4	North	100	45	
5	South	156	50	
6	East	109	78	
7	West	124	82	
8	Midlands	96	24	

	A	B	C	D
1	Sales by region for March			
2				
3	Region	Used cars	New cars	
4	North	120	320	
5	South	178	309	
6	East	107	245	
7	West	135	209	
8	Midlands	110	210	

Figure 3.95

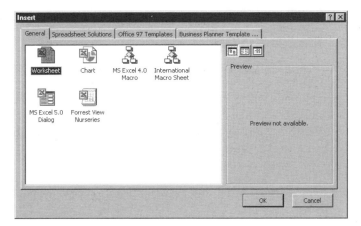

Figure 3.96

	A	B	C	D
1	1st Quarter Sales by Region			
2				
3	Region	Used cars	New cars	
4	North			
5	South			
6	East			
7	West			
8	Midlands			

Figure 3.97

8 Check that the Qtr 1 worksheet is as shown in Figure 3.98.
9 Because the worksheets are all linked a change in the data in any of the other worksheets will result in a change in the Qtr 1 worksheet. Change a couple of figures in the monthly figures and check that the Qtr 1 worksheet has been adjusted accordingly. ▶▶

	A	B	C	D
1	1st Quarter Sales by Region			
2				
3	Region	Used cars	New cars	
4	North	343	421	
5	South	479	568	
6	East	315	501	
7	West	368	394	
8	Midlands	330	290	

Figure 3.98

What are the main advantages of linking worksheets together?

10 Save the workbook (you do not need to save each worksheet separately) as 'Car Sales Figures 2000'.

Using multiple views or windows

Excel, like all Windows software, allows you to have more than one window open at the same time. You often need this facility if you are making comparisons between two worksheets. It is also useful if you have linked worksheets, because when you make changes in the data in one worksheet it often changes the contents in other worksheets.

Worked example: splitting windows

This exercise shows how to split a window.

1 Open the workbook 'Car Sales Figures 2000' which you created on pages 182–184.
2 Select the January tab to display the sales figures for January.
3 Click on Window and then New Window.
4 Select the February tab to display the sales figures for February.
5 Click on Window and then New Window.
6 Select the March tab to display the sales figures for March.
7 Click on Window and then New Window.
8 Select the Qtr 1 tab to display the figures calculated from the other worksheets.
9 You should now see four buttons at the bottom of the screen. Check that you get the four worksheets. If you have any duplicates, close the relevant window. As you can see, it is impossible to tell which window is for which worksheet without pressing it. We could of course re-name them (e.g. January Car Sales rather than 'Car Sales January'.
10 Click on Window and then Arrange. The box shown in Figure 3.99 appears.
11 Select Tiled and the screen will now be rearranged with all the worksheets displayed (Figure 3.100).
12 Click on Window and then Arrange again but this time select Horizontal. The screen will now look like Figure 3.101. Notice that you can see the headings on each worksheet, which can be useful. To see an entire worksheet on the screen you must drag the edge of the worksheet to make it bigger. Try this out.

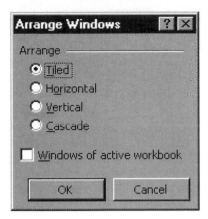

Figure 3.99

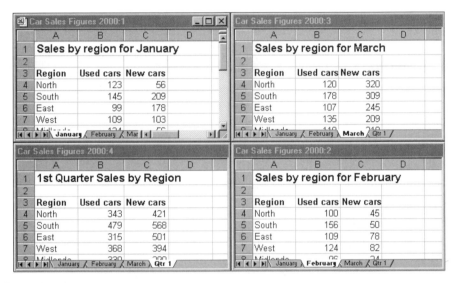

Figure 3.100

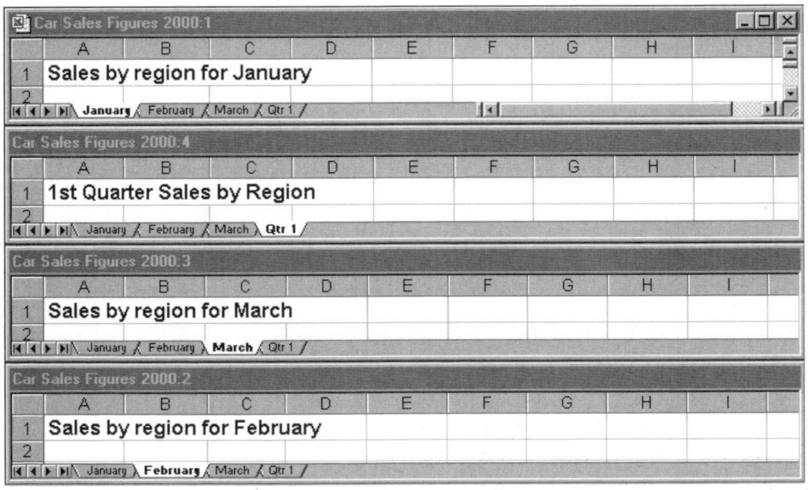

Figure 3.101

> **Activity 22**
>
> **a** Experiment with the other arrangements available.
> **b** Amend some of the figures for the monthly sales and see the effect this has on the quarterly sales.
> **c** Close each window in turn using the document close box.

Macros

Macros are very useful features of software packages because they can save the user a lot of time.

What is a macro?

A **macro** is a software utility found as an addition to most application packages. Using a macro you can instruct the computer to perform a large number of tasks with the press of only one button.

Basically a macro is a set of commands that you can play back whenever you want to perform a simple task. Simple tasks include inserting your name and address at the top of the page when you use your word processor or automating the tasks in loading your spreadsheet software as soon as the operating system has been loaded.

A complex task might be to load a program, copy some data from it, start a different program and paste the data into the new program.

When you are looking for things to automate using macros, you should think of all those tasks that take time and are repetitive in nature.

There are two ways you can produce a macro:

- Using the macro recorder
- Using the Visual Basic Editor.

Only the first method will be looked at here.

Using the macro recorder is the easiest method of recording macros, but the sort of things you can do is limited. You use the macro recorder to record commands and options made using mouse clicks. You can then play back the commands or options in the order they were made simply by accessing the macro from a menu – you can assign the macro to a toolbar.

Using the Visual Basic Editor is a more advanced way to create powerful macros that can not be created using the macro recorder.

When a macro is recorded, *all* the steps will be recorded – even the wrong ones if you make any – so before starting the macro you need to be clear about what you are trying to do. It is often worth going through the steps in a dummy run before you actually record the macro. You'll also find it useful to write the steps down.

Always save your work before starting a macro because you can always go back to the saved version if something unexpected happens. You cannot use the Undo command with a macro so stopping it and going back to a previously saved version of your worksheet is your only option.

What are macros and why are they so useful?

Worked example: your first macro

1 Open the worksheet Car Sales Figures 2000.
2 Click on Tools, Macro and then Record New Macro. The screen shown in Figure 3.102 appears.

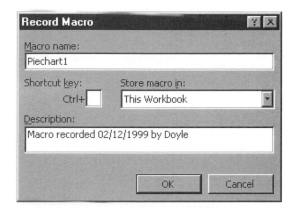

Figure 3.102

3 Change the Macro name to Piechart1 and click OK.
4 Select cells A3 to B8 in the worksheet.
5 Click on the graph icon in the toolbar. The first step of the Chart Wizard will appear (Figure 3.103). Click on Pie in the Chart type list. Select Exploded pie with a 3-D visual effect. Click on Next>.

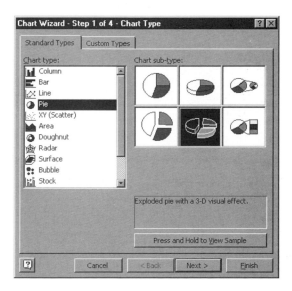

Figure 3.103

6 The next screen appears (Figure 3.104). There is nothing to change here so just press Next>. The next screen will appear (Figure 3.105).
7 Enter the chart title 'Used cars' and then press Next>.
8 We will place the graph in the same sheet as the data so there is nothing to alter in this last screen (Figure 3.106). Just press Finish.

Figure 3.104

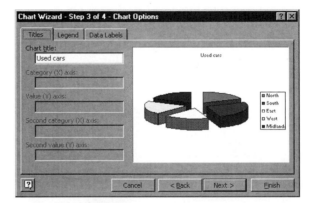

Figure 3.105

Figure 3.106

9 The pie chart will now be drawn and your screen will look something like the one shown in Figure 3.107.

10 Now stop recording the macro by clicking on Tools, Macro and Stop Recording or using the special toolbar. The macro has now been recorded and can be played back at any time.

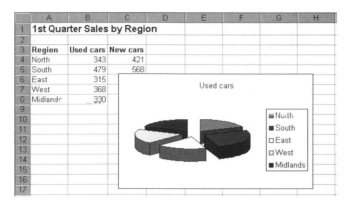

Figure 3.107

11 To test the macro first remove that chart by clicking on the corner handles (these appear as small black squares) and choosing Edit, Cut.
12 Save the worksheet (this will automatically save the macro).
13 From the menu bar choose Tools, Macro and then Macros. Notice the name of the macro you have just recorded appears (Figure 3.108).

Figure 3.108

14 Click on Run to run the macro.
15 The chart will be drawn automatically. If you change any of the data in the worksheets the data in the first quarter worksheet will change, which would obviously mean you have to draw a new chart. By using the macro you no longer have to go through all the processes to draw the pie chart.

Buttons

Rather than go though the usual procedures for running a macro (Tools, Macro, Macros) you can set up a special button for the purpose.

Worked example: creating a button for the pie chart macro

1 Load the worksheet 'Car Sales Figures 2000'.
2 Click on View, Toolbars and then Forms. The toolbar that appears is shown in Figure 3.109.

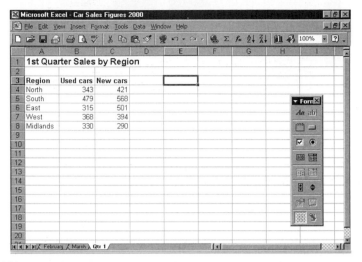

Figure 3.109

3 Click on 'Button' in the toolbar. You can now draw a rectangle on the worksheet to set the position and size of the button by clicking and then dragging the mouse. On releasing the mouse button, a button will be drawn on your worksheet (Figure 3.110).

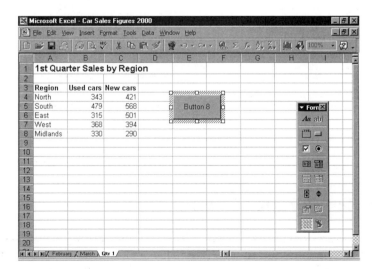

Figure 3.110

4 Change the Macro name to 'Used Car Piechart' (Figure 3.111).
5 Click on the text on the button in the worksheet and type in 'Pie Chart for Used Cars'.

Figure 3.111

6 If you click on the button the macro will run and the pie chart is produced automatically (Figure 3.112).

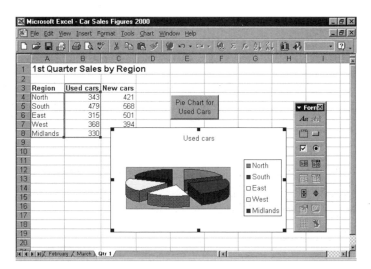

Figure 3.112

A word about macro viruses

There are a number of computer viruses that are recorded using macros – these are called macro viruses. For this reason many people are wary about using macros. If you have created the macro yourself, then you will know that the macro is safe to use, but bear in mind that other people may not be so sure.

When a workbook/worksheet is opened containing macros, Office displays a message warning the user that it contains macros. To use any macros you have created you will have to enable the macros. Do be careful about doing this if you are not sure that the macros are innocuous. It is safer to open the worksheet with the macros disabled.

What is a macro virus?

How are macro viruses different from other viruses?

3.4 Presenting spreadsheet information

Any computer output needs to be presented in the most appropriate layout and form for the user. To ensure that this is done, the purpose to which the information will be put must be established at the start – results could be presented on screen or on the printed page; it might be necessary to import the data into other packages (e.g. an Excel worksheet could be imported into a PowerPoint presentation). You need to consider making use of suitable cell formats, page layout, charts and line graphs.

Page layout

The **page layout** in Excel is determined by the default settings and any alterations are made in the Page Setup screen. Using Page Setup you can alter the page orientation, margins, set up headers and footers and alter the way the actual sheet is shown on the page. To see the Page Setup screen click on File and then Page Setup. By clicking on the tabs at the top of the screen (Figure 3.113) you can move between the layout tools.

Here is a summary of the main parts of each tab on this important screen.

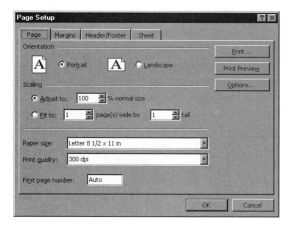

Figure 3.113

Page

Orientation
The **page orientation** is by default portrait orientation. If a worksheet is wider than it is long then it would be better to print it out using landscape orientation. Change the orientation by clicking on Portrait or Landscape.

Scaling
You can alter the percentage of the normal size using scaling, but the most useful feature is the ability to fit the worksheet to a single page. By clicking on Fit to and then making sure the page counter is set at 1 the spreadsheet will adjust the worksheet to fit on a single page. Use this with care because reducing the size of a large worksheet can render it illegible.

Paper size
Clicking on the paper size menu shows all the paper sizes you can use with the worksheet (Figure 3.114). Most of the time you'll use A4.

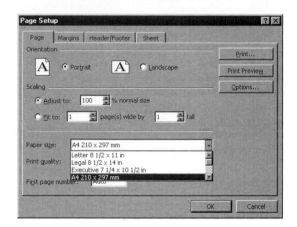

Figure 3.114

Margins

The Margins screen is shown in Figure 3.115. Using this box you can alter the page margins (i.e. the space to the right, left, top and bottom of the working area) and the distance that the header/footer is positioned from the top and bottom edges of the page. Notice the diagram showing the margins and headers/footers and working area on the screen – this is a general diagram and the grid will not show your *actual* worksheet. You can position the worksheet on the page horizontally, vertically or both. To see how your own worksheet looks after making any adjustments, press Print Preview.

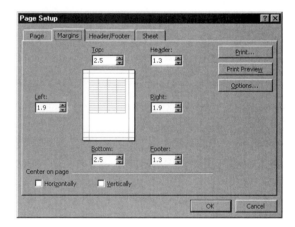

Figure 3.115

Header/footer

Using this screen (Figure 3.116) you can set up headers and footers (a header is text that appears at the top of every page; footers appear at the bottom). Some headers and footers are already set up for you to use (Figure 3.117), but you can also make up your own headers/footers by choosing Custom Header or Custom Footer. When you do this a screen appears to allow you to enter the text or click one of the buttons to add such things as page number, date, time, filename or tab name (Figure 3.118). You can decide where you want to position the various parts of your header and footer.

Sheet

In this box (Figure 3.119) you can select just a range of cells to print or the entire worksheet. You can also choose to display the row or column headings (or both) so that they appear on multiple pages.

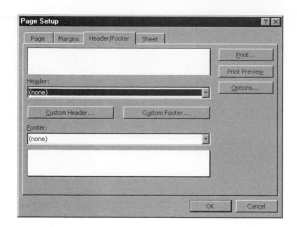

Figure 3.116

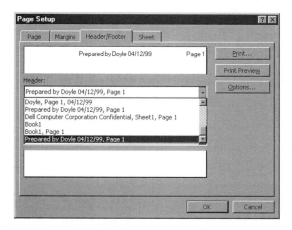

Figure 3.117

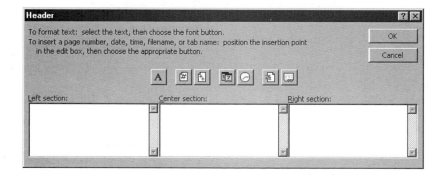

Figure 3.118

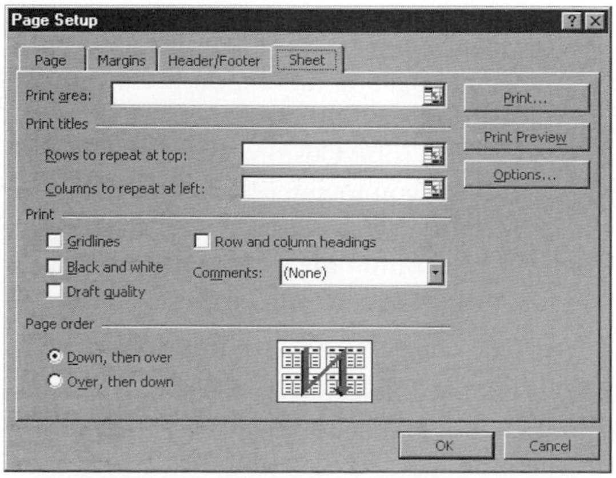

Figure 3.119

In the Print section of the box you can choose:

- to show the gridlines
- to print in black and white
- draft quality (this means that it is printed more quickly, but usually without the gridlines or any graphics)
- whether any comments are printed with the worksheet
- whether to print the row and column headings (i.e. the letters for the columns and the numbers for the rows)
- the order in which the pages are printed.

What is the difference between portrait and landscape orientation?

Printing formulas

When you print out a worksheet, it will always print out the results of any formulas along with any text or numbers. In order to check the spreadsheet it is useful to be able to see these formulas so that you can check them. Go back through the worked example on pages 133–134 to refresh your memory on how to display the formulas on the screen and then print them out.

Printouts of spreadsheets with the formulas shown are an important piece of documentation for the technical user. Why?

Presenting results graphically

Results from worksheets can be displayed pictorially in **charts**. Charts are much better at explaining any trends in data than the numbers in a worksheet.

Excel's Chart Wizard

Chart Wizard will guide you through the steps involved in producing a chart from the data in a worksheet. Most of the time you will create the chart from results but you can also set up a worksheet with the main aim of drawing a chart.

A **wizard** is a facility provided by the software package to guide you through a process using an easy-to-follow series of steps. The four steps used to produce a chart using Chart Wizard are summarised here.

Step 1
This step allows you to choose the chart type and sub-type. Think carefully about your choice of chart – ask yourself if readers will want to take actual values from the chart because with some charts (like 3-D bar charts) this is quite difficult.

Step 2
Before you started using the Chart Wizard, you will have selected the data that you will use to produce the chart. This step shows the range of the data from which the chart is to be produced. If it is wrong, change it.

Step 3
The tabs at the top of the box give you the following options:

- Titles – you can type in a main title as well as give names to the different axes.
- Axes – you can select the axes to be shown.
- Gridlines – you can select the gridlines to show (e.g. major or minor gridlines).
- Legend – this box shows what the parts of the graph are.
- Data Labels – allows you to set the data labels.
- Data Table – you can attach the data table to the chart if you want.

Step 4
This allows you to place the chart in the same worksheet as the data used to produce it or on its own in a new worksheet.

Graphs that look good on the page are not always useful. Why?

195

Pie charts

One of the simplest charts to produce is a pie chart. These are suitable for showing what s up the whole of something.

Worked example: producing a pie chart

1 Create a new worksheet and enter the data and formatting as shown in Figure 3.120.

	A	B	C
1	The composition of a beefburger		
2	*Nutrients*	*Percentage*	
3	Carbohydrate	40	
4	Protein	45	
5	Fat	10	
6	Vitamins and minerals	5	
7			

Figure 3.120

2 Select cells A3 to B6.
3 Click on the graph icon ( ) on the toolbar.
4 Choose Pie for the chart type and then Exploded pie with a 3-D visual effect. Your screen should now look like Figure 3.121.

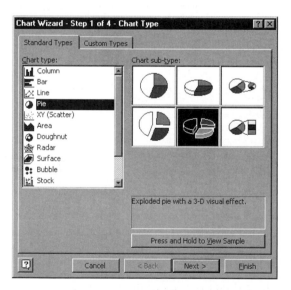

Figure 3.121

5 Now click on the button Press and Hold to View Sample and keep the button held down. This will let you see what the pie chart will look like. You can change your mind at this stage and choose one of the other pie chart designs. Click Next>.
6 Step 2 of the Chart Wizard allows you to choose the data range to be used for the chart. We have already chosen this by selecting the data we are going to use. In the data table, the numerical data is arranged in a column so the Series in columns is chosen. Click Next>.

▶▶

7 The screen in Figure 3.122 appears. Add the title 'Percentage composition of a beefburger'. Using this screen you can change the legend (the legend is the explanation of what each sector of the pie chart represents; this is shown in the box at the side of the chart).

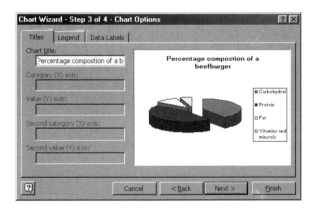

Figure 3.122

8 Click on the Data Labels tab. The box shown in Figure 3.123 appears.
9 Click on Show percent. This will show the data labels (the actual values of the data as percentages). Now click Next>.

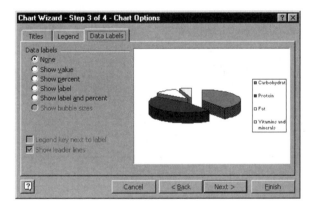

Figure 3.123

10 The screen in Figure 3.124 appears – and you must decide whether to place the chart in the same worksheet as the data or to put it in another sheet. We will put the chart with the data (i.e. included as an object in sheet 1).

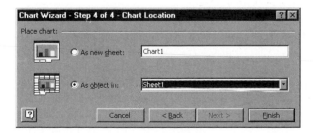

Figure 3.124

▶▶

11 The pie chart will now appear next to the data (Figure 3.125).

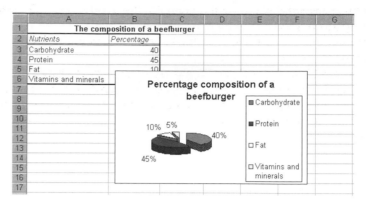

Figure 3.125

12 If the chart covers the data move it by clicking inside the window and keeping the mouse button down (you should see a four-headed arrow). Drag the dotted outline of the window to the required position. To resize the chart window click on one of the corners. The handles will appear, and by clicking on any of the handles you can resize the chart in any direction. Move and size the chart so that it does not cover the data and is as large as it can be on the one page.

13 Investigate for yourself. There are many ways in which the appearance of charts, including this one, can be improved, so do some experimenting. Click once on various sections of the chart to select them, then make the changes by going through menus (double clicking on the selection will open the menu options). By changing options in these menus you will be able to change the appearance. Many of the options refer to the colour (background, bars, etc.). If you have a colour printer see which colours go best together.

14 Save and print a final copy of your worksheet.

Key Skills

You can use Activity 23 to provide evidence for key skills for Information Technology IT2.2 or IT3.2.

Activity 23

Here is some data concerning the method of transport used to get to college.

Transport	Car	Bus	Train	Motorcycle	Bicycle
Number of people	32	38	12	6	2

a Produce a suitable pie chart from this data.
   Make sure that you:
   • Choose the most suitable type of pie chart.
   • Give the chart a suitable title.
   • Use data labels to show the percentage that each sector represents.
   • Use suitable background shading and colours.
b Try to experiment with all the aspects of the pie chart and save each try under a different filename (e.g. pie chart try1, pie chart try2, etc.).
c Produce printouts of your attempts along with a final printout.

Worked example: producing a 3-D bar chart

1 Load the worksheet showing the sales for the Forrest View Garden Centre (Figure 3.34).
2 Select the data in the worksheet (including the row and column headings) from cells A3 to E8.
3 Click on the Chart Wizard icon in the standard toolbar. Step 1 of the Chart Wizard appears. Select a clustered bar chart with a 3-D visual effect (Figure 3.126). You can view the chart by clicking on the button marked Press and Hold to View Sample. Move to the next step using Next>.

Figure 3.126

4 In step 2, click on Series in Rows (Figure 3.127). This tells Excel that you want to group the data according to the rows. The seasons (Spring, Summer, etc.) represent the rows, so the products will be grouped according to the seasons. If you chose 'series in columns', the data would be grouped under department. Move on to the next step.
5 Title the chart 'Department sales by season for 1999'. In the Value (Z) axis box type 'Sales'. These changes appear on the diagram so you can check them (Figure 3.128). Move on to step 4.
6 Instead of putting the chart with the data, we will put it in its own sheet. This means the chart will appear on a new worksheet on its own. Click on As new sheet and enter the title of this new sheet ('Department sales 1999') into the space (Figure 3.129). The chart is now displayed in its own worksheet (Figure 3.130). Notice that the name given to the chart is now on the worksheet tab.
7 Save your chart using the file name 'Department sales 1999' and print out a copy using landscape orientation.

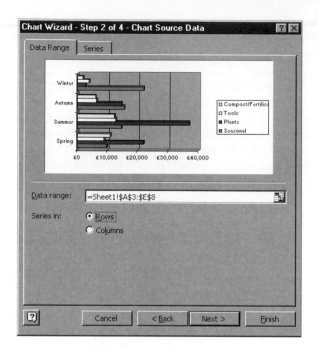

Figure 3.127

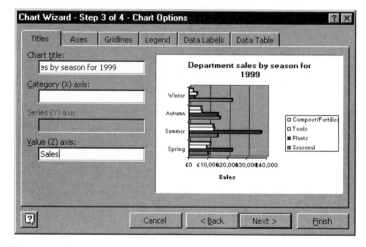

Figure 3.128

Figure 3.129

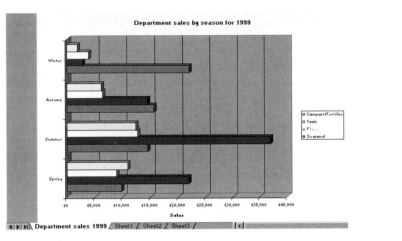

Figure 3.130

Worked example: drawing a line graph

1 Create a new worksheet and enter the data shown in Figure 3.131. Use the Fill feature to fill in the months in column A.

	A	B	C	D	E
1	Average share price per month over two years				
2		99/00	00/01		
3	January	£2.97	£4.60		
4	February	£2.94	£4.67		
5	March	£2.90	£3.78		
6	April	£3.10	£3.99		
7	May	£3.67	£4.78		
8	June	£3.99	£6.70		
9	July	£4.50	£7.00		
10	August	£4.57	£7.23		
11	September	£5.00	£6.99		
12	October	£4.78	£7.78		
13	November	£4.79	£8.67		
14	December	£4.50	£10.00		

Figure 3.131

2 Select cells A2 to C14.
3 Click on the Chart Wizard icon. Choose a line chart and select the chart sub-type shown in Figure 3.132.
4 Skip past step 2 of the Chart Wizard.
5 Fill in the title and the labels for the *x* and *y* axes as shown in Figure 3.133. Click on Finish (you can skip the last step if you just want to put the chart with the data).
6 Move and size the chart so that it appears next to the data on the worksheet. Your worksheet should now look like Figure 3.134.
7 Save this worksheet using the file name 'Monthly share prices' and print a copy of the worksheet on a single sheet using landscape orientation.
8 You are now free to improve the appearance of both the data and the chart on the worksheet using colour, patterns, etc. Produce a printout of your final version.

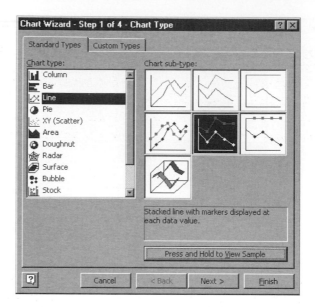

Figure 3.132

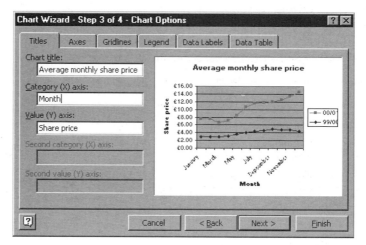

Figure 3.133

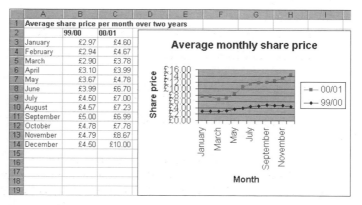

Figure 3.134

Activity 24

The following table shows the amount spent on advertising holidays in the local paper and the number of holidays sold. A graph needs to be drawn to test the hypothesis that the more spent on advertising then the greater the number of holiday sales.

Amount spent on advertising	*Number of holidays sold*
£60	34
£130	56
£210	67
£290	78
£470	98
£500	108
£620	120

A scattergraph shows if there is any correlation between the independent variable (amount spent on advertising) on the *x* axis and the dependent variable (the number of holiday sales) on the *y* axis.

a Set up a worksheet and enter the data shown in the table.
b Using Chart Wizard, prepare a scattergraph from the data with the title 'Scattergraph to show any correlation'. Note the steps you must go through to make it appear in your worksheet.
c Print a copy of the worksheet containing the chart.

Activity 25

The Chart Wizard takes you through the steps involved in creating a chart but you can still make changes to a chart once you've created it. Find out by experimentation and using on-line Help how you would do the following. For each one you will need to explain how to do the task in easy to understand language.

* Explode a pie chart (i.e. move one of the sectors in the pie chart away from the others to emphasise it).
* Rotate a three-dimensional bar chart. This is useful if more than one data set is plotted on the same graph because the bars on one graph can hide the bars on the other.
* Size a chart element (a chart element is part of a chart).
* Use your own colours and patterns for the sectors of a chart.

Using drawings in a worksheet

The choice of drawing tools available in Microsoft Office is the same, whether you use Word, Excel or PowerPoint.

In the later units, you will be expected to produce diagrams called data flow diagrams (DFDs) and entity relationship diagrams. Both of these may be produced using the drawing tools in Excel. Although the drawing tools in Excel are also available in the other packages (Word, PowerPoint, etc.) it is easier to draw them in Excel because you can use the row and column lines to help you line things up. You can remove these lines once you're happy with the drawing.

Why is it sometimes easier to draw diagrams using Excel rather than Word or PowerPoint?

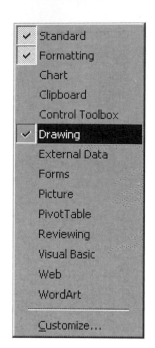

Figure 1.135

To see what drawing tools are available in Excel, click on View and then Toolbars. The menu shown in Figure 3.135 appears.

Click on Drawing and the drawing toolbar appears at the bottom of the screen (Figure 3.136).

Figure 3.136

Each item that can be drawn using the drawing toolbar is called an object. To draw one of these objects click on the object on the toolbar and drag the mouse over the worksheet. For example, to draw an arrow click the arrow button, then click on the worksheet where the end of the arrow is to start and drag until it is the length you require. As soon as you remove your finger from the mouse button, the arrow will be drawn.

Figure 3.137 details all the buttons on the Drawing toolbar and gives a brief description of what they do.

Editing drawing objects

Once you've drawn an object, you can do a number of things with it. Here are some examples.

Copying an object
To copy an object select it, click on Edit and then Copy. This copies the object to the clipboard. Position the cursor where you want to place the copy and use Edit and Paste.

Deleting an object
Click on the object to select it and then press the delete key.

Moving an object
To move an object select it and then drag it to the new position. To move an object to a new worksheet copy it to the clipboard and then paste it where you want it in the new sheet.

The presentation of a spreadsheet is important, especially if lots of people have to use it. What things can be done to improve the presentation?

Resizing an object
To resize an object select it and then drag the selection handles until it reaches the desired size.

Formatting drawing objects
Formatting a drawing object means using fill colour, different line thicknesses, dash styles, arrow styles, etc. To format an object select it, choose the formatting button and select the option from the list.

Grouping objects

Diagrams are produced from lots of different objects (lines, arrows, rectangles, etc.), but to move the entire diagram you don't want to have to move all the components separately. By grouping objects you can tell the computer that you want to treat all the component parts of the diagram as one unit, and you can then move the whole diagram in one go.

This produces the list of draw commands.

 This tells the program that it should select the next object that you click or the next group of objects that you drag across.

 This button displays the Autoshapes list of commands.

 Using this button you can draw a line from the point where you press the left mouse button to the point where you release it.

 This draws an arrow from the point where you press the left mouse button to the point where you release it. The arrowhead appears at the release point.

 This draws a rectangle by clicking and then dragging until the required size is obtained. You can edit the rectangle using the handles.

 This draws an oval by clicking and then dragging until the size you want is obtained. You can edit the oval using the handles.

 This produces a text box that you can size by clicking and dragging. You can then enter text into the box.

 This allows you to insert WordArt, which turns text into interesting and colourful graphics.

 Using this you can select clip art to be inserted into your worksheet using the box that appears.

 This button fills a shape or an area with colour. Clicking on the dropdown arrow reveals a palette of colours from which to choose. The colour on the face of the button shows the colour you have selected.

 This button colours a selected object with the colour chosen from the palette. The colour on the face of the button shows the colour you've selected.

 Use this button to change the text colour by choosing from the dropdown palette.

 This button allows you to choose the thickness of line used to draw the selected object.

 Choose the style of a dashed line with this tool.

 This button allows you to choose a style for lines or arrows.

 Adds a shadow to selected objects with this button.

 This button turns a two-dimensional shape into a three-dimensional one.

Figure 3.137 The buttons on Excel's drawing toolbar

Practice in using the drawing tools and grouping objects

1 In a new worksheet draw a house like the one shown in Figure 3.138.

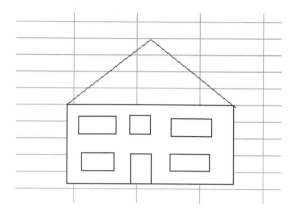

Figure 3.138

2 Make sure that the drawing toolbar is visible. If not, show it.
3 Click on the Select Object button in the Drawing toolbar.
4 Drag the mouse from the bottom left-hand corner to the top right-hand corner so that the entire diagram is inside the dotted rectangle. You have now selected the objects that need to be grouped together. Another way of selecting all the objects is to click on each object in turn while holding down the Shift key.
5 Click on Draw and then on Group. The selection handles are now situated around the entire object rather than each part of it, and the computer will treat the diagram as an entire object.

Activity 26

a Grouping allows the entire object to be moved, copied, etc. Try moving the diagram, resizing it, and making copies until you have a row of these houses.
b By using formatting techniques, colour in these houses and use Word Art to produce an interesting advertisement for an estate agent.
c Save each version using a different file name (in case you need to go back to a previous version) and print out your best attempt.

Worked example: drawing a data flow diagram using Excel

Figure 3.139 is a level 1 data flow diagram for adding a new video to a video library.

Your can produce this diagram using Excel.

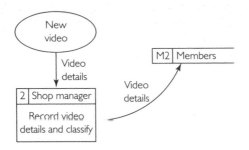

Figure 3.139 Level 1 data flow diagram for adding a new video

1 Open a new worksheet.
2 Make sure that the Drawing toolbar is at the bottom of the screen.
3 Use the Oval button to produce an oval of a suitable size. Use the row and column lines to position the oval on the screen in a convenient place.
4 Place the cursor on the oval and click the right-hand mouse button. The screen shown in Figure 3.140 appears.

Figure 3.140

5 Select Add Text to allow you to enter text inside the oval.
6 Type 'New video' inside the oval.
7 Now use the rectangle button to produced a suitably sized and positioned rectangle.
8 Use the Line button to draw in the extra lines in this rectangle.
9 Now draw the open-sided box. This is best done using separate lines. Again click on the right mouse button and select Add Text. Key in the text.
10 Click on the arrow button in the drawing toolbar. Position the cursor on the point on the oval where the line is to start and click and then drag the line to the correct position on the rectangle. Release the mouse button. The arrowhead appears on the end.
11 We now need to add the text 'Video details' by the side of the line. This is best done by inserting a small text box next to the line. You can alter the style of the text you type into this box using Format Text Box (Figure 3.141) and use the handles to position the box.

12 Double-click on the text box and the dialogue box in Figure 3.141 will appear.

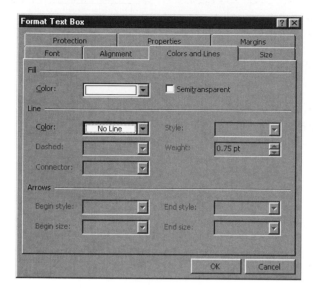

Figure 3.141

13 Click on the Colours and Lines tab and in the Line section click on the drop-down menu. A selection of lines appears. Choose the No Line option, which means that the outline of the text box will no longer be seen.

14 Click on OK and the text will appear next to the line on its own.

15 Now we need to draw the curved line with 'Video details' written on it. To do this click on the AutoShapes button on the drawing toolbar. The menu will appear (Figure 3.142)

Figure 3.142

16 Select Connectors and a sub-menu will appear, from which you should choose the curved arrow connector (Figure 3.143).

Figure 3.143

17 Click on the rectangle and drag the line across to the open-sided box. Release the mouse button to insert the arrowed line.

18 Once you have used the connector, you can move the boxes and the line joining them will follow. In our example it will be necessary to group the objects in each box so that you can move their components together.

19 Now put the text 'Video details' next to this line as described in steps 11, 12 and 13.

20 Save a copy of the worksheet and then produce a printout without the gridlines being shown (you can use the Page Setup menu to get rid of the gridlines).

Why is it useful to group objects together in a drawing to create a single object?

Activity 27

Write a few sentences to explain these terms to a complete novice who knows very little about computers:

*   headers and footers
*   landscape and portrait orientation
*   legend
*   data labels.

3.5 Testing the spreadsheet

There are many ways in which you can make a mistake in a spreadsheet. The most common faults are that the wrong cells have been used in the formulas or the formulas have been constructed incorrectly. Other reasons why the spreadsheet could be wrong include:

*   incorrect use of brackets
*   wrong use of operators (e.g. $+, -, \times, /$ etc.)
*   use of the wrong function
*   incorrect use of logic
*   use of the wrong cells in a formula
*   using the wrong formula (e.g. the area of circle $= 2\pi \times$ radius instead of $= \pi \times$ radius^2).

Never assume that, just because the spreadsheet displays some numbers which appear reasonable, they are correct. You should *thoroughly* test the spreadsheet and fill in a table similar to the one shown below.

Result obtained from the actual spreadsheet	Expected result (the result obtained by manual calculation)	Explanation of difference

By filling in the table you can check all the formulas by comparing the answers from calculations obtained manually or using a calculator with those obtained using the spreadsheet. You must explain and correct any differences.

General rules to follow when developing spreadsheets

Here are some rules to help you when developing spreadsheets. Although these rules will not stop errors occurring, they can cut down on the frequency of errors:

- type labels in cells first
- type raw data in cells
- use formulas wherever possible
- in formulas reference the cell locations rather than the numbers themselves
- use functions where possible instead of formulas
- use the copy commands or fill commands to copy formulas
- do not type anything that the spreadsheet will type for you
- check the formulas using test values
- format the spreadsheet to make it readable and attractive
- use comments to help users know what data is to be put into certain cells
- create validation checks.

Questions to ask when developing a spreadsheet

Testing spreadsheets is not confined to ensuring that the formulas and the ranges to which they refer are correct. You might have developed a spreadsheet that does not do what the user asked for. From the user's point of view this is as useless to them as a spreadsheet containing lots of errors. You need to answer a number of questions as part of the testing.

Does the solution meet the agreed specification?

You should have developed the working specification in consultation with the user and agreed on the requirements of the new system. Without a written user specification there can be conflict over what was agreed. Once the specification has been agreed, the spreadsheet being developed must match it. You might need to make slight changes to the specification as you go along, but these should be agreed by both parties.

Part of the testing stage of the spreadsheet development process is asking the user if the spreadsheet matches the specification. You should make sure that nothing has been overlooked and that there is nothing there the user did not ask for. It is a good idea to use the user specification as a checklist and tick off each part as you complete it.

Do the spreadsheet's results agree with the manual calculations of the same problem?

If a formula has been used to calculate a value, you can always do the calculation manually and compare the two results. You could still get the wrong result – it's quite possible that you have constructed the formulas incorrectly, so any formulas used should be checked. Even if you think you can remember a certain formula you should still check it. It is usually best to show your formulas to someone else as it is always difficult to spot the mistakes in your own work. If you have some input data and some output data from another source that you know is correct you could use this data as the input into your spreadsheet and check that it gives the correct results.

Be careful about any constants used in formulas. For example, the correct value of π is 3.14, not 2.14 or 4.14.

Document your formulas by listing them and explaining the variables and constants used.

Does the spreadsheet cope with normal, extreme and abnormal data?

Before a cell is formatted any type of data can be entered into it, so by formatting cells to match the data type of the inputted data you can perform a simple data type **validation** check.

Try testing the spreadsheet with a variety of data:

* *Normal data*. This data should be accepted into a cell because it meets the data type for that cell and passes the validation check for the cell. Normal data is the general data for which the spreadsheet was designed.
* *Extreme data*. This is data that is outside the range of the normal data and should be rejected by the validation checks. This would include very large and very small values. If designed properly, the spreadsheet will be able to trap these sorts of values.
* *Abnormal data*. This data would not usually be entered into that cell – for example, a negative order quantity or a person's name that starts with a number. This type of data is also classed as *illegal data* and, although an experienced user would know that this type of data could not be used, an inexperienced user might not.

Activity 28

There are many different types of validation checks that can be performed on data. Find out and explain what each of the following validation checks do:

* data type checks
* presence checks
* range checks

* hash totals
* batch totals
* check digits.

Can other people use the solution?

The only way to test if other people are able to use the solution is to allow them to use it and see what they think. After they have used the spreadsheet for the first time give them a questionnaire asking them how they found it. The more people who try the spreadsheet the better, because you will get more feedback. If serious problems come to light during this feedback, you can alter your spreadsheet accordingly.

Is the spreadsheet robust or can it be made to fail?

One of the worst things that can happen to a novice user when using a spreadsheet is that it does something unexpected or gets them into a situation they cannot get out of. By limiting the things the user can do, you can prevent them from doing something which will result in the spreadsheet failing.

What is the best way of ensuring that the spreadsheet you have developed is easy for others to use?

Spreadsheets that crash are more than just an annoyance to the user. What are the dangers when a program crashes?

Part of the testing process is for the user to use the spreadsheet and report back whenever anything unexpected happens. The developer can also try to see if they can get the worksheet to freeze, crash or do something unexpected.

Test specification

The purpose of the test specification is to outline the tests that the spreadsheet will be put through during the testing process. The following should be included in the test specification:

- acceptable data input values (both maximum and minimum)
- unacceptable data values that should be rejected automatically
- a method of checking independently that all formulas and functions work correctly
- a way of checking that the developed system meets the user's requirements. This could be provided in the form of a bulleted checklist created from the working specification.

Sorting out errors

Excel has a variety of error codes to let you know what is wrong. The main ones are discussed in this section.

#########

This means that a formula cannot be calculated because it contains an error or the cell to which it refers creates the error. If there is no formula and this message occurs, it is because the current column width is not wide enough to display the cell contents.

#N/A

This appears when a formula refers to a non-existent value. This can also happen when you are trying to search for data in an unsorted list or are using a worksheet in which the data has not yet been entered.

#REF!

This message appears if a formula contains invalid references – for example, if a cell is deleted and there is a formula in another cell which refers to it. This often happens when cells have been moved to another location or deleted.

#DIV/0

This happens when a formula tries to divide by zero. It could be that you are trying to divide by the contents of a blank cell.

#NAME?

This happens if you do not use the correct syntax when you are using a formula or function. Use the Help facility to check the construction of the formula or function.

#NULL!

This can happen when a formula refers to a cell that contains no data.

#VALUE!

This occurs when the type of data in an argument is incorrect.

#NUM!

This error is produced when the value is too big or too small to be displayed by Excel.

Error checking tools

There is an auditing feature in Excel that can be very useful for checking worksheets and finding where errors lie. To see the auditing toolbar click on Tools, Auditing and then Show Auditing Toolbar. The toolbar shown in Figure 3.144 is displayed.

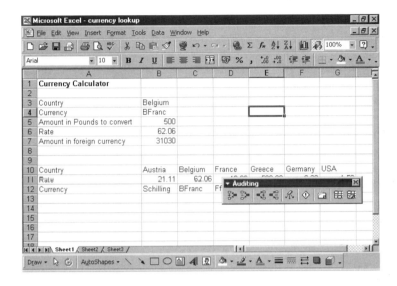

Figure 3.144

There is a variety of buttons on the toolbar to help **trace** mistakes.

Trace precedents
Click on this button to trace all the cells to which the currently selected cell refers. Look at the worksheet in Figure 3.145. The selected cell is cell B6. By pressing the Trace Precedents button you can see that this cell refers to cells B10 to G12 as well as cell B3.

	A	B	C	D	E	F	G
1	Currency Calculator						
2							
3	Country	Belgium					
4	Currency	BFranc					
5	Amount in Pounds to convert	500					
6	Rate	62.06					
7	Amount in foreign currency	31030					
8							
9							
10	Country	Austria	Belgium	France	Greece	Germany	USA
11	Rate	21.11	62.06	10.08	506.26	3.02	1.59
12	Currency	Schilling	BFranc	Ffranc	Drachma	DMark	Dollar

Figure 3.145

Remove precedent arrows
This removes the arrows (i.e. the traces) created by the Trace Precedents button.

Trace dependents
Click this button to trace any cells that depend on the currently selected cell. In the worksheet in Figure 3.146 the selected cell is C11 and cells B4 and B6 depend on this cell.

Figure 3.146

Remove dependent arrows

This removes the arrows (i.e. the traces) created by the Trace Dependents button.

Remove arrows

This removes all the arrows in one go.

Trace error

If a cell contains an error message, select it and then click on the Trace Error button. Each time you click this button you can trace the error back one more step.

New comment

Allows you to add comments to cells to help explain what is happening in each cell.

Circle invalid data

Data which fails any of the validation checks will be circled.

Clear validation circle

This removes the circles created by the Circle Invalid Data button.

3.6 Documenting the spreadsheet

Documenting a spreadsheet solution is essential if other people are to use it. A solution will be easy to document if it is easy to use: if it is obvious to the user what they have to do to use the spreadsheet, then you won't need much documentation to explain it.

You should specify where and how data is to be entered and outline any formulas that can be used. Try to make use of on-line documentation as far as you can by making it clear using cell labels what it is that the user has to enter. In those cells where data is to be entered, use comments showing instructions when the user moves to the cell to enter data.

Bear in mind that, as well as users of the worksheet, there will be people who will want to alter your worksheet or develop it further. Developers will need more information about the development than the users.

There are two types of documentation you should include as part of your spreadsheet solution:

- technical documentation
- user documentation.

Technical documentation

Technical documentation is for computer specialists who will already have a good understanding of spreadsheets or will be able to work things out for themselves if they need to. The idea of the technical documentation is to allow someone to quickly understand how the worksheet was constructed and how it works, so they can develop the solution further or make adjustments to the original sheet. You should remember that you or the person who originally developed the spreadsheet might no longer be around to help – the documentation and the spreadsheet itself is all there is and should therefore be self-explanatory.

Technical documentation should include the following:

* A copy of the agreed design specification.
* Details of the hardware, software and other resources required. Hardware would include the technical specifications of the type of computer needed to run the spreadsheet software. The user might have a very old computer that may not have enough power to run the spreadsheet package you've used. Remember to include the operating system as well as the applications software.
* Instructions for opening and configuring the spreadsheet. The user will need to know where the worksheet is situated and if it is a worksheet or a template they are opening.
* Details of all the calculations, formulas and functions used. Always try to use proper variable names rather than just letters in formulas, as this will help in explaining them.
* Details of validation and verification procedures. Here you would describe what validation checks were incorporated and the types of error they have been designed to trap. Verification will typically include checking that the data on a source document is entered correctly into the worksheet. You could also have two people type in the same data and compare their results.
* Details of input and output screens and printed designs. Rough designs (hand-drawn ones are OK) showing screen designs and designs for any results printed out should be included. It is a good idea to include the draft versions, no matter how 'rough' they are.
* Copies of the test specification. This will contain details of how you intend to test the spreadsheet.

User documentation

User documentation is aimed at the end user, who may have only limited knowledge of spreadsheets and will simply be using your solution to do their job. Your custom spreadsheet should therefore include instructions that are easy to understand. These instructions should include the following:

* how to start the spreadsheet program. Use screen dumps to help explain the processes to the user
* how to navigate through the spreadsheet menus. Use screen dumps and step-by-step instructions
* examples of screens and data entry forms
* instructions about data entry
* advice on how to respond to error messages or conditions
* examples of data output screens and printed copy.

Two types of user documentation are used when a solution to a user's problem has been developed. What are they?

Producing screenshots

It is useful when producing documentation to be able to take a copy of what appears on the screen so the reader can see exactly what is happening. The easiest way to produce a copy of the screen is by using a screen dump or **screenshot**.

Screen dumps produce a copy of the current screen being displayed. If a dialogue box has been opened, then when the screen dump is performed, only that dialogue box will be pasted to the clipboard.

Worked example: producing a screen dump

Being able to produce a screen dump is an extremely important part of documenting a spreadsheet. Here is how to do it.

1 Load Excel and open an existing worksheet. The spreadsheet 'Forrest View Nurseries' created in an earlier exercise has been used in this case.
2 Keeping the Alt key on the keyboard pressed down, press the Print Screen key on the keyboard. Nothing visible happens but a copy of the screen will be pasted to the clipboard.
3 Load your word-processing package (Word has been used here) and create a new document (you do not have to exit Excel first). Move the cursor to the position on the page where you want the picture of the screen to start.
4 Click on Edit and then Paste. The screen dump will appear (Figure 3.147).

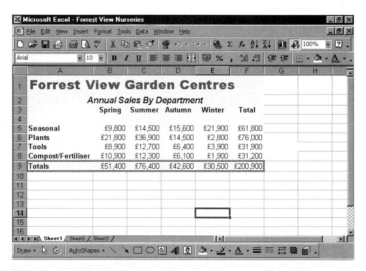

Figure 3.147

5 By clicking on the picture you can use the handles to resize it. You can also move it by clicking and dragging to a new position.
6 Now go back to the spreadsheet and click on File and then Print. The print dialogue box will appear.

7 If you press Alt and then Print Screen this box will be copied to the clipboard. Go back to the word-processed document containing the first screen dump and paste this new screen in.

8 Save your document and print out a copy.

Specialist screen dump programs

There are many specialist screen dump programs available, and you will find it very useful to be familiar with at least one of them. They are easy to use and you can just select part menus as well as show those menus that usually disappear as soon as a key is pressed. Some of these packages are shareware or public domain and you can get them off the Internet. The one I prefer to use is called GrabIt Professional and it is a shareware program.

Activity 29

Download an evaluation copy of a screen dump package or utility off the Internet and use the software to produce a range of different types of screen dumps. Produce printouts of the screen dumps and briefly evaluate the package.

3.7 Standard ways of working

As part of Unit 3 you will have to show that you have used standard ways of working. The whole point of having a standard way of working is to help you to manage your work more effectively. You can save a lot of time and effort if you think logically about how you intend working. All of the standard ways of working are looked at in detail in Unit 1. Here is a list of the ways of working you should apply to this unit.

- Edit and save work regularly, using appropriate names for your worksheets/workbooks.
- Store your work where others can find it easily.
- Keep dated backup copies of your work on another disk and in another location.
- Keep a log of the ICT problems and how they were resolved.
- Make sure that confidentiality (i.e. Data Protection Act 1998) and copyright laws are obeyed.
- Prove that you have avoided bad posture, physical stress, eye strain and hazards from workplace layout.

Key terms

After reading this unit you should be able to understand the following words and phrases. If you do not, go back through the unit and find out, or look them up in the glossary.

Absolute cell reference *Cell referencing*
Autofill *Chart*
Cell *Data capture*
Cell format *Data entry message*

217

Font

Function

Integrity

Logical operator

Lookup table

Macro

OCR (optical character recognition)

OMR (optical mark recognition)

Operators

Page layout

Page orientation

Protecting cells

Relative cell reference

Screenshot

Template

Text concatenation

Toolbar

Trace

Validation

Wizard

Worksheet

Review questions

1 **a** It is essential to find out and analyse the user's requirements before developing a spreadsheet for others. Why?

 b As part of this analysis, name six things the developer will need to find out about in order to produce the detailed design specification.

 c It is important to find out about the 'scope' of the task. Explain briefly what this means.

2 **a** Explain, using sketches of worksheets to help you, the difference between relative cell referencing and absolute cell referencing.

 b Explain the term mixed cell referencing.

 c Give the names of six built-in spreadsheet functions and briefly explain their purpose.

3 During the design and development of a spreadsheet, the developer must make it easy for the user to use. What features can be included in the design to ensure that:

• the user can enter their data into the worksheet easily
• the user is not left wondering what data they have to enter
• the user does not enter the wrong type of data
• the user interface is presented in a thoughtful way
• the user can use a couple of keystrokes to do lots of things and save time?

4 Headers and footers are useful when presenting spreadsheet information. What information is likely to be included in:

• a header
• a footer?

5 Jane, the sales manager of a large cosmetics company, needs to present the sales figures to the representatives in her region. Last year she produced a handout on paper showing comparisons of the sales of various products over the four quarters. This year she would like to produce them as charts using spreadsheet software. What advantages are there in using the spreadsheet to produce the charts? Are there any disadvantages? What particular features of the spreadsheet's charting facilities might she make use of?

6 Explain the processing you would need to go through in order to thoroughly test a spreadsheet you have just finished developing.

7 After a spreadsheet has been developed, it needs to be fully documented. There are two types of documentation.

 a Explain the difference between user documentation and technical documentation.

b What would typically be included in each of these two types of documentation?

8 During the development of spreadsheets the developer needs to employ standard ways of working. Explain what this means and describe some things the developer should do to manage their work effectively.

9 Spreadsheets often contain a feature called a Wizard. What is a Wizard and how does it help the user?

10 A person is completely new to computing. You have the job of explaining to this person what a spreadsheet is and how it might be useful to them. You only have a piece of paper and a pen. Explain the concepts to them carefully and use diagrams wherever possible to help in the explanation.

Assignment 3.1

Most people have some need for a spreadsheet – after all, it is the second most widely used piece of application software next to word processing.

For this assignment you need to identify a real user with a problem to solve using spreadsheet software. The problem must be of sufficient complexity to need at least six of the most complex spreadsheet facilities.

You need to produce a spreadsheet to meet the user requirements and the user and technical documentation for this spreadsheet including a test report.

Here is some detail regarding the specific requirements to be produced.

Produce a design specification which meets the user requirements

You need to include, as part of the specification, the following:

*   an explanation for the choice of the more complex facilities provided by the software
*   details of the sources of data (i.e. where the data comes from before it is processed by the spreadsheet)
*   an outline of the data entry forms used to help with the input of data
*   details of the calculations used (i.e. what formulas have been used and what they do)
*   what user aids have been included to make using the worksheet easy for the user (e.g. macros to reduce the number of keystrokes or mouse clicks and movements needed to perform an operation or data entry to let the user know what sort of data they are required to input)
*   suitable printed or screen output making appropriate use of cell formats, charts or graphs, page or screen layout and graphic images (e.g. drawings, WordArt, clip-art, imported photographs, etc.)

Provide clear technical documentation identifying:

*   formulas and functions used
*   screen and printed report layouts.

Provide clear user documentation including:

- copies of menus and screens
- examples of input and output.

Provide evidence of testing including:

- documentation showing that you have tested the spreadsheet against the design specification
- testing the accuracy of the data used and the output generated.

There is a formula that the police use after a serious traffic accident to work out the speeds the cars were travelling at. They can do this by measuring the length of the skid mark and noting the type of road surface and the weather conditions at the time. A car's speed can be worked out using the following formula:

$$S = \sqrt{(30fd)}$$

where S is the speed of the car in miles per hour (m.p.h.), d is the length of the skid in feet and f is a number that depends on the material the road surface is made from and the weather conditions at the time. The value of 'f' to be used in the above equation can be found from the following table.

Weather	Type of road surface	
	Concrete	Tar
Wet	0.4	0.5
Dry	0.8	1.0

The traffic police would like you to design a system for them which will be easy to use and foolproof. There are over 100 traffic police in the force and all of them will need to use the system you will be developing.

On the basis of the information from your spreadsheet, a criminal conviction could result so it is imperative that the spreadsheet is tested meticulously.

Before you start the design, you will probably need to ask the user (the person who has asked you to develop the application) what they would like the system be able to do.

As a team of 3 or 4, produce a series of questions that you can use to provide further information about the required system. The production of a questionnaire is part of the systems analysis process. If you need guidance about the production of a questionnaire, then look at Unit 5.

Tasks

1 Produce a minimum of 20 questions that can be used to find out from the user what they are looking for in the new spreadsheet system. The questions should ask about:

- inputs to the system
- the processing steps
- how the output is to be presented and used.

Discuss with the other teams, and in conjunction with your tutor, the questions you have produced.

2   Here is a list of things the user has said they want you to consider and include in your design for the new system.

  • Each worksheet is for a particular case and the printout of the worksheet must have on it the following details: an incident number (a six-digit number), the number of the officer who is investigating the accident (a four-digit number), the time and date of the accident and the registration number of the vehicle to which the data refers.

  • The user needs to be able to enter the length of the skid, the type of road and the weather condition. All the data entered must be carefully validated using appropriate and rigorous validation checks. The user should be informed via suitable messages if the data they have entered is not valid. Examples of valid data should be included so that they have an idea of the data they need to enter.

  • The worksheet must be protected from unauthorised access by passwords and the user should be unable to alter anything except where the variable data is entered. Comments to guide the user on the data they have to input should be included.

  • The worksheet needs to have the appearance of an application rather then a spreadsheet. Use colours, borders, backgrounds, etc. to make it easy to use.

  • Police officers do not like spending time on administration so any help in this area – such as reducing the number of keystrokes they have to make to perform certain actions – would be very much appreciated.

Use these requirements, along with any others you can think of, to produce a clear design specification.

3   Implement your design and start producing the actual spreadsheet. Look at what documentation/evidence you need for this in the assessment guidelines. Provide evidence of thorough testing.

4   Produce clear, high-quality user documentation.

Get the grade

To achieve a grade E you must meet all of these criteria.

To achieve a grade C you must also:

  • Show a good understanding of spreadsheet design and attention to detail by creating an imaginative customised spreadsheet that makes good use of design and layout facilities.

  • Produce detailed test specifications together with examples of a full range of acceptable and unacceptable input, associated expected output and any associated error messages.

  • Show that you are able to work independently to produce your work to agreed deadlines.

To achieve a grade A you must also:

• Show that you have a good understanding of the purpose and value of the more complex facilities by using them effectively in your spreadsheet design.
• Provide customised data input using facilities such as forms, dialogue boxes and list boxes that are clear, well laid out, suitably labelled.
• Validate all data inputted.
• Produce comprehensive records of spreadsheet drafting, testing and refinement that show how the spreadsheet was developed and how any problems were resolved.
• Produce high-quality and clear user documentation, making good use of graphic images in detailed instructions for use.
• Provide as part of the user documentation examples of menus and data input screens, types of output available and possible error messages.

Key Skills

Opportunity

	You can use this opportunity to provide evidence for the Key Skills listed opposite.
Communication C2.3	Write two different types of documents about straightforward subjects. One piece of writing should be an extended document and include at least one image.
C3.3	Write two different types of document about complex subjects. One piece of writing should be an extended document and include at least one image.

System installation and configuration

What is covered in this unit:

4.1 **Hardware**
4.2 **Software**
4.3 **Documentation**
4.4 **Standard ways of working**

This unit will give you an understanding of ICT system components and their purpose, and how to specify the components of a system to meet user needs. You will also install, configure and test new hardware and software. It is important to be able to configure systems (hardware and software) to meet the needs of a user. You must also understand and implement safety and security procedures.

As part of the evidence for the unit you will need to set up a computer system to meet a specification. You will also have to demonstrate that you can make modifications to the hardware and software of existing systems and create and maintain records of all your practical activities.

This unit is assessed through your portfolio of evidence and the grade you will get for this will be your grade for the unit.

Materials you will need to complete this unit:

* computer hardware and software
* as many peripheral devices as possible
* access to the Internet in order to perform research work
* access to popular computer periodicals/magazines (most of these are available over the Internet)
* a toolkit (screwdrivers, pliers, earthing strip, etc.).

There are two parts to all computer systems: **hardware** and **software**. Hardware is the name given to those parts of the computer that you can touch or handle, and to the peripheral devices that make up the system. Software is the name given to the programs that allow you to use the hardware – without software, the hardware is useless. Software is made up of a series of instructions that have been written in a computer language.

To understand the difference between hardware and software, think of a tape recorder and a blank tape. The tape recorder and the tape are the hardware: we can actually touch them. If we recorded some music onto the tape, then the music would be the software.

The components of an information system include the hardware, software, information, staff and accommodation, which together enable IT operations.

How do you distinguish between hardware and software?

Activity 1

Classify each of the following as either hardware or software.

- Windows 98
- A scanner
- A word-processing package
- A hard disk drive
- A printer driver
- A floppy disk
- A joystick
- A DVD disc
- Office 2000
- A monitor

4.1 Hardware

This section deals with the hardware associated with a typical ICT system and involves understanding the purpose of significant pieces of computer equipment and their links with other components. As this is a very practical subject, you will need to practise choosing and setting up different combinations of components for a range of different purposes and potential users.

As the demands placed on the hardware by the software increase it is necessary to be able to upgrade the computer hardware by adding more memory, sound cards, hard drives, CD-ROM drives, etc. In order to choose the correct components, you need to understand their purpose and their capabilities. For example, a quick look through any of the computing magazines will reveal the huge choice for the components, all with differing specifications and prices. You will need to base your choices on the user's needs.

As well as deciding which hardware and software to choose, you will need to install them (appreciating the problems in doing this) and test the installation.

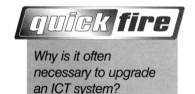

Why is it often necessary to upgrade an ICT system?

Your teacher/lecturer will give you practice in connecting and disconnecting components. From the variety of components available you must pick the ones out that meet a particular user's requirements.

You will need to be able to specify, using reference materials, any of the following components to create an ICT system to meet specified user requirements:

- main processing unit
- keyboard
- mouse
- VDU (monitor)
- connectors
- video card
- sound card
- network card
- disk drives
- optical drive
- printer
- scanner
- serial port

- parallel port
- microphone
- speaker
- SCSI controller.

Most IT systems consist of the following components:

- main processor unit
- input devices
- storage devices
- output devices.

The following sections will explain each of these parts of the system in turn.

Main processor unit

The main **processor** of a computer is often called the **central processing unit** (**CPU**). It acts on instructions – sorting them, filtering them and performing mathematical operations on them. It is called the main processor because many computers now have more than one. With PCs, the main processor unit is normally taken to mean all the components inside the casing, excluding the power supply and the storage devices such as hard and floppy disk drives and CD-ROM/DVD units (Figure 4.1). A typical PC main processor unit would include the following components:

- CPU
- **motherboard** (the board where all the built-in electronics are situated, including the CPU)
- controller boards (e.g. video and disk drive controllers) which are attached to the motherboard)
- expansion cards (e.g. video and sound cards)
- **buses** (the electronic transportation system for sending signals between the components of the computer)

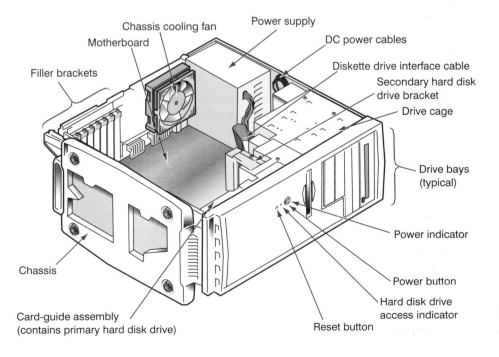

Figure 4.1 Inside a typical computer. Notice where the motherboard is situated

- additional special processors (e.g. maths/graphics **co-processors**), which are usually located on the motherboard
- input and output **ports** (serial, parallel, etc.).

Activity 2

PCs are usually available in four types of casing: desktop, mini-tower, midi-tower and tower (Figure 4.2). There are also different types of casing for portable PCs: laptop, notebook, palmtop, etc.

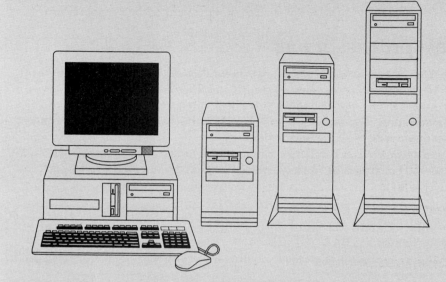

Figure 4.2 Desktop, mini-tower, midi-tower and tower PCs

Imagine you are going to equip 40 users in a company with stand-alone machines (i.e. the machines would not be connected together on a network). Write a list of questions you could ask the users to help them decide which type of casing their PCs should have.

Processor

The **CPU** is the brains of the computer and there are several different manufacturers and many different types to choose from. The main chips are:

- AMD K6
- AMD Athlon
- Intel Celeron
- Intel Pentium

Each is available in a range of different **clock speeds**. It is important to note that each chip has its own particular advantages and disadvantages when compared with other chips and it is impossible to compare the clock speeds of different processors. For example, a Pentium 1000 MHz and an Athlon 1000 MHz have identical clock speeds, but because the internal circuitry of the chips is different one will be faster than the other.

To find out more about the capabilities of each chip you should contact the computing press or the chip manufacturer's Web site.

Activity 3

Have a group discussion about the latest processors available (your teacher will tell you how many people should be in your group). Each member of the group will be expected to contribute to the discussion and will therefore need to be knowledgeable about the subject matter.

Before the discussion you will need to thoroughly research the latest processors on the market from the different manufacturers. This is best done by referring to recent magazine articles. You should be able to search for the back issues and articles of interest using the magazines' Web sites. You should also use the chip manufacturers' Web sites – but they will obviously be biased towards their own chips.

In particular you will need to find out the following:

- the main strengths of the particular chips
- the speed of doing a particular task using the processor
- a price comparison for the various chips
- what the magazines say (summarise the findings that other people have found)
- if it is worth paying the money to buy the latest top-of-the-range chip
- why it is important to find out exactly what the user is going to do with the machine before deciding on the processor.

Motherboard

The **motherboard** is the main circuit board of the computer and contains a series of connectors into which other circuit boards – such as video cards and sound cards – can slot. In most cases the motherboard will contain the processor (i.e. the main chip), **BIOS**, memory (both ROM and RAM), interfaces for the disk drive, serial and parallel ports and all the controllers that control the standard peripheral devices, such as keyboard, monitor and disk drive. There will also be some spaces for extra memory to be slotted in if needed.

Activity 4

There are lots of technical terms used in the above section, and you need to understand what they are. Although they will be explained later in this unit you should look up any unfamiliar terms in the glossary.

Controller boards (also called the input/output controller)

The **input/output** (I/O) **controller** is the link between a microprocessor and its surrounding components, allowing communication with any device connected to it. Devices that are connected to the processor may also be controlled by it. For example, a printer is linked to a CPU and can be turned on and off by it as required. The I/O controller can be a card slotted into the motherboard or may be built into the motherboard. The main purpose of the controller is to translate the signals that pass along the system bus, and the signals that the input and output ports use. The system bus is a series of wires along which the data and control

What is the main purpose of the I/O controller?

signals are able to pass. A port is a connector, usually on the side or the back of a computer, to which input devices (e.g. a keyboard or mouse) and output devices (speakers, printers, etc.) can be connected.

Expansion cards

Expansion cards are circuit boards that fit into the slots on the motherboard of a computer to improve the capabilities of the computer such as the screen performance or the sound capabilities. You can also add extra memory to improve the speed of the system. Adding these circuit boards into the expansion slots is important because it upgrades a computer to cope with the increased demands placed on the system by new operating systems and applications software.

Buses

A **bus** is an electronic transportation system that passes data to and from its different locations within a computer system. A single wire is used for each **bit** of data transferred as a single unit. A computer that is able to transfer 32 bits at a time will have a bus with 32 wires so that all the bits can be transferred in parallel.

There are three types of bus (Figure 4.3).

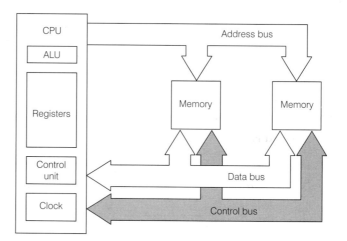

Figure 4.3 How the three buses are used, with each carrying a particular type of information

Data bus
A data bus transfers data to and from the memory locations, so the number of wires in it will greatly influence the speed of the computer. Where several devices are connected to a data bus, they must all take their turn in using it.

Address bus
An address bus carries binary signals (groups of 1s and 0s) for selecting the memory storage locations to be used when data is read from or written to the memory.

Control bus
A control bus carries the signals that tell the devices when they may use the bus and sets up the circuits on the bus for data toward or away from the memory. Because the data and address buses are shared by all of the system components, it is necessary to synchronise their actions.

Industry Standard Architecture (ISA) bus

The **ISA bus** is the type of bus that was originally used in the IBM PC/XT and the PC/AT (older versions of PCs). The AT version of the bus is still used in many computers for slow devices, with a PCI bus being used for the faster devices. There is a new version of the ISA specification called 'Plug and Play ISA' that enables the operating system to configure the expansion boards automatically so that users do not need to alter the DIP switches or the jumpers (see page 246).

Peripheral Component Interconnect (PCI)

This is a local bus developed by the chip manufacturer Intel and is often included with an ISA bus.

Co-processor

A **co-processor** is a separate processor that is designed to take some of the load off the main processor and therefore make the computer run faster. The most common type of co-processor is the maths co-processor, which is designed to perform complex calculations very quickly. The chips in most modern computers come with a maths co-processor already incorporated into the chip. The software must be written to make use of one, otherwise the maths co-processor will be unused. You can also get graphics co-processors for improved and faster graphics capabilities.

Clock speed

The **clock speed** is the speed with which a processor executes instructions. All computers have an internal clock inside them and this clock is used to regulate the rate at which the instructions are executed. It also synchronises all the various components of the computer.

Clock speeds are measured in **megahertz** (MHz): 1 MHz is one million cycles per second. Clock speeds are rising rapidly all the time and 650–1000 MHz is the norm.

Although the clock speed is an important measure of the speed of a computer, the internal architecture of the chip is important. This means it is not possible to use the clock speed to decide which of two different makes of processor runs faster. Only if the chips have identical architecture can you use the clock speed to identify which is fastest.

Ports

Ports are the external connection points on a computer to which the peripheral devices are connected. If you follow the lead from a peripheral (keyboard, mouse, monitor, printer, etc.) it will end up at a port. There are two main types of port.

Serial port
The **serial port** is an interface that can be used for serial communication. With serial communication the series of bits representing the data are sent one after another through the serial port and along the wire connecting the port to the device. Serial ports conform to the RS-232C or RS-422 standards, and allow most devices to be connected to them. Mice, keyboards and modems are usually connected to a serial port. You can also connect a printer to one, although most printers are connected to a parallel port.

Parallel port
A **parallel port** is used for connecting a device such as a printer that is able to make use of the parallel transmission of data (this means that groups of bits can be transmitted at the same time). Parallel transmission is thus much faster than serial

229

What is the difference between serial and parallel transmission of data?

transmission and is the method used for sending data between the computer and an external device, such as a printer, that needs a high throughput (also called bandwidth) of data.

A parallel port makes use of a 25-pin connector of a type called DB-25. All PCs have at least one.

The newer type of port, called a SCSI port, is also a parallel port.

Small Computer System Interface (SCSI)

SCSI (pronounced 'scuzzy') is a parallel interface that is frequently used for attaching peripheral devices to computers. The SCSI interface has a big advantage over the normal parallel interface because it is able to transmit the data much quicker. You can also attach more than one peripheral device to the same interface. Not all computers come with SCSI but you can upgrade a computer to have this interface by installing a circuit board into one of the expansion slots in the computer.

What is the main advantage of a SCSI over an ordinary parallel or serial interface?

Input devices

Input devices are used to get data into a computer system and act as an interface between the computer system and the outside world. The ideal input device would collect all the required data accurately, without human intervention, and be relatively cheap. The input device chosen is usually a compromise, and many systems use more than one device – for instance, a home computer may use a mouse, a joystick and a keyboard.

There are many different input devices and each has its advantages and disadvantages. The most popular ones are:

- keyboard
- mouse
- trackball
- joystick
- light pen
- touch screen
- graphics tablet
- magnetic strip reader
- barcode reader
- optical character reader
- optical mark reader
- punched card reader
- magnetic-ink character reader
- scanner
- video camera
- digital still camera
- microphone.

The keyboard

The keyboard is the oldest input device and is still the most popular. It is an appropriate input device when small quantities of data need to be input. If large quantities of data need to be input, then alternative data capture methods are quicker, less costly in terms of time and are less prone to errors – such as optical mark recognition (OMR), optical character recognition (OCR) or magnetic ink character recognition (MICR).

The mouse

A mouse is considered an input device because it can be used to make selections and, depending on the selection made, the data chosen is then input into the system. Commands are often selected from pull-down menus in Windows-based software and this is used as an alternative to typing in commands using the keyboard or using the cursor keys.

Scanners

Pictures can brighten up the dullest of documents and they are almost essential if you need to make up an interesting and attractive Web page or Web site. Using a scanner, you can scan in previously printed diagrams and photographs into your word-processing, desk-top publishing, Web page/site design software. The most popular scanners are the flat-bed scanners, which have the appearance of a rather thin photocopier. In other scanners the paper is fed through – these have the main advantage that they do not take up as much space on your desk. These paper-feed scanners have a big disadvantage in that you cannot copy from books without taking a photocopy of the page first, because only a single sheet can be fed in.

Flat-bed scanners illuminate the sheet of the item to be scanned by passing a bright bar of light over the page. The reflected light is detected by photosensitive cells and assembled into a picture. Both black and white and colour scanners are available but always bear in mind that if you can't print in colour there isn't a lot of point in scanning in colour – unless you are using the picture for Web page design, etc.

Scanners are very useful if you have a large amount of text that needs to be copied into your word processor. You could, of course, type it in, but this takes a lot of effort and is slow. It is much quicker to scan it in and use OCR, software that is able to read and understand each individual character. However, for OCR to work well you need good originals to scan.

/ **Activity 5** /

For each of the input devices in the list on page 230 that are not discussed separately above produce a short description, mentioning:

- the name of the device
- how it works
- what its advantages and disadvantages are compared to other input devices
- the type of applications it is used for.

Storage devices

Computers store data either in chips inside the computer (**primary storage**) or on other media such as magnetic disk, tape or strip (**backup storage** or **secondary storage**). Primary storage includes ROM (read-only memory) and RAM (random access memory). Data stored in these chips is immediately accessible to the central processing unit, unlike data stored on secondary storage where there is a delay while the data is loaded into memory from the storage medium.

Examples of storage devices include RAM, ROM, magnetic disk, CD-ROM, re-writable CD, magnetic tape and magnetic card/strip.

Primary storage/memory

Another name for the chips inside the computer that store information needed immediately by the computer is memory. There are two types of memory: RAM and ROM. The differences between them will be discussed later.

Terms relating to memory

Many specialist terms are used when the memory of a computer is being discussed. It is important that you understand and use these terms.

Bits and bytes

Bit is short for binary digit, the smallest piece of information that the computer can understand. A single bit can hold only the values 0 and 1. Normally, computers deal with more than one bit at a time.

A series of eight consecutive binary digits is called a **byte**. Computers are normally classified by the number of bits they are able to process at one time or by the number of bits that they use to represent addresses.

A byte is a unit of storage capable of storing one character – any letter, number, punctuation or other symbol that you can type from the keyboard. On most modern computers, a byte is equivalent to 8 bits. As a byte is an extremely small unit, memory is usually expressed in kilobytes (1024 bytes) or megabytes (1 048 576 bytes). These are accurate conversions but it is easier when doing quick calculations to consider 1 kilobyte (Kb) to be 1000 bytes and 1 **megabyte** (Mb) to be 1000 kilobytes. A normal floppy disk has the storage capacity of 1.44 Mb. As 1 byte can be used to store one character, this means that about 1.4 million characters could be stored on such a disk.

A **gigabyte** is approximately one thousand megabytes or, more accurately, 1024 megabytes. One gigabyte is equivalent to 1 073 741 824 bytes.

A **terabyte** is approximately one trillion bytes (1 099 511 627 776 bytes).

RAM and ROM

There are various memory locations in RAM where coded data and instructions can be stored. Each byte of data (letter, number, punctuation mark, etc.) is stored in RAM at an address with a specific address number. The computer uses this address number to locate the data. This type of memory is called 'random access' because the computer is able to move to any address in the memory to retrieve the contents with equal speed.

ROM is slower than RAM and is used to hold data and instructions that the computer needs in order to boot up. '**Booting up**' is a term used to describe the process of the computer starting itself up when the power is switched on. The contents of ROM are stored in the electronic circuitry of the chip and are not erased when the power is turned off, and for this reason ROM is often called non-volatile memory. The contents of RAM are erased when power is turned off, and this type of memory is therefore called volatile memory.

The data stored in RAM can be read from as well as written to the memory (called read/write memory), whereas data stored in ROM can only be read (that's why it's called read-only memory).

/ **Activity 6** /

If you look at catalogues for computer components (these can be found in magazines such as *Computer Shopper*) you will notice that there are lots of different types of RAM. As the prices for the same amount of RAM vary considerably between these different types, there must be some significant difference between them.

Research the different types of RAM and identify the advantages and any disadvantages of each.

Here are some other terms about memory that you will need to be familiar with.

Volatility

RAM is volatile memory because it loses its contents when the power is switched off. Data held in ROM is non-volatile, which means that the data stored in it is not lost when the power is switched off.

Cache

A **cache** consists of high-speed memory in a computer where data can be held temporarily, rather than in slower memory or on a much slower hard disk drive. Its effect is to speed up the computer. Only a relatively small amount of cache memory is needed for it to have a dramatic effect in speeding up a computer. Typical sizes of cache memory are 256 Kb or 512 Kb.

Buffer

A **buffer** is a storage area, usually in RAM, that acts as temporary storage for the data before it is passed to a device. The buffer area allows the computer to manipulate the data before it is passed to the device and also allows the data to be passed to the device at a speed that the device can cope with. For example, the data transfer speed onto and from a disk drive is low compared with the data speed in the CPU and the buffer is used almost as a 'waiting room' for the data.

It is interesting to note that, when you type in a document using a word processor, the document changes that are made are stored in a buffer. When the document is changed, the file on the disk is updated with the contents of the buffer.

Buffers are also used when a print instruction is issued by a program. When the instruction is given, the operating system copies the data to the print buffer, which is either in memory or on disk, and the printer draws the data out of the buffer at its own pace. Print buffering is useful because the printer can operate in the background so that other tasks can be performed on the computer while it is printing a long document.

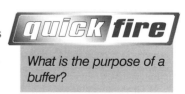

What is the purpose of a buffer?

Printer driver software often contains a buffer so that instructions that have been typed in can be edited before sending the instruction to the computer.

Secondary storage/backing storage

Secondary storage is storage outside the processor. The most common backing storage devices for personal computers are the floppy disk drive and the hard disk drive.

Floppy disks

Floppy disks are an important and popular storage device but their low storage capacity (1.44 Mb) has meant that they are really only suitable for transferring small files between computers or providing a backup for word-processing documents, spreadsheets, etc.

Floppy disks, as the name suggests, are flexible plastic disks coated with a magnetic material and encased in a plastic sleeve. When the disk is placed in the drive, it rotates at 360 revolutions per second, which is a lot slower than hard disk drives; this limits the speed at which the data can be read from or written to the disk, called the data transfer speed. Another limiting factor is the fact that the disks are not continually rotating like a hard drive, so the disk has to accelerate up to the operating speed before the data can be read off or written to it. Floppies are the most widely used backing storage medium and, as nearly all computers have a floppy drive, there is rarely a problem in transferring the data from one machine to another using this medium.

Activity 7

Many inexperienced users have problems with data stored on floppy disks, mainly because they do not take enough care of their disks. Produce a list of the precautions a user should take when using floppy disks.

Hard drives

Hard disks are not usually a single disk but are composed of a series of metal disks on a plinth. Each surface of the disk is able to store data and therefore has its own read/write head (Figure 4.4).

Because the data for a particular file is stored over several disk surfaces, the read/write heads can operate independently and this means that high data transfer speeds can be achieved. The disks are rotating continually, so no time is wasted in waiting for the disks to reach their operating speed.

Most hard drives are fixed inside the computer case but removable drives are available and these are useful for backing up an internal hard drive because they can be locked in a fireproof safe.

A number of factors need to be considered when comparing the performance of hard disk drives.

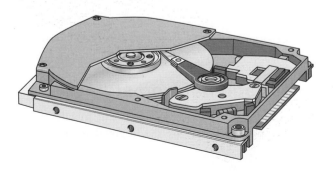

Figure 4.4 The inside of a hard disk drive showing the arrangement of the disks and the read/write heads

Rotation speed

The rotation speed of a disk drive is measured in revolutions per minute (rpm): the faster the disk rotates the less time it takes to locate the data on the disk. Faster disks take less time to transfer the data either to the disk (copying) or from the disk (reading). A typical hard drive rotates at 7200 rpm.

The access time for a disk drive

The speed of a hard disk drive is expressed in terms of the access time, which is measured in milliseconds (thousandths of a second). Hard disk drives have a typical access time of about 9 milliseconds. The access time for a disk drive can only ever be an average value as it includes the seek time (the time taken for the read/write heads to locate the data stored on the surface of the disk). This will obviously depend on how far away the heads are from the data that is required.

Backup storage devices

Hard disks have such large storage capacities that it no longer feasible to take backup copies of the hard drive using floppy disks. Instead one of the following backup storage devices can be used.

Zip drives

These are cheap storage devices that plug into the parallel port of a PC. They consist of a single removable hard drive with the same diameter of a standard floppy disk (3.5 inches). The zip disk looks like a standard floppy disk except it is much fatter. Typical storage capacities are around 250 Mb, with transfer rates of roughly 1.7 Mb per second.

Jaz drives

Jaz drives are similar to zip drives but they have a much larger storage capacity (about 2 Gb on a single cartridge). The data can also be compressed using 2:1 compression, which means that 4 Gb can be stored on a 2 Gb cartridge. The cartridges on which the data is stored are removable and they have a data transfer

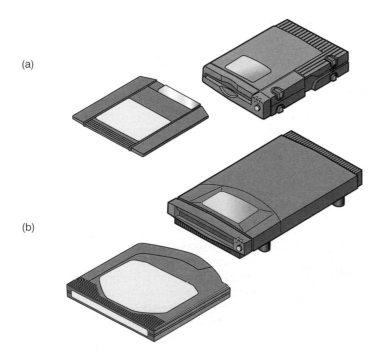

(a)

(b)

Figure 4.5 The zip (a) and jaz (b) drives are popular backup devices for PCs

rate of around 7 Mb per second, which makes them faster than most hard disk drives. Like zip drives, jaz drives are light, and are ideal backup media for most PCs.

Ditto drives

These are tape drives which store the data on a tape in a removable cartridge and can typically store 2 Gb of data with a data transfer rate of 10 Mb per minute. They are easy to install and use and have a facility to back up data as you work.

CD-ROM drives

CD-ROM discs are the best media for the distribution of software because of their high storage capacity, typically 600 Mb, and the uniformity of their use (most PCs are supplied with a CD-ROM drive). Floppy disk drives have a storage capacity of only 1.44 Mb and many packages now need storage of tens or even hundreds of megabytes so installing such packages using floppies is a laborious process.

There are many different CD-ROM drives available, each with a different multiplier (e.g. 40-speed, 48-speed, 50-speed). This is the speed at which the CD is able to transfer its data, called the transfer rate. When CD-ROMs first came out they had a speed of 150 Kb of data per second and then, when a faster speed CD-ROM came out (with a transfer rate of 300 Kb), it was called a 2-speed CD-ROM. This means that a 40-speed CD-ROM would have a data transfer rate of 40×150 bits per second.

To write onto a recordable CD, a special and more expensive CD drive is needed. This type of drive is particularly useful if you are a software producer or if you are developing multimedia applications.

DVD (digital versatile disc)

DVD-ROMs are set to replace CD-ROMs soon and all computers will probably come with DVD supplied as standard. The DVD is a high-capacity optical disc, which is similar in many ways to a standard CD-ROM. At present the capacity of a DVD is about 17 Gb (compared to the 600 Mb of a normal CD-ROM). Although a new reader is needed to read DVDs, the DVD reader can also read standard CD-ROMs.

The main use to which DVD is put is to store high-quality digitised video – up to 8 hours' worth can be stored on a single disc. DVD is likely to replace the standard video recorder and videotapes for higher quality digital video. Its uses in computing will probably be for the storage of data and for the storage of video-based multimedia packages. Soon the domestic television and the home PC will come together to provide everyone with Internet access, video conferencing, video on demand and digital video recording from a camcorder.

At present DVD-RAM is being developed, and it will soon be possible to record data to a disc by using a simple inexpensive drive.

CD-R and CD-RW drives

CD drives are gradually going to be phased out as DVD becomes universal but there are two more modern CD drives that are becoming popular. **CD-R** drives use disks that can be written once and never overwritten. They can be read by any CD-ROM drive. The **CD-RW** drives are read/write, which means that the information written on them today can be overwritten with new information tomorrow and they can (in most cases) be read by an ordinary CD-ROM drive. However, if you have an old CD-ROM drive you might find that it is incapable of reading CD-RW disks.

CD-RW drives are very useful because they allow you to share large amounts of data with other people. You could use zip or jaz drives to do this but the other person would also need one of these drives to be able to read the files. With CD-RW, once you have stored the data on the disk, the disk can be read by anyone with a CD-ROM drive (almost everyone).

Activity 8

Using recent copies of computer magazines as your source, produce a document that compares and contrasts a range of the most popular backup devices that can be used with a stand-alone computer containing around 200 Mb of very important and valuable data files. Include details such as typical storage capacities, prices, speed of data transfer and any other details you feel are important.

Data compression

Data compression means storing data so that it takes up less space on the storage medium (hard disk, for example). There are many different **utility programs** that enable files to be compressed. You need to know how to use these to compress files because files containing graphics/photographs are often too large to store on a normal floppy disk without first being compressed. The most common file compression utilities are PKZIP and PKWARE, copies of which may be obtained free off the Internet.

Compression is also used when sending files over the Internet. Many files that you download will be compressed because they download faster, which in turn costs less. In order to use a compressed file you will need to decompress it. The communications software provided by most **Internet Service Providers** (**ISPs**) does this automatically to the file so that it is ready to use immediately.

Under what circumstances would it be desirable to compress data?

Write protect

It is possible to set magnetic media so that it is read only. For example, by closing the write protect slide on a floppy disk you can store data on the disk but when it is open you can only read the data off the disk. Using the write protect facility you can make sure that important disks are not copied over.

Integrated drive electronics (IDE)

IDE is a type of hardware interface that is used to connect hard drives, CD-ROM drives and tape drives to a PC. Its main advantage over the other types of interfaces is that it is the cheapest way of connecting peripherals. The low cost of IDE in the computer is achieved because the computer itself has only simple connections with all the main interface electronics in the actual device you are connecting to it. A lot of modern computers have two enhanced IDE sockets built into the motherboard and these can be connected to devices using a 40-pin ribbon cable.

Hardware devices used for networking

There are several hardware devices needed for networking, such as the **modem**, **IDSN** adapter and the **network card**. Many other devices are used with local area networks and wide area networks but their use is quite technical and complex and we will not look at them here.

Network card

In order to use PCs as terminals for a network a piece of hardware called a **network interface card**, sometimes just called a **network card**, must be included inside the PC (see Figure 4.6). This is a small circuit board that slots into one of the connectors on the motherboard. Network cards can be bought separately and their main purpose is to reduce the cabling between each terminal to a single wire.

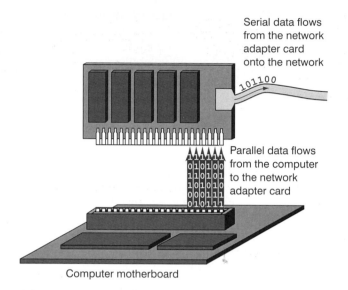

Serial data flows from the network adapter card onto the network

Parallel data flows from the computer to the network adapter card

Figure 4.6 A network interface card/adapter

Computer motherboard

Modems

Telephone lines were not originally designed to transfer data – they were developed to transfer sound signals from one place to another. In a telephone, the varying frequencies of the human voice are converted into electrical signals which are then passed along the telephone wires to an earpiece that converts the electrical signal back into sound. Sound signals are **analogue** signals; their waveforms vary continuously with time. As the electrical signals generated are replicas of the sound, the electrical signals are also analogue.

Analogue and digital signals

Analogue signals have an infinite number of 'in-between' positions, whereas digital signals jump from one value to the next. For example, a watch with a dial and hands can show an infinite gradation of times, whereas a digital clock jumps from one minute (or second) to the next.

Digital signals consist of pulses of voltage or current which represent the information to be processed by a computer. The digital voltages can be 'high' or 'low', the high voltage being used to denote a '1' and the low voltage a '0'. Data is thus represented as binary pulses (1s and 0s) consisting of high and low voltages. It is cheap to produce electronic circuitry to deal with digital signals, and noise is not too much of a problem with binary signals because if the signal becomes distorted it can usually be regenerated and the noise removed. This is not possible with analogue signals because of the complex nature of the waveform, which easily becomes confused with the noise, and thus is continually degraded.

A **modem** (abbreviation for modulator/demodulator) is used to convert the digital signals from a computer into analogue form so that they can be passed along the telephone lines. The receiving modem then changes the analogue signals back to their digital form. Most external modems are small boxes with several indicator lights which flash on and off. Internal modems are situated inside a PC and often show what is happening by flashing indicator lights on the screen. Modems plug directly into the telephone socket, although the lead provided is seldom long enough and generally requires an extension. Figure 4.7 shows two modems connected at either end of a telephone line.

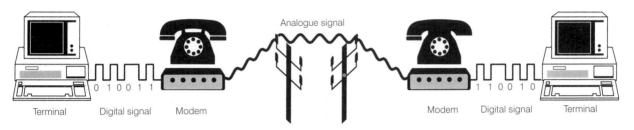

Analogue signal

Terminal   Digital signal   Modem   Modem   Digital signal   Terminal

0 1 0 0 1 1   1 1 0 0 1 0

Figure 4.7 Using a telephone line to carry data

Your modem determines the speed at which you can interact with other computers over the telephone line. When a modem makes a connection to another computer it will negotiate with the receiving modem a speed for the transmission and receipt of data. The speed of transmission is important because this determines the time it takes to transfer the data and hence the cost of the phone call, and possibly the on-line charges levied by the service provider.

Some of the latest modems are able to deal with both voice and data at the same time using a single telephone line with a system called digital simultaneous voice and data (DSVD). This is important during a video conferencing session because it allows participants to speak and exchange data (in the form of an on-screen whiteboard) at the same time.

Integrated services digital network (ISDN)

Although all data (from camera images, music recordings on CD and videos to computer data) now seems to be stored in digital form, the technology used to transmit the data (the telephone system) is over 100 years old and was originally developed for the transmission of voice signals. To transmit data along telephone lines the digital material has to be changed to analogue form and then converted back again at the other end, but the telephone system is not really up to the job. Files are becoming much larger and it is not just textual data which needs to be transferred.

There has been a gradual development in the UK towards a completely digital telephone system; known as **integrated services digital network** (ISDN). Many other European countries are also developing ISDN and countries such as America and Japan already have all-digital systems. The main advantage of ISDN is that it delivers data communication at speeds up to eight times faster than a modem, even without compression. If data compression is used, ISDN is able to transmit data at speeds up to 512 000 bits per second; without compression a respectable 128 000 can be achieved. Surfing the Internet or transferring multimedia files can be done almost instantly using ISDN.

Other advantages of ISDN include the facility to conduct 'virtual' meetings using desktop video conferencing and being able to use the telephone as a sophisticated business telephone with a wide range of new facilities. It is now possible to do several things at once on the same line, so you could be transferring a file to a person while holding a conversation with them.

To use ISDN you either need terminal equipment (TE) or a terminal adapter (TA) in place of a conventional modem. Remember that no modem is needed because ISDN is itself a digital service so there is no need to convert digital signals to analogue, and back again.

Activity 9

Pettifog & Scrimp, a medium-sized business specialising in the publishing of historical journals, is thinking of converting from the slow world of analogue communications to a fully digital system. The IT manager has asked you, one of the systems developers, to investigate ISDN and report your findings.

Your report should be written in simple, non-computer language with all specialist terms explained, because it will be handed to the board of directors, most of whom have only a limited knowledge of IT.

You will need to include the following information in your report:

- what ISDN is and what is new about it
- how ISDN works
- all the advantages of ISDN over the traditional telephone system (make sure the advantages you state are relevant to a publishing company)
- the extra equipment needed to set up the ISDN connection
- how to obtain ISDN from a telephone company and the costs involved.

Can devices which use analogue signals such as modems, faxes and answering machines still be used with ISDN? Explain your answer.

Network connectors

Network connectors connect the wires between terminals to the input socket situated at the back of each terminal. The type of connections used depends on the type of wire/cable used for the network.

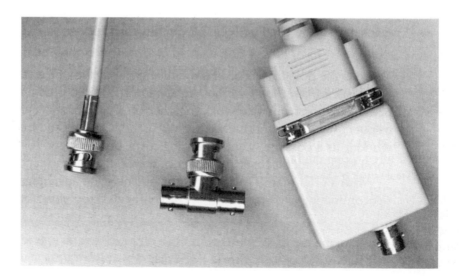

Figure 4.8 Network connectors

Figure 4.8 includes a T-piece connector which connects two network wires to a network interface card via a BNC socket at each end. This type of connector is used mainly with thin cables such as coaxial which is used in token-ring, Ethernet and ARCnet systems.

Output devices

Output devices are used to show the results from a data processing system, often in the form of hard copy (another name for a printout) or a screen display. Data can be passed readily from one computer to another as electronic signals, so the output data from one computer can be the input data for another.

Here is a list of the main output devices:

- monitor or visual display unit (VDU)
- graphical display unit
- laser printer (black and white or colour)
- ink-jet printer
- dot matrix printer
- plotter
- speakers for voice output or sound.

In control applications the output could be used to control an electric motor, switch on a heater or control the movement of a robot arm.

Terms used when talking about output devices

In order to be able discuss output devices, it is important to understand a few terms.

Resolution
Resolution can refer to a printer, scanner or a VDU. With a printer it is the number of dots of ink per inch (dpi) – a laser printer having a resolution of 300 dpi means that in a line one inch long the printer is capable of printing 300 separate dots. This means that 300×300 dots (i.e. 90 000 dots) could be printed in a square inch.

The resolution of a VDU is the number of **pixels** (dots of light) on the whole screen. VDUs have rectangular screens, which means that there are more pixels in the horizontal direction than in the vertical. A typical resolution for a monitor is 1024×768, meaning that there are 1024 pixels in the horizontal direction and 768 in the vertical direction. Multiplying these two figures gives the total number of pixels on the screen. For the 1024×768 screen there will be a total of 786 432 pixels.

The resolution of a scanner is also measured in dpi, but in this case this is the number of pixels along a line one inch long of the image when an picture is scanned using the scanner. A typical scanner resolution would be 1200×2400 dpi.

Monitors, printers and scanners can also be classified according to their resolution – as high, medium or low resolution.

Refresh rate/vertical scan frequency
The **refresh rate** or vertical scan frequency is the number of times an entire display screen is refreshed or redrawn per second. This frequency is typically above 70 Hz. The greater the refresh rate, the less the strain placed on your eyes when using it. Good monitors have a refresh rate of above 100 Hz.

Video card
Video cards are also called video adapters, graphics cards or graphics adapters. A video card is a circuit card that plugs into the motherboard of a personal computer to give it its display capabilities. How good or bad the display is does not just depend on the video card: it also depends on the VDU. The RAM of the

processor is not used for storing the display for the monitor as this would decrease the amount available for programs and slow the computer down. Instead, some RAM (called video RAM) is included as part of the video card. The resolution of a monitor is a measure of the number of pixels and the greater the resolution required for the monitor, the greater the amount of video RAM needed. The amount of video RAM required is also determined by the number of colours needed on the screen (see the table below).

The video RAM required for different resolutions and number of colours

Resolution (pixels)	256 colours	65 000 colours	16.7 million colours
1024 × 768	1 Mb	2 Mb	4 Mb
1152 × 1024	2 Mb	2 Mb	4 Mb
1280 × 1024	2 Mb	4 Mb	4 Mb
1600 × 1200	2 Mb	4 Mb	6 Mb

As well as containing memory in video RAM, most modern video cards also contain their own graphics co-processors for performing graphics calculations. These calculations are needed particularly if three-dimensional graphics are used – for CAD work or games, for example. Calculations need to be performed very quickly and graphics cards often contain their own graphics co-processors. Such graphics cards are frequently called graphics accelerators.

Interlacing

A monitor builds up the image on the screen by scanning an electron beam horizontally across the screen, starting from the top and moving down during each horizontal scan until the bottom is reached. Each horizontal line scanned during this building-up process is numbered so there are even and odd lines. The image on the screen is built up in two stages called fields. In an interlaced screen, the even-numbered lines are displayed in the first field and the odd lines are displayed in the second. The effect of this **interlacing** is to reduce screen flicker because the screen is redrawn twice as often as with a non-interlaced screen.

Monitors (VDUs)

Monitors are the most popular output device and are used in nearly all applications. They are ideal for on-line enquiries, where the results of a query can be displayed on the screen. Monitors come in many sizes but, with the introduction of Windows, there is a lot of information on the screen when working with most packages and this has meant that standard-sized screens are now 17, 19 or 21 inches. Note that the size of the screen is measured across the diagonal, not vertically or horizontally, so a 21-inch screen has a *diagonal* length of 21 inches. Monitors are often neglected when choosing computer systems – people tend to spend their money on faster processors or larger hard drives rather than on a better quality monitor. Yet if you looked at monitors in a line showing the same image in the same way as you view television sets when you are going to buy one, you will see that they vary widely and – as always – you get what you pay for. The monitor's screen resolution determines the maximum number of pixels that can be drawn on the screen at any one time and is quoted using two numbers (the first being the number of horizontal dots, the second the number of vertical dots). The resolution refers to the sharpness and clarity of the image on the screen. A typical screen resolution is

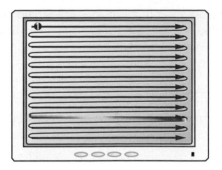

A line of pixels is drawn, then quickly returns (without drawing) to do the next line. The scan frequency, or refresh rate, is the number of times per second the complete screen is drawn.

Figure 4.9 How a picture builds up on the screen of a non-interlaced monitor

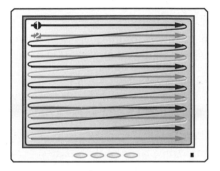

Pairs of lines are scanned first; the monitor does another scan to draw the lines it has missed out.

Figure 4.10 How a picture builds up on the screen of an interlaced monitor

1024 × 768 pixels. Flat panel liquid crystal display screens are becoming very popular because they take up much less space on a desk than a conventional cathode ray tube monitor.

Printers

Printers are used to produce hard copy output which can be taken away and studied or posted. This section looks at the main types.

Dot-matrix printers

Dot-matrix printers are impact printers. They work by hitting pins against an inked ribbon, leaving a character on the paper. The main disadvantage is the low quality of the text and graphics produced using such printers and the amount of noise created. They are very cheap to buy, reliable and cheap to run because the only consumable (other than the paper being used) is the inked ribbons, which last a long time and cost little to buy. Dot-matrix printers are useful if you just want to see a draft result or if you need to use multi-part stationery (an impact printer is needed for this). Dot-matrix printers are being replaced by ink-jet or laser printers because of the higher quality of the printouts they produce.

Ink-jet printers

Ink-jet printers work by spraying black or coloured ink onto the paper. Because there is no impact, they are relatively quiet.

Although the initial cost of an ink-jet printer is set attractively low, the cost of the consumables, which are mainly the ink cartridges, can be high and many manufacturers make more money out of selling the consumables than from the sale of the actual printer. For perfect copies using an ink-jet printer, it is necessary to use special paper that has a whiter look and a glossy surface coating. This paper is more expensive than normal photocopying paper and obviously adds to the running costs.

What are the main advantages in having a large screen with a desktop PC system? Are there any disadvantages?

Most people buy an ink-jet printer because most of them are capable of printing in colour – but the snag with this is that the colour ink cartridges use three different coloured inks, and as soon as one of the colours is exhausted the cartridge is no longer of any use and will need replacing.

One big disadvantage is that when the paper comes out of the printer the ink is often still wet and the image is easily smudged. They do, however, provide a low-cost alternative to a laser printer. If you require colour printing on a limited budget then an ink-jet printer will be your only realistic choice because colour laser printers are very expensive to buy and run.

Laser printers

Laser printers make use of photocopier technology. A laser marks out an image over the paper, the toner sticks to parts of the paper and is then fused to the paper by a high temperature. This process is quite fast (typically 10 pages per minute), so laser printers are ideal if you want quick, high-quality printouts. Operating costs are relatively high because toner cartridges will need to be bought and the metal drum will need to be renewed periodically.

Most laser printers are black and white – colour ones are available but they are very expensive. The main factors to consider when deciding on a laser printer are:

- whether you need to print in black and white or colour
- the print resolution, which determines the clarity of the image
- the speed of the printer, which is usually expressed in pages per minute (ppm). You have to be careful about the speed – a nominal speed of 15 ppm would only be exact if the same line of text were printed on all 15 pages. Obviously this never happens, so the actual print speed is less than that quoted.

Activity 10

Suggest a type of printer (dot-matrix, laser or ink-jet) that can be used as a network printer by five terminals in a small business. Write a list of questions you would ask the business owner to make sure that you make the right choice. For each question, place in brackets after the question your reasons for asking it.

Graph plotters

These are used to produce accurate line diagrams such as maps, plans and three-dimensional drawings. Diagrams produced in this way look more professional than those produced by a printer. A **graph plotter** is often the only choice if the drawing is very large. There are two types of plotter: the flat bed plotter for small drawings and the drum plotter for large drawings.

Configuring the system

If you need to connect a new printer, DVD drive, mouse, etc. it is not always simply a question of connecting the cable and it working. Sometimes you have to configure the device.

To get the hardware to do a useful job you must install both the systems and applications programs and then get them to work with the hardware by configuring the system. A stand-alone computer system means one which is not connected to a network or other communications device.

/ **Activity 11** /

Some input and output devices are used with PC systems whereas others are more specialised and are used only for certain applications.

For each of the input and output devices listed on pages 230 and 241, produce a set of notes covering:

- examples of applications to which the device is most suited
- a brief description of what the device does
- some indication of the cost of the device (if the device is very specialised then it may prove difficult to find out the price; in these cases say that the price is unknown)
- the advantages and disadvantages of the device over other similar devices.

Include a diagram or picture of each device. You could draw these yourself or cut out pictures from computer magazines.

Connecting the hardware components

When you buy a computer you need to connect the components. Because you need to do this every time you transport a computer or buy a new one you need to know how to connect the components in the correct way.

On a basic (non-multimedia) system, the hardware will, as a minimum, consist of:

- main processor unit
- keyboard
- mouse
- VDU
- printer
- ports (serial and parallel)
- connection cables (power and data).

/ **Activity 12** /

A friend borrows your computer system, which has the basic components listed above, and to get it into her car she has to disconnect all the components. She later telephones, saying that she doesn't know how to connect it up again because there is no manual, and asks you to fax some instructions to help her.

Produce a series of instructions, accompanied by diagrams, that your friend could use to connect up the computer. It would be a good idea to study the different leads and connectors at the back of the machine so that you can tell her exactly what goes where.

Terms related to connector plugs and sockets

Computer systems consist of lots of parts that need to be connected and there are many different types of cables and connectors. These connectors will need to be connected into the plugs in the back or side of the computer, and there is also a range of different plugs. Once you start networking the computer the numbers of cables, connectors and plugs increases dramatically. Here are a few of the ones you may come across when connecting up systems.

USB

Universal Serial Bus (or USB) is a very fast external bus that supports very fast transfers of data (typically in excess of 12 million bits per second). A large number of peripherals – keyboards, mice, modems, DVD drives, etc. – can be connected to the computer using USB. USB also supports '**plug and play**' and is likely to replace serial and parallel ports shortly.

BNC connectors

The **BNC connector** is a connector that is used to connect coaxial cables. A special type of BNC connector, the BNC T type connector, is used to connect two cables with normal BNC connectors to a network interface card.

DIP switches

These small switches are built into circuit boards to allow you to configure the circuit board for a particular computer (Figure 4.11). **DIP switches** have only two states: on or off. New developments such as 'plug and play' mean that the use of DIP switches to make configuration changes is becoming obsolete.

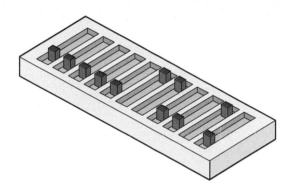

Figure 4.11 What DIP switches look like

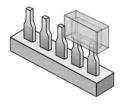

Figure 4.12 How jumpers work

Jumpers and settings

A **jumper** is a small piece of metal that is placed across two pins to create a bridge, thus allowing the current to flow (Figure 4.12). To make alterations in the setting simply select two different pins across which to place the jumper. Jumpers can be used to configure expansion boards.

Configuring a new system

When you first buy a system it will probably come pre-configured, which means that it will have been set up to use whatever optional equipment (e.g. disk drives, memory, sound card, display adapter) that was originally installed. However, if you plan to change any part of the system, for instance to install some extra memory or add a peripheral device such as a scanner or modem, then you will probably need to reconfigure the system. (Peripheral devices are devices that are connected to and under control of the central processing unit.) Reconfiguring a modern Windows-based PC is much easier than it was in the past, when extra instructions had to be included and some existing ones changed in a **configuration file** called the config.sys file. Most hardware devices now come with special software, called installation programs, whose purpose is to reconfigure the system automatically, and most users aren't even aware that it has been done.

Config.sys

If you have DOS (Disk Operating System) on your computer you may like to look at the config.sys program by entering the directory where DOS is to be found and then issuing the command 'Type config.sys'. Remember not to change anything without doing a printout of the original file first. The instructions in a typical config.sys file are shown in Figure 4.13.

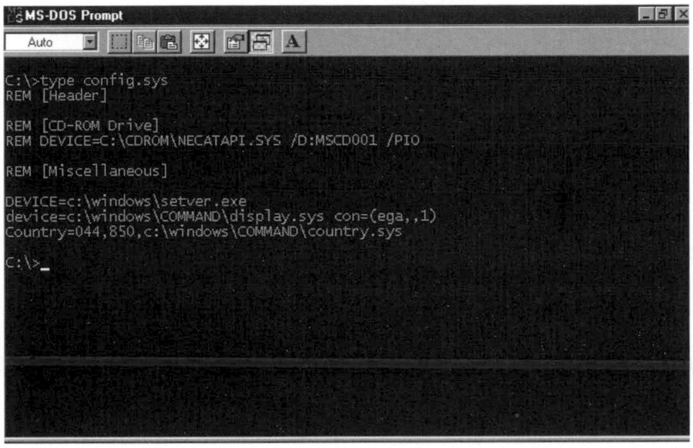

```
C:\>type config.sys
REM [Header]

REM [CD-ROM Drive]
REM DEVICE=C:\CDROM\NECATAPI.SYS /D:MSCD001 /PIO

REM [Miscellaneous]

DEVICE=c:\windows\setver.exe
device=c:\windows\COMMAND\display.sys con=(ega,,1)
Country=044,850,c:\windows\COMMAND\country.sys

C:\>_
```

Figure 4.13 The instructions in a typical config.sys file

Activity 13

As part of the evidence for this unit you should be able to

- configure the display
- configure the memory
- configure devices
- set the date and time.

Your tutor will be able to show you how to do each of the above.

The config.sys file is a DOS or OS/2 (another operating system) configuration file that is loaded when the machine boots up. When programs are installed, they often alter the config.sys file in order to customise the computer to behave in a certain way for the program. Windows will execute the lines in the config.sys file if there no drivers are found in the windows driver folder.

Autoexec.bat

Autoexec.bat is short for Automatic Execute Batch, which is a DOS batch file that is executed when the computer is booted up. It may be used to load programs that stay in memory and pop up whenever they are needed. You can also use it to start an application as soon as the computer is turned on.

Win.ini

Win.ini is short for Windows Initialisation. It is a Windows configuration file that is used to describe the current state of the Windows environment. It is read on start up and contains hundreds of entries. The purpose of the win.ini file is to tell Windows what files to run automatically and what the screen, mouse and keyboard settings are. The settings for the desktop such as the borders, titles, icon positions and spacings, fonts used, etc. are all stored in this file.

System.ini

System.ini is short for System Initialisation and is a Windows configuration file that describes the current state of the computer system environment. Like win.ini, it contains hundreds of separate small files and is also read by Windows on start up. It is used to identify the hardware drives to the system and contains a lot of detail about the Windows settings. When you change any of the default values to improve the performance of your system, the changes can be made in the system.ini using either a text editor or a word processor that can import ASCII files.

Configuring the hardware to work with the software

For the computer to know what hardware is connected, and that the operating system in use is capable with communicating with it, it is necessary to configure the hardware. The way this is done depends on the operating system. Some systems use 'plug and play', which is discussed in the next section.

A useful Web site for helping with installation of equipment, configuring hardware and software and fixing bugs can be found on www.zdnet.com/zdhelp/ (Figure 4.14).

Figure 4.14 The opening screen of the ZDNet Help & How To Web site

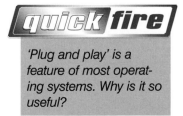

'Plug and play' is a feature of most operating systems. Why is it so useful?

'Plug and play'

The Macintosh operating system has had a simple intuitive system for installing new peripherals for a number of years. The Windows 95/98/2000 operating system now provides similar ease of installation with **'plug and play'**, which means that if you want to add new peripherals such as graphics adapters, CD-ROM drives, sound cards or new hard drives the operating system will automatically detect them and reconfigure the system so that the new devices can be used immediately.

Consumable materials

ICT systems need consumable materials, such as:

* paper
* transparencies
* toner
* ink-jet cartridges
* floppy disks
* printer ribbons.

It is essential to make sure that, after installing the computer system, you know what consumables are needed. It is also important to know how to put the paper into the printer, insert toner, ink-jet cartridges or ribbons, and the correct direction for inserting headed stationery into the printer. Most IT departments will keep a stock of regularly used items (paper, toner, etc.) because it is important not to run out. It is a good idea to also know the difference in quality between products, the prices of the various consumables and where they can be bought most cheaply.

Activity 14

A cheap printer can end up being more expensive than a more expensive one because of the high cost of the consumables that need to be bought regularly. Some companies can sell printers really cheaply, knowing that they will make their profit on the ink-jet/toner cartridges.

Verify this statement by doing some research. Select printers in the cheap, medium and expensive price ranges and then determine the costs associated with running each one.

In your comparison you should work out the costs associated with printing 2000, 6000 and 12 000 pages per year, both in black and white and in colour. Include the cost of the toner or ink-jet cartridges.

You should also work out the times for printing 100 pages using each of the printers that you are considering.

Remember to give the names of the printers in your evaluation as well as the costs.

4.2 Software

Computer hardware is useless without the software required for it to operate – the software is the programs that give the hardware the instructions on how to operate. Some software operates the computer system as soon as it is switched on and there is a minimum amount that must be loaded for the user to communicate with it.

There are several categories of software:

* operating systems software
* user interface systems software
* network operating systems software
* communications software

- programming languages, which can be used to write applications software
- applications software.

Other categories of software, such as games, are outside the scope of this book.

Systems software/operating system

Systems software is the program used to control the hardware of the computer directly and without such programs the hardware would be useless. The systems software is often said to form the bridge between the applications software and the hardware. The purpose of the system software is to take control of the inputs, outputs, interrupts and storage and it does this in most cases in an efficient manner without us knowing. Figure 4.15 shows how the **operating system** 'sits' between the hardware and the applications programs.

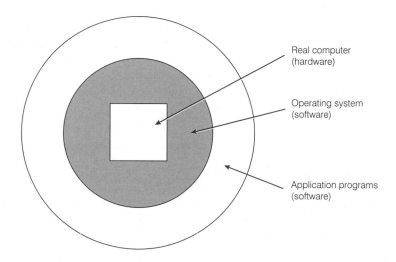

Real computer
(hardware)

Operating system
(software)

Application programs
(software)

Figure 4.15 The operating system 'sits' between the hardware and the applications programs

Systems software performs two groups of functions:

1 It ensures the efficient management of the computer's resources such as internal memory, input and output devices and files. Operating systems are frequently **multi-tasking**, enabling the user to do one task whilst another is being performed. For instance you can work on one document using your word-processing software whilst printing another.
2 It protects the user from the complexities of the hardware. This means that the user need consider only the application package being used. For instance, the user does not have to worry whether a file is being stored on a vacant part of the disk; the operating system does this automatically.

Some of the detailed tasks performed by an operating system:

- Performing certain diagnostic tests when the computer is first switched on ('booted up'). The operating system checks the peripheral devices attached to the computer and checks the memory by writing data to the memory locations and then reading it back to see if the data matches. If there are any problems, an error message appears on the screen.
- Scheduling and loading jobs that are to be executed, thereby maximising the computer's power and time.
- Selecting and controlling the operation of any peripheral devices.

- Dealing with any errors that arise and keeping the system running should any faults occur.
- Maintaining system security by checking passwords.
- Loading program files and data files into memory from the backing storage devices.
- Handling the interrupts that occur when there are problems with the applications software due to bugs or problems with the computer itself.

Systems software is taken to include the following groups of software:

- operating systems
- utility programs
- file management programs
- virus detection software.

Textual/command-based interfaces

Textual or command-based interfaces are a bit like programming languages because you have to type in carefully constructed instructions. One problem with these interfaces is that if you make only a simple spelling error or a mistake in the way the command is constructed the computer will not be able to understand the instruction. Another problem is that there is a lot to learn just to do relatively simple tasks. Most modern PC-compatible computers still allow the user to use MS-DOS (see below) and you should be familiar with some of its simple commands.

Microsoft Disk Operating System (MS-DOS)

MS-DOS was the original PC operating system developed by Microsoft and it required users to type in commands at a prompt on the screen. Users of MS-DOS need to have a fair amount of knowledge of the system.

Commands in MS-DOS are of two types: internal and external. Internal commands are always immediately available to the user because they are held in internal memory, external commands refer to larger programs that are held on backing store and therefore take longer to execute. The command DELETE is an internal command used to erase a file or a group of files, whereas FORMAT is an external command used to format a magnetic disk.

MS-DOS was developed as a single-stream operating system, which means that it can perform only one task at a time. However, software such as Windows 3.1 (not Windows 95, 97, 2000 or NT) is a **graphical user interface** (see page 254), which enables the user to run several tasks simultaneously (this is called **multi-tasking**).

MS-DOS uses a hierarchical or tree file structure. To understand how this file structure works, it is necessary to understand what is meant by file, **directory** and subdirectory.

A file is a unit of storage on the computer. Some files are program files, which can be run to perform a task, whereas other files are documents, storing data in one form or another. All files stored using MS-DOS are given file names, and the file name ought to bear some resemblance to what is stored in the file. As you store more and more programs and data on a hard drive you can very soon accumulate several thousand files. Keeping these in some sort of organised structure is extremely important and it is here that the operating system can help.

Directories are used to group together files containing similar material. For instance, a directory could be set up for games, another for word-processing software and yet another for letters. When a disk is formatted using MS-DOS, a root directory is created on the disk and from this root other subdirectories can be specified. Look at the directory structure in Figure 4.16.

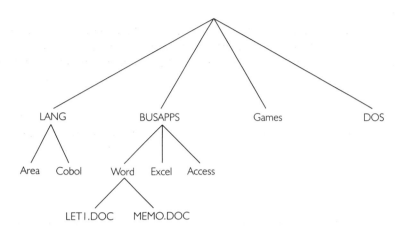

Figure 4.16 The tree directory structure

This tree structure is organised into areas with all the programming languages in a subdirectory called LANG. The LANG directory is divided into two further subdirectories, one for all the Pascal programs and data and the other for Cobol with its programs and data.

In order to locate files in certain directories it is necessary to use a path. For example, to select the letter with the filename LET1.DOC you could use the path:

CD\BUSAPPS\Word\EXT.DOC

A sound knowledge of MS-DOS produces a deeper understanding of how computers work, and so we will take a brief look at some of the commands available and how they can be put together. You need this knowledge to understand some of the tasks an operating system can perform.

In order to use MS-DOS commands you must be at the command prompt which appears as C:\>. If your PC uses Windows you must exit Windows and go to MS-DOS.

As with all modern software, there is an on-line help facility and this can be accessed by typing HELP at the command prompt. If you know the command name with which you require help, it is quicker to type HELP followed by the name of the command. For example, for help on the command to copy disks you type:

HELP DISKCOPY

To write batch files successfully using MS-DOS, it is necessary to understand a little about how certain operations are tackled. For this reason a brief guide to MS-DOS is provided here.

Viewing the contents of a directory
Type dir at the command prompt.

Changing to another directory
cd\windows

This changes to the windows directory ('cd' stands for 'change directory').

Changing back to the root directory
cd\

No matter which other directory you are in, this moves you back to the root directory.

Creating a directory
Suppose you want to create a directory called 'games'. You could do this using the command md games. Notice that the command prompt now changes to C:\games>. If you now type dir at the command prompt, you get a list of all the files in the games directory.

Removing a directory
You can remove a directory only if there are no files in it – if there are files in it they must all be deleted first. You must also be in the root directory when issuing a command to remove one of the other directories. The command for removing the games directory is rd games.

Copying files
For this you use the copy command, which has the format copy (source destination). The source needs to specify the drive and directory of the file to be copied; the destination needs to specify the drive and the directory to which the file is to be copied. The following command shows this arrangement:

copy a:task1.doc c:\mydocs

The above copies a file called task1.doc on drive A to a directory on the hard drive called mydocs. Because no name is specified for the file in the destination, the file will be given the same name as the original.

Using wild cards for copying groups of files
Suppose we had ten files with the file extension .doc which we wanted to copy. Doing them one at a time would be tedious, so we look for something the files have in common; in this case it is the .doc file extension. The command to copy a .doc file is:

copy a:*.doc c:\mydocs

The asterisk means that the computer will recognise any group of characters before the .doc file extension, so all such files will be copied, in this case into a directory on the hard drive called mydocs. If you wanted to copy *all* of the files you could use:

copy a:*.* c:\mydocs

Activity 15

Use any of the above methods to answer the following questions.

a Explain how MS-DOS organises files in directories and subdirectories.
b What is meant by the 'root directory'?
c What are the commands to:

- select the A drive
- select the C drive
- create a directory on the C drive
- remove a directory from the C drive
- change to a directory
- list all the files on a disk in the A drive
- format a disk placed in the A drive
- list all the files in a certain subdirectory on the C drive?

There are various ways you can find out about MS-DOS:

- the MS-DOS manual supplied with your computer
- the on-line help facility
- books available from the library
- notes provided by your tutor.

Graphical user interfaces

A **graphical user interface** (GUI) uses pictures rather than commands in words.

What is a GUI?

GUI software controls the information presented on the screen that helps the user to learn and eventually use the applications software. Computing has moved away from command line systems towards GUIs, which are much easier for both beginners and experts to use. All the applications packages that use the same GUI have the same 'look' to them so, once you have used one application, you will find others much easier to learn and use.

In a GUI windows and buttons, icons, dialogue boxes, toolbars, etc. are all shown on the screen and the user makes selections using a mouse. However, if the user is an experienced typist he or she can still issue typed commands rather than move the mouse if they want to speed up their use of the interface. GUIs need a lot of main memory to be used with reasonable speed (the use of Windows-based operating systems and applications software has pushed up the average memory requirements for a computer). If you are considering writing a program or developing a system with a GUI, you should always bear in mind the extra resource requirements needed, which could limit the use of the software because many organisations have older hardware which may not be able to run your software. One way around the problem is to limit the use of pictures, use fewer graphics and limit the range of colours used.

How does a GUI differ from a command-based interface?

Windows 3.1 is a GUI but it cannot be classed as an operating system because it cannot be run without MS-DOS being loaded first. Windows 95,98, NT and 2000 are all true operating systems that make use of a GUI.

Here are some of the things that make a GUI easy to use.

More ways of performing the same operation
Many GUIs offer more than one way to achieve the same result and the choice is left to the user. For example, a novice user may prefer to use a drop-down menu or click on an icon in order to print out a file, whereas the experienced user may find it faster to issue a command using a sequence of keys such as Ctrl + P.

Multi-tasking capabilities
GUIs support multi-tasking, which makes it easy for the user to switch between applications. Multi-processing, where more than one application is open at any one time, places great demands on the processing power of the chip as well as the main memory requirements. As users demand this facility, there is a need for faster processors and more main memory.

Faster searching of help files
Chips with a higher clock speed are able to search for on-line help faster and display the results more quickly. If the help does not appear almost immediately, the user may be put off using it.

The resource implications of GUIs

Any sophisticated user interface is going to push the existing technology to the limit as GUIs need faster processors and higher main memory storage capacity.

The easier the interface is for humans the more demands it places on the computer and as we move towards speech recognition systems, where the user can simply use natural language to communicate with the computer, the greater the demands and level of sophistication of the hardware and software.

Some user interfaces are quite expensive in terms of hardware and software. For example, touch screen technology enables people who may have never used a computer before to communicate with one to find out about a range of products and services. Banks use touch screen technology to enable the users to point to items on a screen to make selections. CAD work needs to make the maximum use of the screen and so does not want the screen cluttered up with toolbars and menus; these are transferred to a hardware device called a graphics tablet, which increases the cost of a system.

Resource implications for the processor

The greater the demands placed on the processor by the sophisticated operating system and applications software, the slower the programs will run. Processors are continually being redeveloped to cope with the demands placed on them by the software. Graphics-hungry applications will stretch the capabilities of the chip and to run such software quickly requires a processor with a high clock speed.

Resource implications for the immediate access store

Manipulating large graphics files on the screen means that the immediate access store (i.e. the main memory) needs to be high, otherwise the system will be very slow and frustrating to use. Having a lot of main memory (e.g. 256 Mb instead of 128 or 64) means that many windows can be opened at the same time without any appreciable loss in the speed of the system.

Resource implications for backup storage

Somewhere will be needed to store these large files, so a high-capacity hard disk and high-capacity removable storage (not floppy disks) such as zip or jaz drives for backup copies are needed. Also, sophisticated GUIs are themselves stored on the hard drive and because these files are so big a high-capacity hard drive should be used.

Resources for development

If a company is writing software to do a particular task using a sophisticated human–computer interface, more time will be needed to develop and test, which will add to the cost of the project. However, this should be placed in context with the lower training and support costs – users should not get into difficulty as often and can sort their own problems out without needing to call the help desk.

Folders and directories

In GUIs such as Windows and the Macintosh environment, a **folder** is an object that can be used to contain multiple data or program files. Folders are therefore used to keep similar files together and are therefore a way of organising information.

Activity 16

Research, using the Internet, magazines, manuals for the software, etc. the resource implications of making the following upgrades:

a from Windows 95 or Windows 98 to Windows 2000
b from Microsoft Office 97 to Microsoft Office 2000.

Booting up (powering up the computer)

As soon as the computer is switched on it takes some time to run its self-check before the operating system is automatically loaded from disk. This self-check consists of a set of operations designed to make sure that all the components are working correctly. If they are not, the operator is alerted, usually by a few 'beeps' and a brief message on the screen. The self-check is often called the **POST** (**'power on self-test'**). It is usually quite simple to correct many faults, which are often nothing more than a loose keyboard or mouse connector at the back of the machine.

When the power is switched on, some instructions stored in ROM are obeyed and the computer begins its self-test routine. It checks the memory by storing a value in each memory location and then reading it to check that they are the same. It also checks for the presence of essential hardware such as disk drives, monitor, keyboard and mouse. The computer's configuration settings are stored in a type of chip called CMOS RAM, which has a low power consumption and can therefore be powered by a small rechargeable battery. Because the contents of the chip are maintained even when the power supply is switched off, the computer is able to remember that it has certain devices connected to it. During the self-test routine, the computer checks this configuration against the actual hardware connected to see if the system is the same as when it was last used.

Activity 17

To see how the self-test can detect bad connections or missing devices, remove the following peripherals and boot up the computer:

- the mouse
- the keyboard
- the printer.

Write a short description of what happens in each case.

ROM-BIOS start-up software

The BIOS (basic input–output system) reminds the computer that it has a processor, memory and a hard drive and tells it to load this information before loading the operating system. It is part of the operating system and is stored in ROM on a special chip on the motherboard. The BIOS keeps all the information to start up the computer from scratch. Instructions in the BIOS tell the computer which device to look at first in order to find the operating system – most BIOS settings tell it to look on the floppy disk first but this can be changed to the hard drive or even a CD-ROM.

Entering setup to look at the BIOS settings

To enter setup you need to boot up the computer and, during the boot process, press F1, F2 or the delete key. A screen with the heading BIOS Setup Utility should appear.

A word of warning: just fiddling with the settings on the screen could mean that your computer may not work properly. The first thing you should do before anything else is to print out a copy of this screen so that you know what the original settings were before altering them. Alter the settings only if you have done this first.

Creating a boot disk

Usually the computer is booted up from the hard disk but if the hard disk has been damaged (by a virus or an accident for example) then the operating system could be lost and you will need to use a bootable floppy disk. To prepare for these emergencies, you should always make sure that you have a bootable floppy disk handy.

Customising software

Most software packages allow the user to customise them to match their individual work style or tasks. By taking some time to look at the customisation options you can save time in the long run. Such options in Microsoft Word include:

- *Creating new toolbars* by including your selection of icons and getting rid of any you never use. Some software will automatically put those items in the toolbars that you use regularly. In other words, the software is configured on the basis of what the user uses the most.
- You can customise your documents by *altering the templates* that come with Word to suit the document style you generally use. Templates are the framework for a document, and the word-processing software you use will have many already set up which you can configure and customise to your own requirements. This saves a lot of time when creating routine documents such as letters or memos. It also ensures that all the documents have a consistent style.
- It is possible to *change the appearance* of the screen by hiding or displaying screen items such as the toolbars, ruler, the menu bar.
- You can *alter the frequency* with which automatic backup copies are made.
- You can alter the *directory in which Word stores your data files*, thus avoiding having to use the 'Save as' function or waste time having to select the directory/folder being used.

Software is often customised to simplify the interface. For example, if you develop a database using the relational database software Microsoft Access, you will understand its use fully but simplification could be needed when a novice has to use what you have developed. His or her job might just be to input orders or stock details into the system and they will not be required to understand the intricacies of the package. In fact, they need to be kept *away* from the interface supplied by the package because it is really only suitable for experienced users. To do this the developer will have to develop menus for the user to select what they require for their particular job.

Installation of software

Installing new operating systems or applications software is a lot easier than it used to be and demands little specialist knowledge. Most operating systems and applications programs can be installed automatically because most software comes with an installation program which takes the user through the installation process using a series of questions and options. The user does, though, need to have enough technical knowledge to understand and respond appropriately to the options offered. If the installation is being done from floppy disks, the user's main task is to feed the correct disk into the disk drive when requested. As software programs have grown in size, more and more floppy disks have been needed for installation, so it is now common for a CD-ROM to be used instead. CD-ROMs typically have 650 Mb of capacity; a single CD-ROM can easily store all the data needed for a typical applications program. Problems do still occur with program installation: there are so many different programs and versions of them, and so many different hardware configurations, that occasional difficulties are inevitable. It is when these difficulties arise that experience and expertise are needed.

/ **Activity 18** /

Figure 4.17 shows the opening screens from two operating systems: the older Windows 3.1 and the more recent Windows 98.

Just by looking at the pictures of these screens, say why you think the interface provided by Windows 98 as an improvement on the other interface. (Hint: Put yourself in the shoes of an inexperienced user who may have never used a computer before).

Creating folders/directories

To make it easier to keep track of files on a disk, folders/directories are used to store them. All disks have at least one directory, called the root directory. It would be possible (within limits) to store every file in the root directory, but this would make it difficult for the user to know which files went with which programs. In any one directory, no two files can have the same name and if all the files on a computer were in a single directory, conflicts would arise if two programs tried to create files with the same name. Directories other than the root directory are therefore created, into which groups of files with similar properties can be placed.

Directories can be further split into subdirectories. The names of directories and subdirectories should be chosen carefully to avoid any conflicts.

Installing software in a subdirectory

Software that runs under DOS must be installed into certain directories, and these directories need to be created before the data can be copied into them. To do this you need to know quite a lot about the DOS commands. In particular, you need to understand how to:

- change between different disk drives
- list all the files in a directory
- create a new directory
- move to a new directory
- copy all the files from a floppy disk to a named directory/subdirectory
- delete a file from a directory
- delete all the files from a directory/subdirectory
- remove a directory/subdirectory
- copy the entire contents of a directory/subdirectory on the hard disk to a floppy disk.

In Windows you can use the File Manager to create directories, select files and groups of files with certain properties, delete files and copy files. On-line help is available if you encounter any difficulties.

Setting defaults

A **default** is a predetermined setting. You can usually either accept a default setting established by a program or operating system or change it to suit your personal preference. For example, in most word-processing packages there is a default template which you can either use or, if you don't like it, change to a different

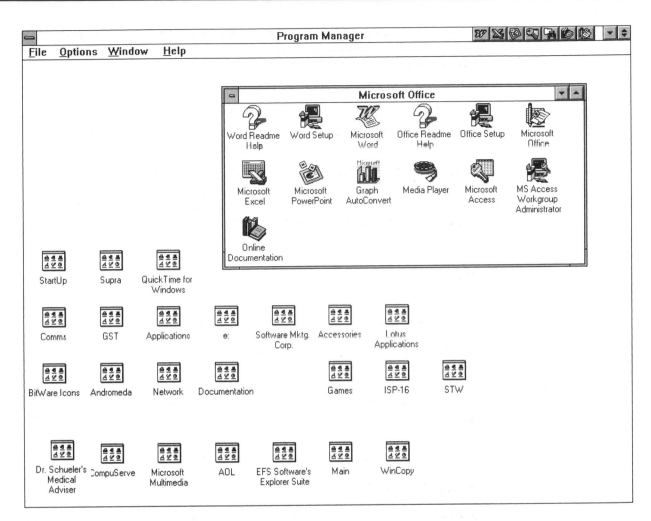

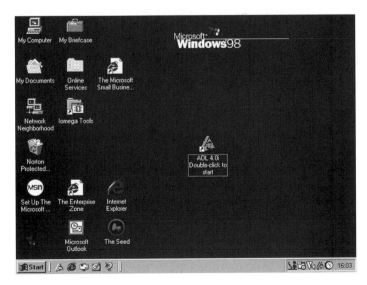

Figure 4.17 The two opening screens from different operating systems: Windows 3.1 (a) and Windows 98 (b)

What is meant by the term 'setting a default'?

default template. Also, when setting up some word-processing software you may need to specify the type of printer from a list. This printer then becomes the default printer. The software automatically assumes that the default printer is the one being used unless it is told otherwise.

Installing device drivers

A device driver is a special file which tells the computer how to use a device like a mouse or a printer. A mouse driver makes the mouse pointer appear on the screen and then translates the mouse-clicks into action.

Testing the system

On adding new software, new hardware or both, you need to make sure that they all work together properly, and to do this a series of system tests must be performed.

Accessing software

Once the computer has self-checked it will load the operating system. When this has been completed successfully it is possible to load the applications software and start using the computer.

Entering data

Software can be tested by running it and then checking that the results are correct. If you are running payroll software, for example, you could enter last month's payroll data and check that the results generated are the same as when the task was performed manually.

Saving data

To check that a system and program are working as expected, data can be entered and saved. The data can then be reloaded into the system from disk to check that it has been saved correctly.

Printing

To make sure that the correct printer driver has been installed successfully, it is necessary to print out some results. If strange symbols appear it usually means that the wrong printer driver has been installed and does not understand the data signals being sent to it by the computer. You need to make sure that the printer can deal with the various fonts supported by the software (fonts, such as Times and Helvetica, are the different styles in which characters are printed). Most computer systems and programs now provide a range of fonts. Some, such as TrueType and Postscript fonts, appear on the screen exactly as they will be printed; others look different. Some fonts can be printed out only using a special output device such as a plotter.

Windows 2000

To make computers as easy to use as possible, a very sophisticated and powerful operating system is needed which 'shields' the user from the intricacies of the hardware. Ideally the operating system should be compatible with all the existing software, easy to use, fast and powerful.

Windows 2000 has all the above features and will become the standard operating system for the latest PC computers with their large memories (typically 128 Mb), large hard disk capacity (12 Gb) and high-speed processors (typically the Pentium with a clock speed of 1000 MHz).

Windows 2000 is not just a replacement for the popular operating systems Windows 95 and 98 – it is also a replacement for **Windows NT** (Microsoft's preferred operating system for businesses who use networks). There are four versions of

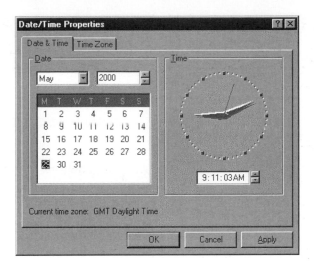

Figure 4.19 The window for altering date and time

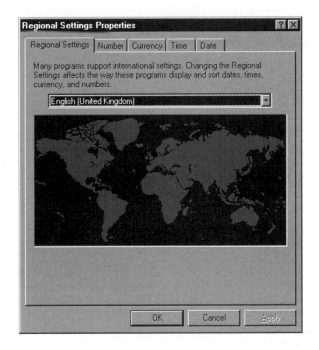

Figure 4.20 The window for altering the regional settings

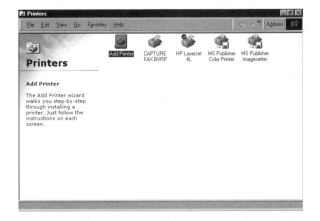

Figure 4.21 The printer settings window

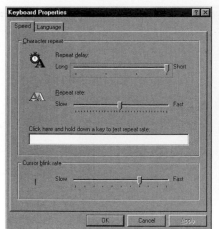

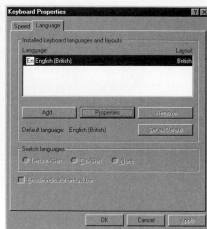

*Figure 4.22 These two
windows are used to alter
the keyboard settings*

*Figure 4.23 The window
used to alter the
mouse settings*

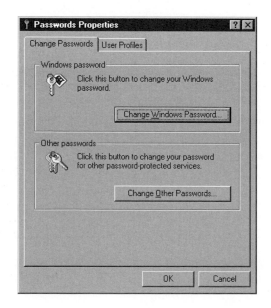

*Figure 4.24 You can set a
password using this window*

If you are printing a job you can see the current status of the printer by clicking on the icon with the name of the printer underneath.

Clicking on the keyboard icon enables you to alter the language and the speed settings for the cursor and the character repeat (Figure 4.22). If your keyboard does not give you a pound sign when you press the £ key then check that the language is set to English (British).

By clicking on the mouse icon you can alter the mouse settings using the menu that appears (Figure 4.23).

Password properties

Click on the passwords icon in the control panel and a screen appears that allows you to choose a password that a user needs to enter in order to access Windows (Figure 4.24). Use this with care.

Virus protection configuration

Click on FindVirus scan settings and a screen appears for you to alter the types of files that should be scanned, what happens if the computer detects a **virus**, whether the computer should disinfect the file automatically and so on (Figure 4.25).

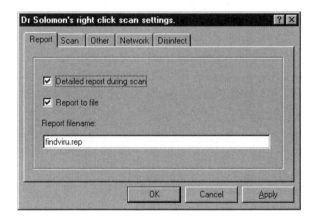

Figure 4.25 You can alter the virus scanner settings using this window

---

Activity 20

Click on each of the tabs in the screen shown in Figure 4.25 to find out what the options are.

Power management

The main purpose of **power management** is to minimise power usage when the system stays inactive for a while. In most cases, the power management system is incremental, meaning that the longer a system stays inactive, the more parts will close down. The most obvious use of power management is to shut down the monitor after the computer has been inactive for a while. As soon as a key is pressed the monitor springs back into life. The disk drive may also be shut down and the processor slowed down as well to save power.

Click on the power management icon in the control panel. The menu shown in Figure 4.26 appears.

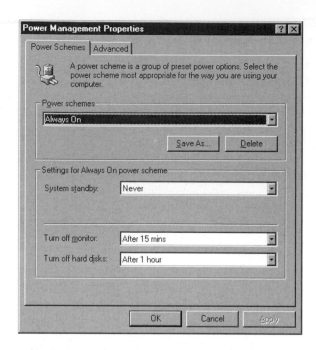

Figure 4.26 The power management properties can be altered using this screen

Notice that you can turn off the monitor and the hard disk drive in order to save power after a set period of time. As soon as you press a key or move the mouse, the system springs into life again.

Multimedia configuration

Click on the multimedia icon to obtain the screen that lets you alter the devices used for multimedia such as the sound card or graphics card (Figure 4.27).

Figure 4.27 You can set the multimedia capabilities of the computer system using settings in the window shown

By clicking on the tabs you can get to other menus where you can see what multimedia hardware is installed.

System properties

Click on the system icon (you might have to scroll down to find it). Using the tabs you can get to screens that allow you to see the devices that are available to the system and also look at the performance of the system (e.g. the amount of hard disk capacity used, the operating system being used, etc.). You can also choose the hardware setup at startup (Figure 4.28).

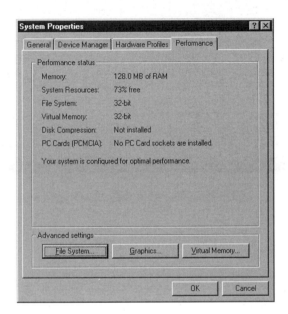

Figure 4.28 The system properties window lets you see information about the hardware and performance of your computer system

Uninstalling software

As well as installing software you have to be able to remove or uninstall it. Uninstalling software is not as simple as deleting all the files in the folder where the program is situated because files belonging to the application sometimes reside in other folders not directly associated with the application. Deleting them from the main folder will not remove these other files and means that there are links back to files that no longer exist.

As well as coming with an install file, many packages come with an uninstall routine to make removing the files easy. Earlier software does not usually contain an uninstall routine so you can either use the add/remove programs icon in the Windows control panel or use a special package for doing the job.

Frequently when you uninstall the software you are left with stray files that may need to be removed in another way.

There is an icon in the control panel in Windows 98 that will remove software that is no longer needed. When you click on the add/remove software icon, the screen shown in Figure 4.29 appears.

You can also install programs using this screen.

Scheduled tasks

Find the **scheduled tasks** button/icon (Figure 4.30). It may be at the bottom of the screen.

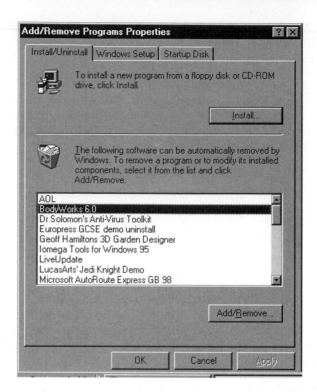

Figure 4.29 The window you can use for adding or removing software

Figure 4.30 You can schedule tasks using the schedule tasks icon in this toolbar

A screen appears (Figure 4.31) for you to schedule tasks, which means get Windows to automatically do these tasks at a certain time. For example, you could get the computer to clean up the hard disk by deleting any unnecessary files while you take your lunch break.

Figure 4.32 is a screenshot of the Scheduled Task wizard, which takes you through the stages of scheduling the time that a particular program should run.

Many of the tasks that you can schedule the computer to perform automatically are best done when you are not working at the computer. Why is this?

Graphic design software

Graphic design software is described in detail in Unit 1 (pages 47–50). Look back at this section for information.

Configuring applications software

All users have different requirements from their applications software so it is important to find out what their needs are and to be able to configure the applications software to suit them.

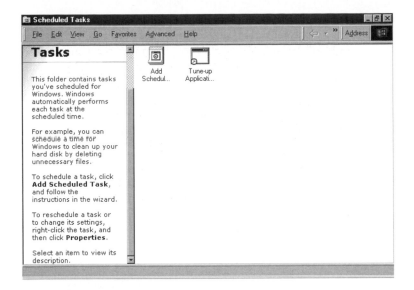

Figure 4.31 Using this window allows you to schedule a task such as run a virus scanning program

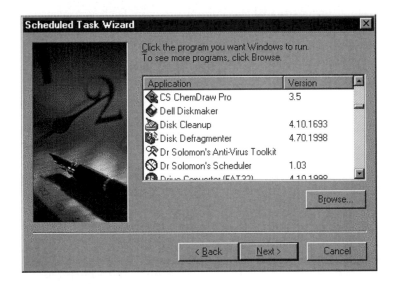

Figure 4.32 The Scheduled Task wizard takes you through the steps involved in scheduling a task

It is important to be able to set:

* preferences (or configuration files)
* macros
* the toolbars and buttons available to the user
* directory structures and defaults
* data templates
* saving and backup security
* menu layout and contents.

Preferences (or configuration files)

One of the annoying things about working on lots of different computers is having to change the settings after the previous person has altered them. For example, the icons may be in different places on the screen or toolbars may be missing.

Many people use computers in a network environment and each time the software is loaded from the server the user will be presented with the same layout. He or she would then have to make the changes to suit them. There *is* a way around this – you can use what is called a user profile to define your customisation preferences and options. For example, in the case of a word-processing package you can use your own custom dictionary, templates, AutoCorrect settings, icons, rulers, toolbars and so on. All these things together make up what is called your user profile. When you log onto the network the operating system knows who you are and will use the user profile file to customise your computer. No matter which terminal you use, you will be able to have your own settings no matter how many other users use the same terminal.

If you are the only person using a stand-alone computer you can just customise your settings. These will be available each time you use the application.

Macros

What is a macro?

A **macro** is a high-level programming tool that can be used to automate a series of tasks or procedures within a program. A macro consists of a series of individual instructions; to start it you need only press a couple of keys on the keyboard or click the mouse.

A macro used within a particular applications program, such as the spreadsheet Excel, will run only when that application has been loaded; such macros cannot be run on their own.

Why are macros useful?

Some macros can be created by users who know nothing about programming. This type of macro will record your keystrokes (or mouse movements) and selections so they can be automatically 'played back' when you need them simply by activating the macro.

Activity 21

Macros can be found as an advanced feature of most software.

a Using the software manuals or on-line help menus, find out how to write a macro for a task that you find tedious when using your spreadsheet or word-processing software.

b Produce an easy-to-understand set of instructions which you could give to a friend who needs to perform the same task.

Toolbars and buttons

Applications software usually comes with many different toolbars and to display them all on the screen would leave little room for the variable data. People generally just show those toolbars they use the most. Any icons that the user rarely uses can be removed from the toolbar. Some software is intuitive; it will remove these icons automatically and display only the toolbars and icons regularly used. If you are setting up a system, it is usually worth making up your own toolbars containing the icons you use regularly.

An experienced typist will probably never use the icons in word-processing software, preferring instead to issue commands using combinations of keystrokes, which are faster than moving the mouse. He or she would usually remove these toolbars so they have the maximum area of screen in which to type.

Activity 22

Using the help menus to guide you, produce a customised toolbar containing the icons that you use regularly. To supply the evidence for this activity you will need to produce a screenshot showing the toolbar before the customisation and after the customisation. You should also provide a brief set of notes explaining how you performed your customisation.

Directory structure and defaults

When you save work that you have created in an application (word processing, spreadsheet, database, etc.) it will automatically be saved in the default folder/ directory unless you tell it otherwise. You can store files in any folder/directory on the hard disk but it makes sense to organise them into folders and subfolders (or directories and subdirectories).

The word-processed files in Word (normally called documents) are by default stored in a folder called My Documents. If you have lots of these documents to store then it is better to put them into subfolders within the My Documents folder.

Data templates

In word processing, a **data template** is a framework for a document (fonts, sizes, headings, formatting, etc.) that is already stored – all you have to do is to fill in the variable information. Microsoft Word has a large number of templates for documents such as letters, memos or reports.

Data templates can also be found in DTP packages where their use can save an enormous amount of time. All you have to do is to decide on the type of document you want to produce and look for a suitable template.

If there are no suitable templates to suit your needs you can produce your own template for them and save it so that it may be re-used.

Activity 23

a Investigate the templates available in the word-processing and DTP packages that you are familiar with. Produce a report outlining your findings.
b Produce a template for a document, using word-processing software, for a holiday company. The company would like to use this template to reply to people who have been dissatisfied with their holiday. The template will need to be tailored to their complaints so there needs to be quite a bit of space for the variable information to be inserted.

After completing the template prepare a short user guide to explain how a user should use it.

Saving and backup security

It is a fact of life that computers can crash unexpectedly for no apparent reason, usually when you're in the middle of a piece of work. Even if you have not saved your work for some time you might still not lose any of your work – AutoRecover may have saved your work automatically and it's possible that you will be able to recover the work using this saved version. It is important to set the AutoRecover to a suitable time interval so that your work is saved as a recovery file every five minutes or so.

Another precaution that can be taken is to create a backup copy automatically. Using this method, as soon as one file is saved a copy of the previous version is also saved. This means that there are copies of the old and new versions – which is handy if you make changes to a file and then prefer the previous version of the file.

Menu layout and contents

Some applications software is very configurable and users can tailor the package by using specially designed menus of choices. Databases are particularly configurable and the user can create menus containing choices such as create a new record, append a record or delete a record.

4.3 Documentation

ICT systems often do not behave as expected or even crash – the crash might be caused by installed hardware conflicting in some way with the operating system, or perhaps the operating system or application software contains bugs. Unfortunately, bugs are fairly common, especially in the earlier versions of software. Later versions of the software usually contain fewer bugs. If there is a serious problem with the software then the company may have a patch (i.e. a piece of software code that can solve the problem), which you can download from the software manufacturer's Web site.

To be able to investigate the problems with ICT systems it is necessary to be systematic about recording exactly what went wrong. Records about such occurrences provide useful reference material to show others – at least the IT manager can see the scale of the problem.

In many organisations a computer is not used by just one user: several different people will use it during the day. Because of this, it is important to make sure that all the computers in the company are set up and configured in the same way. If all the computers are configured differently the users could be confused by a different set up on each computer they used.

When configuring computers it is important to produce a document saying exactly what has been done so that the same settings can be applied to any other computers the user may need to use. Typical information you could include in these records is:

- dates work is undertaken
- specifications used
- components installed
- configuration tasks undertaken
- faults and problems experienced
- solutions applied
- support services accessed
- diagnostic software used.

Records of installation, configuration and associated problems need to be indexed so that information about particular problems or events can be accessed easily.

Why is it important to keep records when new hardware or software is added to a system?

4.4 Standard ways of working

Standard ways of working were looked at in detail in Unit 1. Here we will look at the standard ways of working that are applicable to Unit 4.

You need to show in your assignment that you have used the following standard ways of working:

- Edit and save work regularly, using appropriate file names for your documents.
- Store your work where you and others can easily find it.
- Keep dated backup copies of files on another disk and in another location.
- Keep a log of ICT problems you met and show how you solved them.
- Avoid bad posture, physical stress, eye-strain and hazards from workplace layout.

Working safely to protect yourself and others

There are many potential risks to yourself or others when you are moving, connecting, disconnecting or installing hardware.

Just by moving a heavy VDU from a box on the floor to the desk you can easily injure your back unless you use the correct procedure for lifting heavy objects. When lifting heavy objects you should always keep your back straight and bend your legs so that you are using your legs rather than your back to bring the object up to desk height. This reduces the strain on your back and can prevent back damage such as a slipped disc.

People can trip over wires trailing across walkways. If wires need to go across a walkway, they should be laid under the carpet or through a special piece of trunking. This trunking has a gentle slope on either side that prevents people from tripping.

Modern offices have special areas (called plenums) between the ceiling of one floor and the floor of the next. Wires are run through this area and fed down special hollow pipes to the computers on the desks.

Because you will be inserting circuit boards into the computer and connecting and disconnecting ICT components, safety is very important. Before connecting or disconnecting any ICT components you must make sure that all the devices are switched off by their own switch, at the wall and the plug removed from the wall socket.

Important advice when working inside your computer

Before working inside your computer you must observe the following important steps:

1. Turn off the computer and any peripheral devices attached to it using their switches.
2. Disconnect the computer and all the devices connected to it from the electrical supply. Also disconnect any telephone or telecommunication lines from the computer.
3. Pull all the plugs out of the mains socket. Do not just assume that the ones nearest to the computer are the ones being used – follow the power supply cables from the back of the computer to their plugs and remove them.
4. Always wait at least 5 seconds after turning off the computer before disconnecting devices from the system board to avoid damage to the system board.

5 To ensure that you do not damage the delicate chips inside the computer, you need to Earth yourself by touching an unpainted metal surface such as the power supply bracket. You could become charged up with static electricity and if you touch a component when charged you could damage it. Touching the metal allows the electricity to flow to Earth and you can then touch the components without damaging them. If you go away from the computer, always Earth yourself again before touching components.

Good ergonomic habits

As part of the evidence for this unit, you will need to show that you have set equipment up with regard to good ergonomic practice.

- The system should be positioned so that the keyboard and monitor are directly in front of the user as they work.
- The monitor should be situated at eye level or slightly lower when the user is sitting in front of it.
- The monitor should be about 24 inches from the user's eyes.
- The tilt, contrast and brightness of the monitor should be adjusted to be comfortable.
- There should be plenty of space around the monitor so that there is room for the user to place paperwork, etc. There should also be space in front of the keyboard to allow the user to rest their hands.
- Make sure that there is enough room for the user to use a mouse.

These points are illustrated in Figure 4.33.

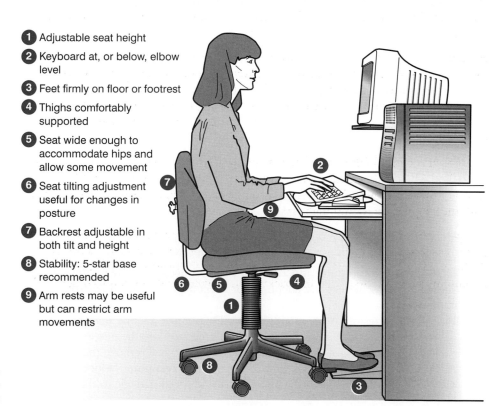

1 Adjustable seat height

2 Keyboard at, or below, elbow level

3 Feet firmly on floor or footrest

4 Thighs comfortably supported

5 Seat wide enough to accommodate hips and allow some movement

6 Seat tilting adjustment useful for changes in posture

7 Backrest adjustable in both tilt and height

8 Stability: 5-star base recommended

9 Arm rests may be useful but can restrict arm movements

Figure 4.33 Alleviating strain when working with computers

Security

You must also be able to implement or recommend proper security procedures, including those that ensure:

*   data and software backup is maintained
*   confidential information is protected
*   passwords are used
*   virus checking is undertaken
*   copyright is protected
*   theft (of data, software or equipment) is avoided.

Virus checking

It is essential that virus scanning software is used to protect the computer against **viruses**. As new viruses are being created all the time, a virus checker soon becomes out of date so it is essential that you update the virus scanning software frequently. Some virus scanning software comes with monthly or quarterly updates, which are either sent out on disk or downloaded from the software manufacturer's Web site.

It is essential, before any programs are installed on the computer, particularly those obtained from unlicensed sources, that the program disks are scanned for viruses. Again, regular backups should ensure that if a virus infection occurs any damage to data and programs is minimal.

Data and software backup

There are many ways in which files containing data or software can become corrupted – perhaps a virus attack or a hard disk malfunction. Software loss is not as devastating as data loss because a sensible user will have made sure that they copied all the software or kept the original disks. This means that any lost software can be reinstalled. However, data files will be lost completely unless backups have been made.

Hard disks should be backed up at least once a week and a daily backup kept for those files that have changed. This ensures that no more than one day's work will be lost if the worst were to happen. To further ensure against data loss, duplicate copies of the daily and weekly backups should be kept off-site. You will therefore lose no more than a week's work even if the off-site backup is corrupted or destroyed.

Because of the very high storage capacity of modern hard disk drives, floppy disks are no longer a feasible medium for the backing up of program and data files. Instead, high-capacity backup devices such as tape, zip, jaz or CD-RW can be used with their high capacity and high transfer speeds.

It is possible for the backups to be taken at certain times of the day automatically by the computer. You could for example, get the computer to schedule a backup during your lunch hour on a certain day. As long as the drive has the suitable backup medium inserted, then the backup can be taken without you having to remember.

Key terms

After reading this unit you should be able to understand the following words and phrases. If you do not, go back through the unit and find out, or look them up in the glossary.

Analogue signal	Jumper
Backup storage	Linux
Bit	Macros
Bit-map graphics	Megabyte
BNC connector	Megahertz
Booting up	Modem
Buffer	Motherboard
Bus	Multi-tasking
Byte	Network card/network interface card
C	Network operating systems software
C++	Operating system
Cache	OS/2 Warp
CD-R	Parallel port
CD-ROM	Pixel
CD-RW	Plug and play
Clock speed	Port
Cobol	POST (power on self-test)
Configuration file	Power management
Co-processor	Primary storage
CPU	Processor
Data template	Refresh rate
Default	Resolution
Digital signal	ROM-BIOS
DIP switch	Scheduled task
Directory	SCSI (Small Computer System
DVD (digital versatile disc)	Interface)
Folder	Secondary storage
Gigabyte	Serial port
Graph plotter	Software
Graphical user interface (GUI)	Systems software
Hardware	Terabyte
IDE (integrated drive electronics)	Unix
Immediate access store	Utility program
Input/output controller	Video card
Interlacing	Virus
ISA bus	Visual Basic
ISDN (integrated services digital	Windows 2000
network)	Windows NT
ISP (internet service provider)	

Review questions

1 **a** While setting up ICT equipment for others to use you are exposed to some health and safety risks. Describe three such risks that you could be exposed to, and explain the steps that can be taken to alleviate them.

   b Once the hardware has been set up and the software installed, the user can use the system but, on using the system, the user can be exposed to certain health and safety risks. Describe three such risks and explain the steps that can be taken to avoid them.

2 To take advantage of the latest software developments, a user has upgraded their computer.
   a Explain the term 'upgraded'.
   b State two changes that you would expect to see as a result of the upgrade.

3 The table lists some of the devices found in a typical computer system, and gives a brief description of each. However, they have been muddled up. Rearrange the table, matching the device to its correct description.

Device	Description
RAM	The electronic transportation system used to carry data and instructions from one part of the CPU to another
Video card	A store that is volatile
ROM	A circuit board that plugs into the motherboard to give a monitor its display characteristics
Bus	The memory used to hold the 'boot' program
Motherboard	A temporary storage area used as a 'waiting room' for the data to allow for different speeds in the data
Co-processor	The main circuit board in the computer
Expansion slot	The interface that allows data to be sent in a line one bit at a time
Buffer	An interface that allows more than one device to be attached to it at a time
SCSI	The interface that allows data to be sent a group of bits at the same time
Serial port	A processor that is used to take some of the load away from the main processor
Parallel port	The link that switches devices such as disk drives on and off by the processor
Input/output controller	The vacant locations in the motherboard where extra circuit boards or memory can be added at a later date to improve the characteristics of the computer

4 **a** Explain the difference between a textual (command)-based interface and a graphical user interface (GUI).
   b Name one command-based interface and one GUI.
   c Describe three different functions of an operating system.
   d Describe three features of a GUI that make it easy for an inexperienced user to use.

5 Read the following paragraph carefully and answer the questions that follow.

   When an image is scanned, there is a lot of data that needs to be sent back to the computer. The serial port is not fast enough for this so most scanners use the parallel port. When you need to scan in colour, as well as storing all the pixels for the picture, the scanner also needs to store data about the colours of the pixels. Colour scans therefore need a very fast connection between the scanner and the computer.
   A SCSI interface is ideal for this. The trouble is that many computers do not come with SCSI as standard because they are relatively expensive. You can add a SCSI card to the motherboard; some scanners have them built in to the scanner.

    a Name three different uses to which a scanner may be put.

    b When scanning in images the picture is turned into pixels. What is meant by pixels?

    c Explain the different ways in which the data is transferred using a serial interface and a parallel interface.

    d What is the main advantage in using a SCSI interface rather than a parallel port for transferring data?

    e A SCSI card can be added to a computer that is not equipped with one. What precautions should be taken when inserting one of these cards into the motherboard?

6 **a** Graphic files may be classed as 'vector' or 'bit-map'. Explain the main differences between these two types of graphic file.

    b When application software is installed various folders and links are created. Explain why special programs are needed to remove programs once they are no longer needed.

    c Application software is usually installed from a CD-ROM rather than from floppy disks. Why?

7 **a** Most personal computers come with a power management facility. What is meant by power management and why is it useful?

    b A user is having difficulty printing when they are using a unfamiliar computer system. Produce a list of the things you would investigate (with the obvious and easy ones first) to try to find out what is wrong.

8 Windows 98 makes use of a system called 'plug and play'. Explain what this means and how it is useful to someone who needs to attach a new peripheral to their computer.

9 In order to meet the needs of the user, application software can be configured in many different ways. Choose a popular piece of applications software and explain the ways in which it can be configured to meet the needs of the user.

Assignment 4.1

Your friends decide to buy a PC system which they hope to use for the work they do at home (they are both teachers – one teaches all subjects in a junior school and the other teaches physics in a secondary school). Their daughter, who is taking her GCSEs, will also use the computer. Ideally they want a system that will last them several years, and they intend to use it for running multimedia software, word processing, desktop publishing and games software. They would like to learn more about the Internet so they will need a modem as part of their system. They don't know much about computers but have seen an advertisement in a magazine. The advert outlines the following system specifications:

- Intel Pentium III processor with a clock speed of 1 GHz
- 128 Mb RAM expandable to 768 Mb
- 40 Gb 7200rpm hard disk drive
- 10-speed DVD with movie playback software
- CD-RW drive ($4 \times 4 \times 24$)
- 64-bit stereo sound card
- Powerful 150 W stereo speaker system
- 1.44 Mb floppy drive
- mini-tower case
- 32 Mb high-performance graphics card
- 2 USB high speed ports plus parallel and serial ports
- 19-inch colour monitor
- Windows keyboard
- Microsoft PS2 Intellimouse
- Microsoft Windows 98 with manuals and CD
- anti-virus software
- I56k V90 PCI Modem

Tasks

1 Your friends have sent you the following note:

> Dear ....
>
> Please find enclosed an advert for the system we are thinking of buying. This system has been recommended by a friend, but the system specification, as we think it's called, means nothing to us and we were wondering what it all means. Could you put it into English that a non-technical person could understand? For instance, what is a cache and what effect does it have on the performance of the overall system?

Write a reply to their note, explaining what each item in the system specification means and its relevance to the system performance.

2 Your friends have heard of Windows 2000, Windows 98 and MS-DOS.

   Produce a brief report on each of these software packages, comparing and contrasting their use. You should also mention their main strengths and weaknesses.

3 Your friends' neighbour, who lectures in IT at the local college, has suggested that they build their own system from the various components. He says that the main advantages in doing this are, that it will be cheaper and they can make sure all the components are of the highest quality.

   Your friends have asked you to investigate this possibility, pricing all the components they would need, and to present the details in the form of a spreadsheet. To do this you will need to determine the components they would need, which components to choose from a range to ensure good quality and price, and where you would buy them.

   As well as the spreadsheet showing the components and their prices, produce a short word-processed recommendation saying how easy or difficult it would be for your friends to build their own computer and whether or not you think it is a good idea.

Assignment 4.2

You need to prepare a short talk explaining to a small group of users how they should tackle the tasks listed below. Because there is a lot for them to take in and remember you must provide a set of instructions on how to perform these tasks. The users are fairly new to computing but are keen to learn so you will need to make these instructions as easy to use as possible.

Tasks

1   The clock is wrong on my computer. How do I adjust the time?

2   As I work in the Human Resources department, there is a lot of confidential material on my computer. I need to make sure that there is no unauthorised access to this material. I would like to be able to use a password system so that users can only get access to my stand-alone machine if they type in a correct password. Is this possible? How do I do it?

3   I'm a bit lax in taking backup copies, especially on a Friday afternoon when I am in a hurry to go home. I have heard that it is possible to let the computer do this automatically. Is there any way of getting the computer to schedule a task like taking a backup copy and how do you set it up to do this?

4   How do I do the following?

   • Create a new folder.
   • Format a floppy disk.
   • Delete some of the files on a floppy disk without formatting the whole disk.
   • Copy a group of files on one floppy disk and place a copy onto another floppy disk.

   In each case to supply the evidence you will need to write out a set of correct instructions.

Assignment 4.3

It is impossible to give you full details of this assignment because they will depend on the hardware and software resources of your school/college. Your teacher/lecturer will give you the precise details of what you have to do.

As the assessment evidence for this unit, you will be required to produce:

1   A specification for a complete ICT system to meet user requirements, together with an operational system.

2   A specification for a user upgrade to an ICT system that requires installation of at least two items in the processing unit and configuration of software, together with an operational system.

You must also show that you can remove the installed items and use uninstall procedures to install the system to its original state.

3   Records of set-up, installation, configuration and test activities.

Your configuration of software must include setting up a toolbar layout, a menu a template and a macro.

Your teacher/lecturer may allow you to work in teams when performing the installation.

Get the grade

To achieve a grade E you must:

1   Define user requirements and produce clear specifications for the ICT system and the upgrade, including for each full details of hardware, operating system, applications software and configuration.

2   As part of the above, you will select suitable hardware and software and correctly. The tasks involved in doing this are:

   • connect hardware
   • install items in the processing unit
   • install software
   • set ROM-BIOS parameters.

3   You must also design and implement a suitable toolbar layout, menu, template and macro to meet the user requirements and must be able to restore the upgraded ICT system to its original state.

4   To supply evidence for this, you must produce clear records of work done that includes suitably annotated printed copy or screen prints of your toolbar, menu, template and macro, together with details of a suitable system configuration check, test procedures, problems experienced and solutions implemented.

To achieve a grade C you must also:

5   Show, through your records of practical work, a systematic approach to specifying and constructing an operational ICT system.

6   Clearly define and implement test procedures to check each task undertaken and show how you overcame problems or limitations found as a result of using the test procedures.

7   Show that you can work independently to produce your work to agreed deadlines.

To achieve a grade A you must also:

8   Show a good understanding and imaginative use of options for customising applications software, such as keyboard configuration, toolbar layout and menu design, by providing users with facilities that improve efficiency.

9   Demonstrate an imaginative use of design and attention to detail in the creation of a template and macro that clearly enables users to improve their efficiency and effectiveness.

10   Make effective use of system diagnostics, system monitoring procedures and de-install routines, implementing adjustments as necessary to ensure correct system operation.

11   Keep records in an organised way and index them to enable easy reference to the problems experienced and the solutions implemented.

12   Show that you can work independently to produce your work to agreed deadlines.

Key Skills — Opportunity

You can use this opportunity to provide evidence for the Key Skills listed opposite.

Communication C3.1a	Contribute to a group discussion about a complex subject
C3.2	Read and synthesise information from two extended documents about a complex subject. One of these documents should include at least one image.
C3.3	Write two different types of document about complex subjects. One piece or writing should be an extended document and include at least one image.
Information Technology IT2.1	Search for and select information for two different purposes.
IT2.3	Present combined information for two different purposes. Your work must include at least one example of text, one example of images and one example of numbers.
IT3.1	Plan, and use different sources to search for, and select, information required for two different purposes.
IT3.3	Present information from different sources for two different purposes and audiences. Your work must include at least one example of text, one example of images and one example of numbers.

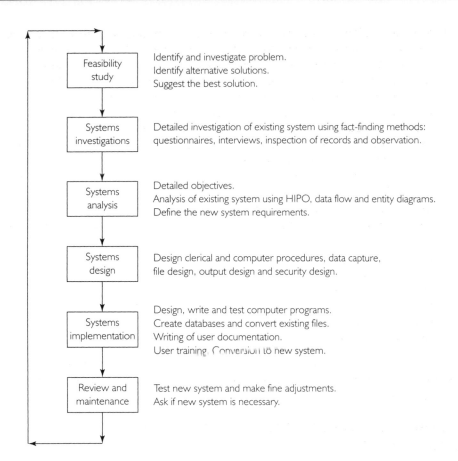

Feasibility study	Identify and investigate problem. Identify alternative solutions. Suggest the best solution.
Systems investigations	Detailed investigation of existing system using fact-finding methods: questionnaires, interviews, inspection of records and observation.
Systems analysis	Detailed objectives. Analysis of existing system using HIPO, data flow and entity diagrams. Define the new system requirements.
Systems design	Design clerical and computer procedures, data capture, file design, output design and security design.
Systems implementation	Design, write and test computer programs. Create databases and convert existing files. Writing of user documentation. User training. Conversion to new system.
Review and maintenance	Test new system and make fine adjustments. Ask if new system is necessary.

Figure 5.1 The systems development life cycle

specification for the new system is agreed between the management/users and the analyst.

System design

After the requirements have been agreed, the design of the new system begins. Appropriate input devices, computers and output devices should be chosen in this design. Any communications links should also be mentioned. A detailed specification for each piece of hardware should be given.

Software specifications should also be given so that a programmer can write the program code. An unambiguous statement outlining what the program should do must be provided at this stage. Various diagrams are included in this – such as **decision tables**. **Structured English** should also be used as an aid to program design.

Why is the analysis of a system carried out before systems design?

5.2 The feasibility study

The **feasibility study** starts with an outline of a problem that needs to be solved using information technology. To identify the problem in the first place, **fact finding** will probably be necessary.

Once all the facts have been investigated it is possible to prepare a feasibility study report. This document looks at the likelihood of achieving the stated aims and objectives at reasonable cost. Many projects are rejected at this stage so it is not worth going into too much detail in this report. However, the report does need to give an idea of the costs and timescale associated with the project, as outlined in the project brief or problem statement. One of the main problems you are likely to face when producing a report is that it can be very difficult to estimate costs and benefits without going so deeply into the system that you incur too much cost and take too long.

The amount of time a feasibility study takes depends on its scope and the experience of the systems analyst.

A feasibility study is also needed to find out if the project is technically possible and if the organisation has the human resources (staff) with the time and qualifications to successfully complete the project.

Is a feasibility study necessary?

A feasibility study is an optional stage in systems analysis and design. Where there is a clear-cut need for a new system and it is obvious that a certain system is required, there is no need for this stage, and once analysis is complete the systems design can start. Much of the work involved in a feasibility study will be performed in greater detail during the analysis stage.

Structuring a feasibility study

Projects are all different, and the following steps are only a guide to a typical feasibility study.

A statement of the purpose of the system

Systems development starts with identification of a problem in the existing system – there is nothing to be gained by changing a system if it works perfectly well and the management is satisfied with it. To develop a new system it is necessary to look in detail at the existing system and identify all its shortcomings. The process of investigating the existing system is called **fact finding**.

After a full fact find you should be able to answer the following questions:

- Is there anything that the existing system does not do or anything extra you would like it to do?
- Could IT be used to improve the timescale for the completion of certain tasks?
- Would it be possible to use new technology to reduce the costs involved when performing certain tasks?

A definition of system scope

This gives some indication of the amount of detail into which the project will go and how far-reaching the project will be. For instance, an order-processing system could just concentrate on the sales department, or it could also look at how information about credit limits comes from the accounts department.

It is often difficult to look at part of a complete system in isolation because other parts of the system affect it, even if indirectly. Figure 5.2 shows how the various parts of a typical system overlap.

If we were replacing just part of a system it would be necessary to make sure that links to other parts of the system are not removed. If part of a system is computerised and other parts still manual, links between the computerised and manual parts would need to be considered and a description of any problems envisaged mentioned in this section of the feasibility study.

Figure 5.2 How the parts of a typical system overlap

To summarise, the scope would normally consist of the following two parts:

- Define the range or extent of the project.
- Set the boundaries of the project (it is as important to outline the things the project will not cover as well as the things it will cover).

A list of deficiencies of the current system

It may be that the current system has been used for a while and that any changes have been 'bolted on' to it, creating an inefficient system. Such 'old' systems need a complete overhaul from time to time and a new streamlined system must be designed and implemented.

Part of looking at an existing system will be to examine it critically to discover its main deficiencies. A system that worked perfectly well a few years ago might not be able to cope with the increase in demand by customers over time.

Typical deficiencies would include:

- different departments holding separate databases with much of the same data
- information systems which still depend to some extent on paper-based systems.

A statement of the user requirements

This must be a clear statement of what is required by the user. For instance, an order-processing system could have the following requirements:

'A new order processing system will enable three-quarters of the existing staff to process the orders in around half the present time. The new system must also be able to cope with an expansion in the number orders of around 10% per year over a period of five years.'

The statement of user requirements is an important document because it enables the developer and the customer to agree on exactly what the system has to do.

The costs and benefits of development

It is easy to see the benefits of a system that costs £20 000 to develop and saves £30 000 in the first year, but not all benefits are this clear cut. For example, how do you measure, in terms of money, a more satisfied customer or fewer customers queuing at the tills? Many of the staff who decide whether the feasibility study should be accepted and the go-ahead given are no computing specialists, and will need to be convinced that the benefits of the new system will outweigh the costs. It should be borne in mind that the development of a new IT system may be in competition with non-IT projects such as a new production facility or a new fleet of delivery vans and the directors will choose the project with the greatest apparent return.

Benefits of development

To help users (and management) accept a change in a system it is necessary to convince them that it will be an improvement. They therefore need to be told about the likely benefits, which may include:

- the job can be performed more quickly
- fewer people are needed in administration and so can be transferred to more productive work, such as creating more sales
- smaller offices are needed, reducing rental, heating and lighting costs
- better security of information will be provided
- customers/clients will be more satisfied with the service, leading to fewer complaints
- management information can be obtained more quickly (e.g. the value of the sales made one month compared with the same months over previous years)
- it is more environmentally friendly (an electronic mail system would save trees by saving paper)
- it provides more accurate management information.

It is important to note that benefits to the management may not always be benefits to the user. Many of the benefits will be tangible and easy to measure in money terms but others will be intangible and very difficult to measure.

The costs of development

It is very difficult to anticipate the likely cost of completing a project, but to obtain approval it is important to get some idea. The costs are not simply those of the hardware and software. Some of the likely sources of costs are:

- *Creating a new working environment* – this might include rewiring, decorating the office, fitting blinds or new carpets.
- *Staffing* – new staff might be required because there is no one with existing expertise in the area. This might apply if a network manager is required. If fewer staff are needed redundancy needs to be considered.
- *Training* – staff will need to be trained to use the new system, either in house or at a college/private training centre. Either way, training costs need to be taken into account and you must bear in mind that while staff are being trained their work still needs to be done, incurring overtime or temporary staff costs.
- *Development costs* – these are the staffing costs involved in developing the new system and should include the wages/consultant fees of the systems analyst. There may be some cost incurred for the conversion of manual files to computer files.
- *Hardware and software* – these are more obvious costs. Remember to include items such as magnetic media and backup devices such as tape streamers.
- *Costs associated with running the system* – these would include consumables such as pre-printed continuous stationery, printer ribbons or toner cartridges. With wide area networks the costs of data transmission also need to be included.

quick fire

The cost of a new system is not simply the sum of the costs of the hardware and software needed. Why?

Activity 3

An old-fashioned company has been taken over by a more progressive competitor that has decided on complete computerisation of all its administrative procedures. Write a list of the disadvantages to the staff of this computerisation.

Conclusion and recommendations

The final part of the feasibility study will look at the conclusions reached and what the systems analyst recommends. In some cases alternative systems will be proposed, and these systems will differ in their complexity and corresponding cost. The systems analyst will recommend one of the proposed systems.

The feasibility report

The feasibility study culminates in a document called a **feasibility report**, which provides a summary of the findings from the study and is the document that will be discussed by the senior management or directors of the organisation and used to make the decision to go ahead with the project or not.

In the early stages of a systems development project, an estimate of the amount of time, effort and money involved needs to be made, though this can be difficult. There may be serious consequences if the project runs over the allotted time or exceeds its budget.

Once a project has been identified, the person or team responsible for system development has the job of selling the benefits of the proposed system to the management. A summary of all the activities outlined as part of the feasibility study should be contained in the feasibility study report. This can be a written report or an oral presentation. A feasibility report should contain all the headings outlined above.

5.3 System investigation

Investigation takes place throughout all the stages of systems analysis, though most of the investigation needs to be performed before the analysis.

In order to improve a system it is essential first to understand how it works. This process is called fact finding and can be performed by various methods, as shown in Figure 5.3.

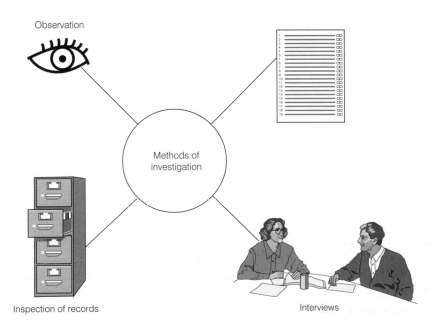

Observation

Methods of investigation

Inspection of records

Interviews

Figure 5.3 Fact-finding methods

Some of the many ways you can find out about an existing system are outlined below.

Interviewing

Interviewing is one of the best ways of finding out how a business operates. It is usually best to start at the top to gain an overview of the organisation's activities and find out about the company's overall strategy (staff restructuring, development of new markets) and likely developments (mergers and takeovers) in the future. Lower down the management structure, interviews should reveal the problems experienced in particular departments. These managers will be closer to the information flow in the business, and systems analysts need to find out about this. At the lower management end a systems analyst will probably get some suggestions as to how the existing system could be improved on a local scale.

At the bottom of an organisation's structure are the operational staff who deal with the day-to-day processing of information. These people often have strong opinions of what should and should not be done, but their experience of the whole organisation is narrow and often restricted to their own jobs. Their jobs may be threatened by the development of any new system, so a considerable amount of tact needs to be exercised when interviewing them. The company employs systems analysts to reduce the costs of operating the business – which usually implies job cuts. However, although the systems analyst researches and develops the new system, the decision to lay off staff is ultimately a management decision.

Some rules to follow when interviewing

A systems analyst needs to be prepared for staff interviews. Have a crib sheet of the questions you are going to ask and make sure you are asking questions at a level appropriate to the person you are interviewing – it's no use asking a person who processes orders about the future expansion of the company.

Some general rules:

- introduce yourself and mention the purpose of the interview
- allow the person being interviewed to do most of the talking
- make sure you structure the questions so that they start off in general, but lead on to more specific, areas
- keep a record of what the person says – in many cases it is a good idea to tape record it
- before the end of the interview, confirm what has been said
- if you are making notes, tell the interviewee that you will send them a copy for checking once the notes have been typed up
- at the end of the interview thank the person for their time and ask if it would be possible to meet again if the need should arise.

It is sometimes hard to arrive at a true picture of what someone actually does. Why is this?

Observations

This type of information gathering involves a variety of techniques, from watching a person doing their job (which can evoke hostility) to just looking around a building to get a feel for the organisation.

Activity 4

The list below forms part of an observation sheet which could be filled in when making a visit to a company's premises. The document aims to provide a checklist of things to look for during your visit. Complete the design of the sheet ready for your visit, using word-processing software.

Observation	Comments
Working environment	All office furniture old and does not comply with EU regulations. New furniture will be needed. Not enough electrical sockets (building may need rewiring?) Too much light through the windows.
Flow of work	Much of the work comes in the morning. Start of the week much busier than the end.

This may be a good opportunity to learn how to produce tables using your word processor. You will need to produce a variety of information in tabular form in the sections that follow.

Inspecting records

All organisations still use and generate a fair amount of paper-based documentation. By examining these documents you can get a feel for the way an organisation operates.

Documents can be divided into those giving specific information and those giving general information:

Specific information	General information
Order forms	Organisation charts
Invoices	CVs of relevant staff
Dispatch notes	Job descriptions
Picking lists for warehouse staff	Policy/procedure manuals
Files	Results of previous feasibility reports

You have been asked to inspect the records kept in a school. What documents might you want to look at?

All of these documents provide information about the existing system and should be used in conjunction with other methods of fact finding.

Questionnaires

At first sight, questionnaires seem ideal for collecting information about a company. You do not have to spend time interviewing people and a questionnaire sticks to the important points without digression, which can occur in an interview.

However, questionnaires have drawbacks:

- many people forget to fill them in, which can result in an incomplete picture of a system
- respondents may misunderstand some of the questions if the forms are simply posted to them and no personal help offered.

Nevertheless, questionnaires *are* useful when information needs to be collected from a large number of individuals as they consume a lot less staff time than interviews.

When compiling questionnaires, bear the following points in mind:

- make sure that the questions are precisely worded so that the users do not have to interpret the questions
- it is best for respondents not to have to put their name on their questionnaire – otherwise you might not get honest answers
- structure the questionnaire so that general questions are asked first, followed by more specific ones. It is also worth dividing up the questionnaire into functional areas – so, for example one part could deal with sales order processing, another with stock control. Obviously the approach will vary depending on the type of organisation
- avoid leading questions (questions which suggest a preferred answer)
- at the end of the questionnaire, always add the question 'is there anything I have missed that you think I ought to know about'?

Recording the results of the investigation

During the fact-finding procedure, the systems analyst will accumulate a large amount of detail about the existing system and all this material will need to be assimilated during the analysis phase. For instance, during the fact-finding stage the analyst could have collected:

- questionnaires
- interview notes
- observation sheets
- samples of forms used (invoices, order forms, stock lists, payslips, etc.)
- charts and diagrams to illustrate the existing system (flow charts, organisation charts, etc.).

To make sure that all the details are available for analysis, the data derived from these must be collected systematically.

Some facts might be recorded using narrative (i.e. a description in words), others using a variety of formal diagrams available to the systems analyst. It is said that a diagram is worth a thousand words, and this is certainly true when describing systems. Many different types of diagram and chart can be used during investigation. These will be studied later in this unit and also in Unit 6.

What do we need to collect information on?

All the techniques of finding out about the existing system can be used to collect information on:

- flow of information
- types of data
- sources of data
- decisions taken
- data capture methods
- documents used
- personnel involved
- manual operations
- automated operations

- types of processing
- storage methods
- types of output.

When you are performing your own systems analysis, you can use this list as a checklist to make sure you have found out everything.

Activity 5

A group of businesswomen have set up a CD music club offering a selection of the best chart-topping CDs along with some 'golden oldies'. At present they have around 8000 members, although this is likely to increase steadily over the next few years. The members get five free CDs when they join but have to pay postage and packing on them. Over a two-year period they must buy a minimum of eight CDs at full price.

A catalogue is sent to the members every month containing around 50 CDs, some at full price and others at half price. Members fill in an enclosed order form, calculate the total (including the carriage charge) and then send a cheque with their order.

A small warehouse is attached to the offices of the club and the CDs are placed in boxes on shelves. The CDs are picked from the shelves and orders made up. The parcels are then dispatched by either of two parcel firms.

This is all you know about the system. You have no information about whether it is a manual or computerised system. The existing system was set up when the company started and had only a few members; it is now having problems dealing with the workload and you have been asked to design a new system to process the orders.

Before you can design a new system you need to find out more about the present arrangements by asking the managing director of the club a series of questions. Design and word-process a list of questions that will reveal more about the existing systems used

Here is some advice before you start composing your list of questions:

- Try to arrange your questions in a logical sequence. General questions about the business should come first.
- You might be tempted to write the questions and leave a space for the answers but, because you do not know the length of the answers it is better to list question reference numbers along with the answers on a separate sheet.
- Make sure that your questions aim at producing answers that reveal important facts about the business. You will use these facts when creating a new system, so make sure that all your questions are relevant.
- It is a good idea to split the questions up into sections. For instance, you could have a section about how orders come in and are processed; another section could be about stock control and so on.

Flow of information

When explaining an existing system it is important to look at the information flows within an organisation. This is often done diagrammatically using **data flow diagrams**.

When investigating the flows of information around an organisation, why is it a good idea to follow or track a document, such as an order form, around the system to see what happens to it?

Types and sources of data

You need to look at the types and sources of data used by the existing system. The term 'sources of data' refers to how and where the data originates and whether it comes from an internal or an external function. Both type and source of data are needed when drawing the series of data flow diagrams (level 0, context, level 1, level 2).

Decisions taken

It is important to discern from the user information any critical decisions that are made during the running of the operation. Such a decision might be 'If the total invoice cost is more than a certain amount, delivery charge is zero'.

These decisions, and the context in which they are made, need to be noted so that decision tables summarising the rules and corresponding actions can be produced (the construction of decision tables is covered later in this unit). Decision tables should be included in the systems analysis documentation.

Data capture methods

You need to look at how data enters the organisation. For example, orders for goods might come in by telephone, fax, paper documents such as order forms, or even electronically using e-mail. Make sure you have covered all the possible ways that data can enter. The study of **data capture** methods is concerned with how the data entering the organisation is entered into the internal system. For example, a customer may send an order form and the details are typed into a database using a keyboard; faxes may be scanned into a word processor, and are stored on disk so they can be found easily. The paper documents can then be destroyed because they take up too much space.

Documents used

The documents used by the organisation are a useful source of information about the detail that needs to be recorded and the processing needed to produce the output. Such documents could include order forms, invoices, stock pick lists, delivery notes and letters.

Looking at letters (particularly letters of complaints from customers) can reveal useful information about the shortcomings of the existing system. For instance, many complaints about the wrong goods being sent may point to bad stock control with unsuitable alternatives being sent to replace goods that are out of stock.

If you are going to base your project on a real organisation, ask if you can see some of the documents it uses in the course of its business. Look at their design, particularly if they are used for output such as quotations, statements or invoices – it can tell you a lot about the information flows to external organisations. With the organisation's permission, include photocopies of any documents as evidence in your analysis documentation.

Personnel involved

In your initial analysis you should outline the staffing arrangements for the existing system. It is a good idea to include an organisation chart. You need to find out the qualifications of the staff, and in particular their computing expertise because this will tell you about their training needs. It will, for example, be difficult to computerise an organisation if none of the staff has any experience of computers. However, this situation is much less common than it used to be.

If a new computer system is to be introduced, it is essential to find out what computing skills the staff have. You might think that you could obtain this from the personnel department, but their records may not be up to date – people might have only put down their skills in the particular area in which they work, or only if they have a qualification on paper (for instance, computing might be a hobby, of which personnel have no records). Technical staff are computer literate, but seldom have a formal qualification, and often omit their experience when stating their qualifications. You could put together a simple questionnaire to enable staff using the new technology to assess their computing skills and knowledge. This might be particularly useful later on, when you need to work out how much training to provide.

Manual and automated operations

The operations to be carried out on the data need to be investigated. In a manual order-processing system we would look at what information is copied down.

Orders might not always arrive in the normal way on an order form: they will sometimes be phoned through so some record needs to be kept of the transaction. Many companies now record the telephone conversations their staff have with customers, particularly when selling financial services such as insurance, mortgages and loans. In this way if a dispute arises over what was said, a copy of the tape can be played back in the absence of paperwork.

If products are ordered from a catalogue, some check is needed to make sure that the catalogue number quoted on the order exists, and that it matches with the goods' description.

If invoices are issued you need to find out how the VAT is calculated, how the order is totalled and how it is checked.

Records of transactions might need to be communicated to other departments and the way in which this is done might need to be investigated.

Types of processing

The processing needed in the system should be outlined. Remember that processing data is not confined to doing calculations. Merging, searching, summarising are also examples of processing.

An understanding of the system's processing requirements is built up using:

- structure diagrams
- decision tables
- screen reports and printed report layouts.

Storage methods

All the paperwork associated with a manual system needs to be stored somewhere for a certain period of time – filing cabinets, account books, card files, box files, cardboard folders, etc. During the investigation phase you need to find out what information is stored and where it is stored. It is also important to ask *why* it is stored and not disposed of. Sometimes storage is less formal, and small businesses might even keep details of orders in the manager's head, or an old shoe box. However informal the method, it is important that the developer knows where all the information is stored.

One problem with any storage system is knowing how long to keep material before it is disposed of. Computerised storage is useful in this respect because much less space is taken up than in a manual system and it is possible to delete a whole store of data at the press of a button.

Types of output

Sometimes the output is on paper or screen, but it could be in the form of electronic signals transmitted along cables or even using radio signals. You need to include a description of the various forms of output used by your system. Examples of the paperwork developed as a result of processing should be kept for further reference.

The output from one system can also be the input to another system.

5.4 Structured analysis tools

To produce a system that replaces an existing one you need to understand the existing system fully. The process of finding out and documenting the existing system is called analysis and is performed by a systems analyst. A variety of diagrams and charts can be used to document a new or existing system. These include:

- high-level (contextual view) data flow diagrams (DFDs)
- process specifications
- entity-relationship diagrams
- low-level (detailed view) data flow diagrams
- entity-attribute definitions
- data dictionary.

Some of these important tools are described below.

Analysis documentation is used repeatedly in systems analysis, at different levels in the system. The first level usually gives just an overview of the whole system, but as the system is developed more detail is added.

Data flow diagrams (DFDs)

In the feasibility study an initial investigation of the system should look at the inputs, what processes are performed on them and the outputs. The scope of the system is also specified in this phase. To help further analysis, **data flow diagrams** are drawn to consider the data while ignoring the equipment used to store it. They are used as a first step in describing a system.

There is a series of symbols used in these diagrams but unfortunately different authors use different-shaped symbols, which can be very confusing. The convention used in this book is the one adopted for GNVQ although you will see different ones in other books.

A process or action

Figure 5.4

This is a rectangular box (see Figure 5.4) and represents a process that does something with the data (performs calculations on it, for example). The box is divided into three parts, the top left box having a number in it which identifies the box. The main body of the box is used to record a description of the process and the top right is used to record the person or area responsible for the process.

An external source of data (where it comes from) or an external sink of data (where it goes to)

Figure 5.5

This is an oval shape (see Figure 5.5) which is used to describe where, outside the system, the data comes from and goes to. We are not concerned with what happens to the data before it reaches the box (if it is a source) or what happens to it when it goes past a sink.

If a data source is duplicated on the diagram, the oval has a line going through the corner of it, as in Figure 5.6.

Figure 5.6

Data flow

Data flow is shown by an arrow pointing in the direction of the flow (see Figure 5.7). Usually it is advisable to put a description of the data flow on the arrow to aid understanding. By convention, we never use a verb on a data flow.

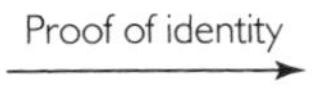

Proof of identity

Figure 5.7

A store of data

The symbol for this is shown in Figure 5.8. A data store can be anywhere data is stored; it could be a drawer where you keep letters, file boxes, folders, books, a filing cabinet (or a certain drawer of a filing cabinet), floppy disk or a hard disk. Again, the symbol bears a number which is used to reference the store when describing it, but there is also a letter placed in front of the number. M is used for a manual store and C for a computer store.

| M1 | Members |

Figure 5.8

Activity 6

The table some of the phrases used in a data flow diagram. Copy the table and put a tick in the appropriate box to indicate whether the statement represents a flow of data, an external entity, a process or a store. You will see that this is not as easy as it first looks and it may be possible to put a tick in more than one column.

Phrase	Flow of data	External entity	Process	Store
Calculate order discount				
Customer				
Invoice				
Make appointment with patient				
Appointment card				
Order acknowledgement				

It is very difficult to do this task without knowing more detail about the system being considered, because an entity in one system could be a data flow in another.

Drawing data flow diagrams

To draw data flow diagrams you first need to understand how a system works. If possible, it is best to split the system into smaller subsystems and investigate each in detail. These subsystems can then be reassembled to give the whole system.

Let us now take as an example the processes involved in drawing data flow diagrams for a video library. To do this we will divide the system into three subsystems dealing with new members, recording new videos and recording loans of videos.

There are, in reality, many more subsystems, but to keep things simple we will consider only these three.

Membership recording subsystem

The first task is to investigate the membership system to reveal the following basic information.

Applicants to join the library fill in an application form and show certain documents to provide proof of identity. If the applicant does not have this documentation, the library manager will refuse membership. After the membership details have been checked the new member is given a membership card and the member's details are recorded and stored in a filing cabinet. If the member borrows a video and does not return it, this file will be accessed to obtain the member's name and address.

New video recording system

We can now look at the system used when a new video from the suppliers is added to the library. This is a simple manual system, with the details of the video (name, price, etc.) recorded and then stored.

System for recording loans of videos

To borrow a video the member needs two things: the video and a membership card. Each of these items has a unique reference number and the borrowing process simply links these two numbers together, recording, for example, that member 34223 has borrowed video 90234.

Notice that the member data and video data are both needed; the 'loans' store of data will contain only the membership number and the video number, so if we need to send a letter demanding return of a certain video the name of the video and the name and address of the member would be recovered from the other recording systems.

We could get an overall view of the system by joining the diagrams together but the system is much easier to understand if the diagrams are drawn separately.

Levels of DFDs

When analysing systems it is usual to draw DFDs at different levels. The level used reflects the depth in which the diagram looks at the system. The high-level DFD is also called the contextual view DFD because it puts the system being investigated into context by outlining the general flow of information internal to a system and the **relationship** between the internal system and the external **entities** such as suppliers and customers. The diagram might show the flow of orders from the customers (an external entity) to the sales department and goods being despatched from the warehouse to the customers.

The simplest diagram showing the system being investigated on its own (i.e. with those other systems not being investigated removed), with the data flows between the system and the external entities, is called the **context diagram**. The context diagram shows the system being investigated in one simple diagram.

High-level DFDs can be broken down into further diagrams showing greater levels of detail. Figure 5.9 shows the decomposition (i.e. the levelling) of the DFD.

Analogy with a map

The different levels of DFDs are like looking at greater details of a map. Suppose you are looking for a particular street in London. If you did not know where

Level 0 (the context diagram)

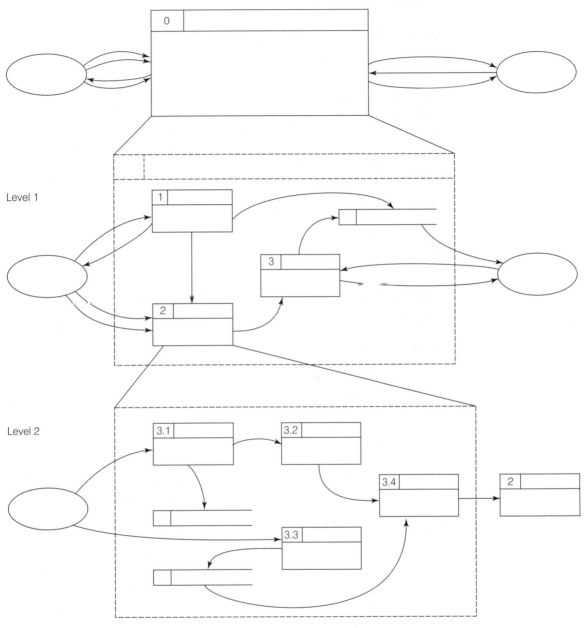

Level 1

Level 2

*Figure 5.9 The decomposition
(i.e. the levelling) of DFDs*

London was, then you might first need to look at the capital cities of the world on a map. Here you can see England in relation to all the other countries (this would be analogous to the level 0 DFD). A map of England on its own would be the context diagram (i.e. we are interested only in England and the main routes in and out of the country). We can then decompose this diagram to show increasing amounts of detail.

We will now look at how to go about drawing a series of DFDs at the various levels for the video library example described before.

Why do we level DFDs?

Level 0 DFD showing the system boundary

This diagram shows the data flows between sources and sinks (or recipients). A dotted line is used to mark the system boundary, the area inside showing the extent of the system being investigated. In a level 0 DFD no processes or stores are shown so the only shape used will be the ellipse.

All the sources and sinks of data are shown in this diagram, inside elliptical boxes with arrows indicating the directions of data flow between boxes. A brief description of the data is added to these arrows.

When drawing the level 0 data flow diagram it is a good idea to fill in a table like the one below, which lists the name of every external source/sink, whether it is a source or a sink, and the name of the data flow.

Name of external source/sink	Source or sink?	Data flow
Potential member	Source	Application details
Potential member	Sink	Membership card
Video suppliers	Source	List of available videos
Video suppliers	Sink	Order
Video suppliers	Source	Invoice
Video suppliers	Sink	Payments
Manager	Sink	Details of members who have not returned their videos
Manager	Source	Letters to members to return videos
Manager	Sink	Lists of number of times each video is borrowed
Video	Source	Returned video details
Member	Sink	Video information
Member	Source	Fines
Member	Source	Card details
Member	Source	Letters saying to return videos
Member	Source	Video requests

Having completed a table like this, it is easier to draw in the sources and sinks and connect them with the correct data flow lines. Figure 5.10 shows the level 0 DFD drawn from the information contained in the above table.

A broken line marks the system boundary, which in this case separates all the activities that go on inside the video store. Some systems analysts do not bother with the level 0 DFD, preferring to start with the context diagram which only looks at the system being investigated.

Drawing the context diagram

A context diagram groups everything inside the system boundary and places it in a single process box. A description is added to the box, which takes in the sources and sinks that it replaces. The sources, sinks and data flow lines are now added using the level 0 DFD as a guide. Figure 5.11 shows the context diagram for the video shop system.

The aim of the context diagram is to define the scope of the system and help examine the system boundary.

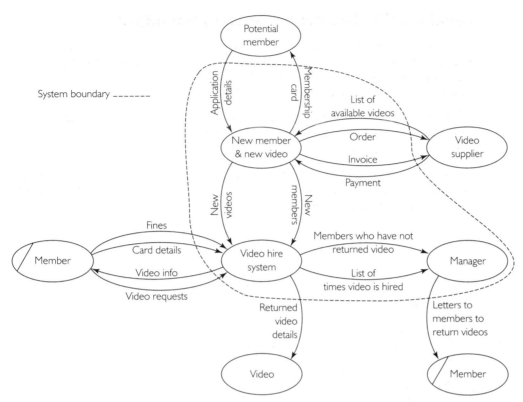

Figure 5.10 Level 0 DFD

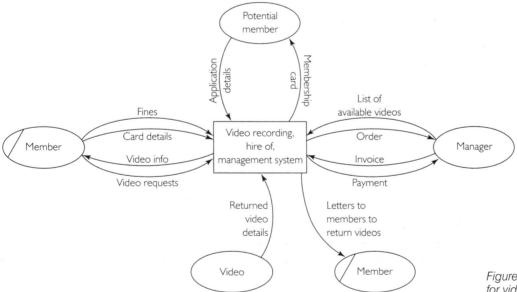

Figure 5.11 Context diagram for video shop system

Level 1 DFD

It is not until the level 1 DFD that the process and store boxes are encountered. At this stage the rectangle at the centre of the context diagram is broken down into processes and stores, and data flows are added. The sources and sinks are also shown, along with their associated data flows. Care must be taken to ensure that not too many processes are included. Try to aim for no more than about six. You can use another layer of diagram (the level 2 DFD) to decompose each of these processes further, if needed.

Figure 5.12 shows the level 1 DFD for processing a member's application to join and Figure 5.13 shows the diagram for adding a new video to the library. Figure 5.14 shows the level 1 DFD for borrowing a video and Figure 5.15 combines the entire system as a level 1 DFD by joining together all the component level 1 DFDs for the subsystems.

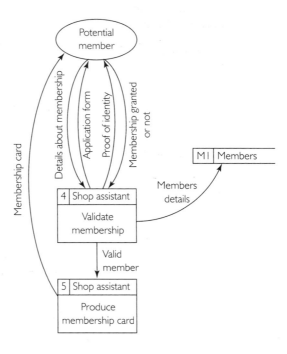

Figure 5.12 Level 1 DFD for processing a membership application

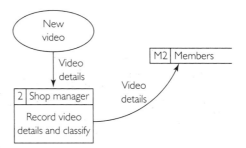

Figure 5.13 Level 1 DFD for adding a new video

Activity 7

A simple order processing system receives orders from customers. Orders are validated to make sure all the items on the order are valid stock items and that they are actually in stock; this is done using the stock file. Customers are checked against a customer file to ensure that they are account holders and credit worthy. The order is processed, the stock file updated and the goods sent to the customer with an invoice.

a   Draw a level 0 DFD for this system.
b   Using the level 0 DFD, draw the corresponding context diagram.
c   Using the context diagram to help you, draw a level 1 DFD.

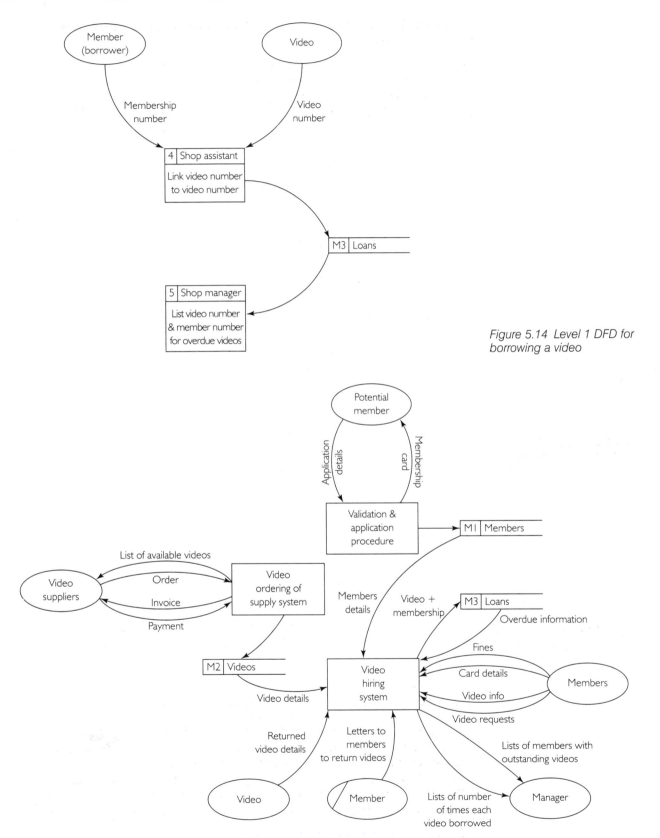

Figure 5.14 Level 1 DFD for borrowing a video

Figure 5.15 Level 1 DFD for entire system

Level 2 DFD

A level 2 DFD breaks down the process boxes in the level 1 DFD. To do this one of the process boxes (one of the rectangular ones) from the level 1 DFD is taken and, if possible, the process it represents is broken down into a series of more detailed subprocesses. The resulting diagram is a level 2 DFD.

The processes can be taken one step further to produce level 3 DFDs by expanding the process boxes to show still more detail. It is very unusual to go beyond three levels of DFDs to describe a system.

The advantage of this 'zooming in' approach is that, by looking at each level starting with the outline DFD, we can gradually build up a complete picture of the information flows through the system. Instead of having a single DFD showing a large number of processes, we end up with a whole family of DFDs showing different levels of detail.

/ **Activity 8** /

When drawing DFDs, it helps to write down in a table the sinks/sources, data flows, processes and stores. This has been done for the CD club encountered earlier:

Sink/source	Data flows	Processes	Stores
Member Orders Credit status Delivery note	Stock details Customers	Process orders	Stock

a   Draw a (very simplified) data flow diagram for this. The diagram you draw will show only the information flows from the club to the members and vice versa.

b   Study the details of the CD club outlined on page 299. Now produce a data flow diagram for the system specified.

The system must:

- process customer orders as they arrive
- check that the CDs ordered are currently in stock
- provide a picking list for the warehouse staff
- update the sales ledger with details of amounts owing
- send an account to each customer every month
- send a delivery note with the order and issue one for the carriers
- remove the items sold from the stock file
- send out the bonus CD provided that the customer has not cancelled it on their last order form.

To make the system easier:

- assume that there is no part delivery of orders
- assume that the initial processes for the free CDs on joining have been sent
- do not be concerned with the purchasing of CDs from suppliers.

You will be unlikely to produce a good drawing on your first attempt. Try a rough one to start with and then look at it carefully to see if anything has been left out or if it could be drawn with the symbols arranged differently so that the diagram is less cluttered.

Using DFDs

Data flow diagrams can be used:

- during system investigation to record findings
- during system design to illustrate how a proposed system will work
- when outlining the specifications of new systems.

Entity relationship models and entity relationship diagrams

Entity relationship models

Entity relationship modelling is a technique for defining the information needs of an organisation to provide a firm foundation on which to build an appropriate system. Putting it simply, entity relationship modelling identifies the most important factors in the organisation being looked at – the entities. Also looked at are the properties these factors possess (their **attributes**) and how they are related to one another (the relationships). Entity models are logical, which means that they do not depend on the method of implementation. If two departments in an organisation perform identical tasks but in different ways their entity models would be identical because they use the same entities and relationships. But their data flow diagrams could be different because their information flows may well differ. Entity models are particularly useful because they are independent of any storage or ways of accessing the data. They are thus not reliant on any hardware or software at this stage.

An accountant often produces a financial model to see the effect of changes to the finances of a business, and the models for this are usually created using spreadsheet software. Systems analysts also produce models, called entity relationship models, which represent a view of the organisation. A model is a simplification of a real system which can be used to understand how the real system works. Good models reflect the real world and bad models do not. When creating a model, it is unusual (or lucky!) to get it right first time, so the process of producing a model is one of continual trial and improvement.

An entity model is an abstract representation of the data in an organisation and the aim of entity relationship modelling is to produce an accurate model of the information needs of an organisation which will aid either the development of a new system or the enhancement of an existing one.

An entity relationship model (ERM) describes a system as a set of data entities.

Entity relationship diagrams

Entity relationship diagrams (ERDs) look at any components important to the system and the relationships between them.

So what is an entity? An entity can be anything about which data is recorded. It could include customers, sales, payments or employees. Each entity has some associated attributes. An attribute is detail about an entity. In the following table the attribute for the entity 'customer' contains further details (attributes).

Entity	Attributes
Customer	Customer number
Name	
Address	
Telephone number	
Credit limit	
Amount owing	

Each entity is represented in an entity diagram by a soft rectangle (a rectangle with rounded corners) with the name of the entity written inside. Relationships between entities are shown as lines between these boxes. The entity is always written inside the soft rectangle in capital letters and always in the singular because the use of plural would imply a certain type of relationship. So instead of CUSTOMERS we would use CUSTOMER.

It is very important to understand the terms 'entity' and 'attribute' so here are their definitions.

Entity

An entity is an object of the real world that is relevant to an information system – a customer, invoice, product, country, course, etc. Entities are written in a soft box (a box with rounded corners), with the name of the entity in the singular and in capital letters.

Attribute

An attribute is a single item of data which represents an individual property of an entity. An attribute can be thought of as something which adds further detail to the entity.

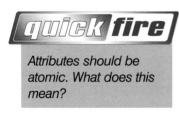

Attributes should be atomic. What does this mean?

For instance, the entity STUDENT could have the following attributes: student number, student address, telephone number, date of birth, tutor, course number. It is important when developing databases to check each attribute to decide whether it is possible to break it down further into more attributes. For example, the attribute student address could be further broken down into the attributes street, town and postcode. When attributes need not be broken down any further, they are said to be atomic – street, town and postcode are atomic attributes.

Breaking down attributes to produce atomic attributes allows flexibility to manipulate the data. For example we could search for students who have a certain postcode or sort students according to the town they are from.

Figure 5.16 shows the properties of an entity box.

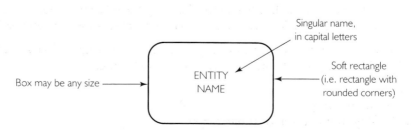

Figure 5.16 An entity box

Activity 9

a Which of the following attributes could be considered to be atomic?

- video details
- date of birth
- membership number
- member name
- enrolment
- quantity in stock
- product number

b For each of the above attributes that are not atomic suggest suitable attributes which *are* atomic to replace them.

c A college library borrowing system has the following entities:

- MEMBER (a person who is a member of the library and eligible to borrow books)
- BOOK (a book which may be borrowed)
- LOAN (a link between a particular book and the person borrowing it)
- RESERVATION (books may be reserved by members, so that when brought back they are kept aside for another person).

The above system is only part of the whole library system.

Identify and list the attributes for each of the above entities. You will need some attributes which uniquely define the entities MEMBER and BOOK. Although the ISBN (International Standard Book Number) is used by bookshops to identify book titles, a library might have many copies of the same title, in which case the ISBN could not be used to distinguish each copy.

Produce your list and show it to your tutor.

Relationships

A relationship is the way in which entities in a system are related to one another. A relationship may be **one-to-one** (1:1), **one-to-many** (1:m), or **many-to-many** (m:m). The three possible kinds of relationship between two entities A and B are shown in Figure 5.17.

The relationships between the entities are drawn as lines between the boxes; a solid line indicates that half of the relationship is mandatory (must be) and a broken line indicates that half of the relationship is optional (may be).

What are the three types of relationship?

Take a look at the simple entity relationship diagram shown in Figure 5.18. The boxes show two entities and the line shows the relationship between them. Notice that on the left-hand side the line has a forked ending (called a crowsfoot), while on the right it has a single one. This relationship is called a many-to-one relationship.

It is important to note that each line in an entity relationship diagram has two ends and to describe these you will need to decide on the following:

- name
- degree (how many?)
- optionality (is it optional or mandatory?).

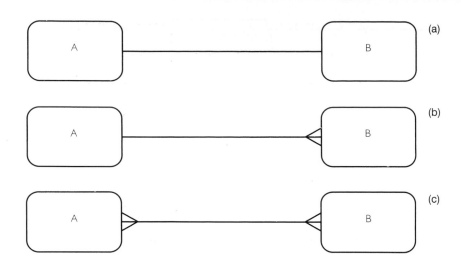

Figure 5.17 (a) A one-to-one relationship; (b) a one-to-many relationship; (c) a many-to-many relationship

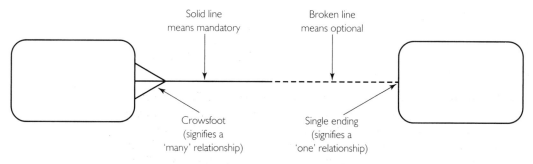

Figure 5.18 Entity diagram for a many-to-one relationship

It is important to make sure that both ends of the relationship are defined. Look at the relationship shown in Figure 5.19.

Figure 5.19

Following the diagram from both ends we can see that:

- each TICKET is for one and only one PASSENGER;
- each PASSENGER might be shown on more than one TICKET.

Let us now consider the CD club we met earlier. It can be described as consisting of the following four entities: member, order, CD and delivery. These could be connected by lines in the way shown in Figure 5.20.

Figure 5.20 is based on the following set of relationships:

- member makes an order
- order consists of CDs

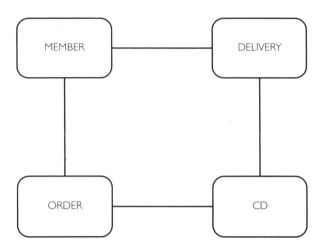

Figure 5.20 Simple entity relationship diagram for the CD club

- delivery consists of CDs mentioned in the order
- delivery is straight to the customers.

The diagram is far from perfect. If an order consists of several CDs this is not indicated. In addition, if the club runs out of a popular CD the part of the order that *is* in stock will be sent, with any out-of-stock CDs to follow.

Let us now look at the relationship between two of the entities: ORDER and CD. We need to examine the relationship from both ends. Looking in the direction from ORDER to CD we can see that one order can be for many CDs. Looking in the reverse direction we can see that a particular CD (that is a particular title) can be in many different orders. In other words, the relationship between ORDER and CD is many-to-many.

Figure 5.21 shows the entity diagram for a many-to-many relationship. We can see from the diagram that many members order many CDs. This would imply that an order is for many CDs but it is impossible to say which CD the order is for. There needs to be a way of linking the CDs to each order so that they can be cross-referenced. We do this by creating a new entity called ORDER LINE. This entity indicates the CD that is on a particular line in a particular order.

Figure 5.21 Entity diagram for a many-to-many relationship

We try to avoid many-to-many relationships because it is difficult to implement them. Instead we use an intermediate stage that is related to the two entities. This intermediate stage consists of intersection data and is also an entity. For example, in Figure 5.22 you can see that another entity called ORDER LINE has been produced. We must now look at the relationships between ORDER and ORDER LINE. This relationship is one-to-many because one order may consist of many order lines. Looking at the relationship between ORDER LINE and CD we can see that this is also a one-to-many relationship because a particular CD can be in many different order lines.

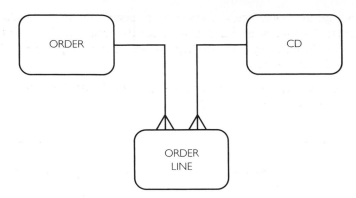

Figure 5.22

The complete entity diagram is the one shown in Figure 5.23.

Entity diagrams are quite complicated and time needs to be spent drawing them; in fact you often need to devote as much time to them as when learning a new programming language. This section presents a superficial and perhaps over-simplified view of entity diagrams – this area can be very complex and you can get whole books devoted just to entity diagrams.

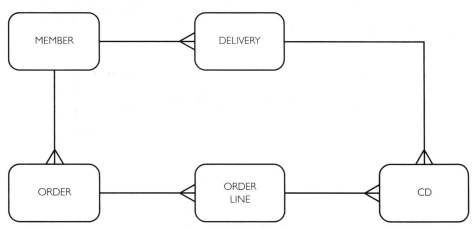

Figure 5.23 Complete entity relationship diagram for CD club

Activity 10

Draw entity diagrams to show each of the following relationships:

- Classes consist of many students.
- One customer has many orders.
- One tutor lectures on many courses.
- Each module is taught by one tutor.
- Many students enrol on many courses.
- Many customers order many products.

Deciding on the type of relationship

In entity modelling it is necessary to decide on the type of relationship between entities. This is best done by looking at the entity from both ends. For example, take the relationship between students and courses in a college. Looking at the relationship from the student end first we can see that a student could take more than one course. Looking at the relationship from the college end, we see that a single course can be taken by many students. So the relationship between COURSE and STUDENT is many-to-many.

We can describe the relationship between the entities COURSE and STUDENT using the diagram shown in Figure 5.24.

Figure 5.24

Notice that the names of the entities are always written inside their boxes in the singular form, hence STUDENT rather than STUDENTS.

As we have already seen, many-to-many relationships cannot be implemented, so in this case we need to create a new entity called ENROLMENT. If a student takes several courses, there will be an enrolment for each course and the attributes of the enrolment record will include the student number (to identify the student) and the course number (to identify the course) as the primary keys. The entity diagram now becomes as shown in Figure 5.25.

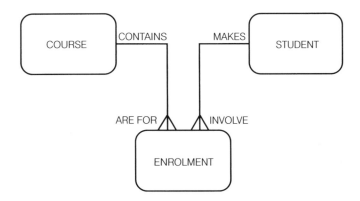

Figure 5.25

Example
An agency employs temporary staff to conduct market research for its clients. The staff are employed only on a 'per job' basis, although as one job finishes another one starts. The agency has asked a consultant to build a database system to store details about staff, jobs, rates and the hours that staff have worked. Initial analysis reveals the following entities:

- EMPLOYEE
- DATE
- HOUR
- JOB
- RATE.

Further investigation revealed the following. At a given moment one employee can work on only one job, with that job having only one rate. The one job can have many hours and can be done over many dates. The one rate can also be paid over many hours and worked over many dates. Finally, one employee can work many hours over many dates.

Figure 5.26 shows the entity relationship diagram for this situation. The situation has been simplified, so optionality is not shown in the diagram, nor are the names of the relationships included.

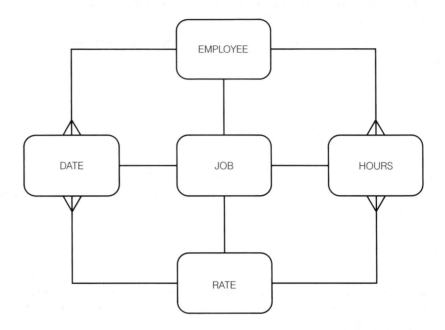

Figure 5.26

Activity 11

a Look at Figure 5.26 and redraw it to include suitable names on both sides of the relationships.

b The above situation applies only to a momentary point in time. Taken over a period of time the situation is slightly different, so the entity relationship diagram will need to reflect this. Modify the diagram to take account of the following:

Over a period of time, one employee can have many jobs which have many pay rates. The many jobs will also be worked over many hours with many rates involved for those hours. The many rates will also be paid over many dates. As before, the one employee works many hours over many dates.

The differences between physical and logical design

You will see the words logical and physical appear many times when talking about systems analysis – and in particular about system design. What is the difference between them?

The **logical design** looks in detail at the system requirements and specifies *what* must take place rather than *how* it is to take place. For example, a DFD is a diagram that shows the logical system. It shows what inputs and outputs there are to the system as well as the data flows, the processes and stores. If you look at any DFD you will see that you cannot tell anything about how the processes are performed or how the data is stored. This is because the data flow diagrams describe the logical system rather than the physical system and you cannot tell from them how anything is actually going to be accomplished.

Physical design looks at how the system is to be implemented and describes the specific components and **system specifications**. If you describe the way something is done, then you are describing the physical system.

Logical data modelling uses many specialist terms. Entity and attribute are two such terms. Here are some more, along with their definitions.

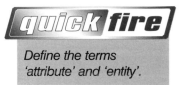

Define the terms 'attribute' and 'entity'.

Key

A **key** is an attribute that has a special significance.

Primary key

A **primary key** is usually one (but occasionally several) attributes that can uniquely define a particular entity. For example, the entity CUSTOMER could be uniquely defined by a unique attribute, such as customer number. The entity PAYROL could be uniquely defined by the entity employee number.

Foreign key

When one of the attributes that is a primary key in one entity also appears in another entity, it means that there is a relationship between the entities. In the other entity the attribute is not a primary key because it may no longer be unique. It is known as a **foreign key**.

Suppose you supplied goods to other companies and you wanted to set up a system to keep the details of these contacts. There are two entities: COMPANY and CONTACT. Here are the attributes for each of these entities:

CONTACT	COMPANY
Contact_ID	Company_ID
First Name	Name
Last Name	Address
Company_ID	Telephone Number

Each contact will be allocated a unique number so Contact_ID is the primary key for CONTACT. In the entity COMPANY, the attribute Company_ID is unique and therefore the primary key. Notice that the attribute Company_ID appears in both entities and that in COMPANY it is the primary key. In CONTACT, Company_ID is a foreign key. Because the attributes appear in both entities we can link them. Hence the entity CONTACT has a relationship to the entity COMPANY.

You may be wondering why we have two entities rather than just one containing the contacts and the company they work for. If each company had only a single contact then it would be appropriate to put the company and contact details together but in reality there will be more than one contact in each company, so if we store contact details the name of the company and its address will appear in the details of many different contacts. Hence the name and address of the company would be a repeating group of attributes. The main problem with this is that if the company changes address it would be necessary to change the address for each contact working for that company. If two entities are used then if the address of the company changes then it will only need to be changed once – in the entity COMPANY.

Making sure that attributes are atomic

As well as making sure that repeating attributes are removed and placed in their own entity, you must make sure that all the attributes are atomic. Atomic attributes are attributes that do not need to be broken down any further.

For example, a beginner to systems analysis and database design might use an attribute called Name to hold the data items

Mr Stephen Doyle

The trouble with this is that if a sort is now performed it will be done in order of title, so names starting with Miss will appear before names starting with Mr. Additionally, if you want to search for the surname Doyle, you can only do so if you know the title and forename. Using a single attribute called Name causes problems because the attribute is not atomic. The attribute needs to be broken down into three attributes, which can be called Title, Forename and Surname.

The attribute Address is not atomic. What needs to be done to correct this?

Data dictionaries

A **data dictionary** system is a tool used during systems analysis, and particularly during database design. It is quite often provided as development software, called CASE tools.

The purpose of a data dictionary is to provide information about the database, its uses and participants in the system. A data dictionary could be said to provide 'data about data'.

The contents of the data dictionary

Data dictionaries usually contain some or all of the following features:

- *Entity names* – Each entity needs to be given a name and these names need to be recorded and described in the data dictionary.
- *Relationships between the entities* – The relationships between the entities can be described and then shown in an entity relationship diagram.
- *Attribute names* – Each attribute should be given a name, and these names should be chosen so that they describe the data as fully as possible without being too long. It is best to avoid leaving spaces in field names, using – or – to separate the words instead.
- *Synonyms* – Synonyms are alternative names for the same thing. In many large organisations, a database is at the centre of the computer system and is used by different departments for various activities. It is therefore common to find users in different departments employing a different name for the same concept; however, this can be very confusing for systems analysts trying to design and build a new system. To prevent confusion, users should list any alternative names in the synonyms section of the data dictionary.
- *Data type* – A data type needs to be specified for each attribute. It is very important that once an attribute's data type is set, it has the same data type wherever it occurs. In most cases, relationships can be made only between attributes with exactly the same attribute names and data types.
- *Format* – Details of formats should be included in the data dictionary for those attributes which have formats set by the user. For instance, a numeric field can be a short integer, long integer, currency, etc. and dates can be given as 12th June 2001, 12/06/01, etc.
- *Description of the attribute* – Each attribute should have a description, and these descriptions can be transferred to this section of the data dictionary.

- *Attribute length* – Here we specify the length of the attribute for those attributes that can have their lengths set.
- *The names of other entities in which the attribute appears* – An attribute may appear in more than one entity. For example, the attribute Order_Number could appear in the entities ORDER and ORDER_LINE. If we wanted to change the name of an attribute then it is useful to know the entities where it is repeated.

Here is a data dictionary entry for an attribute called REORDER_QUANTITY:

Attribute name	REORDER_QUANTITY
Synonyms	None
Data type	Numeric
Format	0 decimal places (i.e. an integer)
Description of the attribute	The number of units of a stock item which can be ordered at any one time
Attribute length	4 digits
Entity Names	STOCK_LIST
	STOCK_LIST_EXCEPTION
	REORDER

Data dictionaries are always created when large commercial databases are being built, especially if the databases are to be used in many different applications. The purpose of the data dictionary is to make sure that data across the whole system is consistent. Large systems are built by large project teams, with each team working on a part of the whole system; in this situation consistent terminology is vital, and the data dictionary is invaluable. The data contained in data dictionaries is often referred to as metadata.

Data dictionaries can be built manually but there are clear advantages in maintaining a computerised data dictionary. Most relational database management systems have software which creates and maintains a data dictionary. It is kept as a separate database and automatically updated as changes are made to the structure of the main database.

Optional features of the data dictionary

In addition to the essential features listed above, some data dictionaries provide the following additional information:

- validation checks
- details of the validation checks performed on data entered for each attribute
- a key
- the type of key specified.

Process specifications

All systems require a certain amount of data processing to be performed and the systems analyst needs to describe this processing.

In this part of the systems specification all the clerical and computer procedures necessary to produce the output must be outlined so that programs can be written for those tasks performed by the computer, and the clerical procedures for the manual tasks are in place.

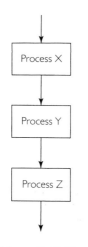

Figure 5.27 Flow chart showing three processes in a sequence construct

The computer and clerical procedures can be described in a variety of ways.

Flow charts

Flow charts are used to show the order in which a series of processes should be performed. There are three main constructs used when drawing flow charts and their purpose is to alter the order in which the processing steps are performed.

Sequence

This is simply a list of processes in which one process is always carried out after another, and always takes the same path. The various processes are placed in the correct order and are obeyed in sequence. A flow chart for three processes in the order X, Y and then Z is shown in Figure 5.27.

Selection

Sometimes it is necessary, within a program, to decide whether to take one particular path or another, depending on a certain condition. If this condition or factor is 'true' then one path will be taken, and if 'not true' then the other path will be put into effect. In other words, depending upon conditions, the computer is able to decide which path to take through a program. If several selections are included the number of possible paths through a program increases.

Take the following English description of a selection: 'If condition X is true then do process B'. Figure 5.28 shows a flow chart for this selection construct.

Another form of selection can be represented by the following English statement: 'If condition X is true then do process B otherwise do process A'. Figure 5.29 shows the corresponding flow chart.

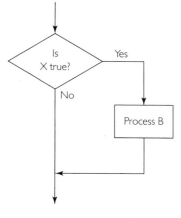

Figure 5.28 Flow chart section showing a selection construct

Figure 5.29 Flow chart section showing another selection construct

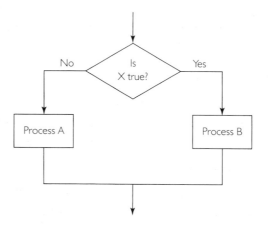

Iteration (repetition)

An **iteration** is a section of a program which is repeated a number of times. The following English statement represents a typical iteration statement: 'While condition X is true, repeatedly do process A'. Figure 5.30 shows the flow chart for this construct.

Iteration of a set of program instructions can take place in one of three ways:

1 a condition is tested for after the first run of the sequence
2 a condition is tested at the start of a sequence
3 an instruction is given for the sequence to be carried out a certain number of times.

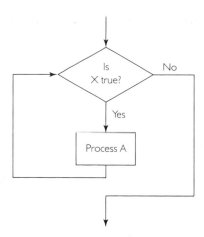

Figure 5.30 Flow chart section showing an iteration construct

Structured English

Structured English is used as a way of describing the processes involved in a system as a series of statements from which a programmer can write the program. It can also be used where diagrams are inappropriate to describe the results of systems analysis.

In some ways structured English looks a bit like the pseudocode that programmers often write on paper as a first step in program development. The main difference is that with structured English there are not as many control structures, although those used are very similar and often interchangeable.

Structured English uses a number of key words. These are: DO, REPEAT, UNTIL, IF, THEN, ELSE, SO. Structured English and **structure diagrams** are used together to outline the system's programming requirements.

Structure diagrams

Structure diagrams can be used to describe information systems. The overall task is taken and broken down into smaller, more manageable tasks, which can then be broken down further if necessary. This is called the top-down approach.

The top-down approach

Let us draw a structure diagram for a task with which we are all probably familiar: doing the weekly shopping. To begin we place the overall task at the top and describe it briefly in the box 'Do weekly shopping', as in Figure 5.31.

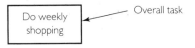

Figure 5.31 A simple structure diagram

The task is then divided into the subtasks necessary to complete the whole, for instance:

- prepare a shopping list
- do the shopping
- put the shopping away.

So we now have something like Figure 5.32.

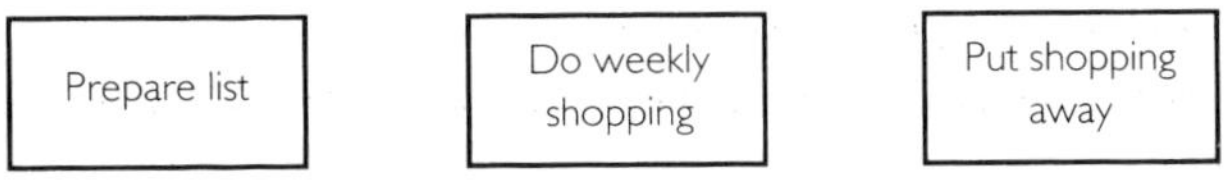

Prepare list	Do weekly shopping	Put shopping away

Figure 5.32

These sub-tasks may again be split up into the stages shown in Figure 5.33.

Figure 5.33

We can now put all the stages together to produce a final structure chart, as shown in Figure 5.34. Do not worry if your structure diagrams look different from this; two sets are rarely the same.

Figure 5.34 Final structure chart

As you can see, the purpose of a structure diagram is to show the tasks in more detail as you move down. The top box is the overall view; hence the term 'top-

Activity 12

a Pick one of the following tasks (preferably the one with which you are most familiar) and draw a structure diagram for it:

- renting a video from a video shop
- borrowing a book from your local library
- programming your video recorder to record a television programme
- getting ready to go for a night out
- washing the dishes.

b Draw a structure diagram for making a roast dinner with turkey, roast potatoes, carrots, sprouts and gravy. Figure 5.35 shows the top part of the diagram to start you off.

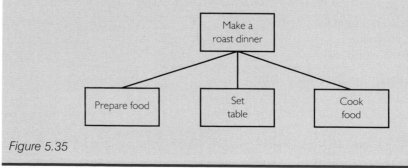

Figure 5.35

Activity 14

a To pass a course in computer studies at college, a student must satisfy the following conditions:

- the student must pass all the computing units
- the student must pass English and maths
- the student must have at least 80 per cent attendance (they must still satisfy this attendance rule even if they pass all the units).

The actions that can happen are:

- pass course
- repeat English or maths or both
- repeat computing modules failed
- fail course.

Unless the student passes the course, they will fail. Draw a decision table to show these rules.

b A mobile phone service provider will allow a person to be a subscriber to their network provided that the following conditions are met:

- they must have a bank or building society account
- they must have lived at the same address for at least one year
- they must be a home owner.

Draw a decision table to show these rules.

c A car insurer who specialises in older drivers will insure only people who are 50 or over, who have been accident-free for at least five years and who hold a clean driving licence. Draw a decision table to summarise these facts.

d An examinations board will satisfy an order for syllabuses and past examination papers, provided that:

- they are in stock
- the correct amount of money accompanies the order
- the correct amount of postage has been sent.

Draw a decision table which could be used to determine if a particular order should be satisfied.

Impossible rules

In some situations there is no point in writing down and considering all the rules because some may be impossible. Suppose the carriage to be paid when ordering a CD from a club is as follows:

- 1–3 CDs carriage is £2.50
- 4–6 CDs carriage is £3.00
- 7 or more CDs carriage is £4.00.

As there are three general conditions there are eight (2^3) rules, but if we look carefully we find that some of these rules are impossible. For instance YYY would be impossible because the number of CDs cannot be 1–3, 4–6 *and* 7 or more. We get impossible rules when the questions in the decision table are related to each other. To take account of these impossible rules, we leave them out of the decision table.

If the impossible rules in the above example are eliminated the decision table shown in Figure 5.43 is obtained.

	Rules		
	1	2	3
1–3 CDs ordered	Y	N	N
4–6 CDs ordered	N	Y	N
Over 7 CDs ordered	N	N	Y
Carriage of £2.50	X		
Carriage of £3.00		X	
Carriage of £4.00			X

Figure 5.43

Activity 15

A CD club offers members the following discounts as part of a pre-Christmas special promotion:

- an order for 3–6 CDs qualifies for a discount of 5%
- an order for more than 6 CDs qualifies for a discount of 10%.

The special offers are available only to those who have been members for 12 months or more. Orders for one or two CDs are acceptable but no discount will be given.

Draw a decision table that will enable a programmer to understand the rules immediately. Your final decision table should not contain any impossible rules.

Redundant rules

In some cases rules may be combined, which results in a much simpler-looking decision table. See how this works by considering the following example.

With a particular insurer, a person under 25 years old applying for car insurance must have a clean driving licence to be accepted. Their premiums will in any case be loaded (increased) to take account of their inexperience. Applicants of 25 and over who do not have a clean licence will also have loaded premiums. Because this insurance is available only to careful motorists, anyone who has had an accident in the last two years which was their fault will be refused insurance. Everyone else is accepted on normal premiums.

A decision table can be drawn up as shown in Figure 5.44.

If you look at rules 6 and 8 you can see that they differ only in the second row of condition entries. So applicants will be refused insurance if they have had an accident in the last two years regardless of whether they are under 25 or not or have a clean licence or not. We can therefore combine rules 6 and 8 to give a single rule and put a dash to represent Y or N. The dash means that it makes no difference if the answer to the condition is Y or N.

Rules

	1	2	3	4	5	6	7	8
Age ≥ 25 years	Y	Y	Y	N	N	N	Y	N
Clean licence?	Y	Y	N	Y	N	Y	N	N
Blame-free accident record for at least 2 years?	Y	N	Y	Y	Y	N	N	N
Normal premium	X							
Loaded premium			X	X				
Insurance refused		X			X	X	X	X

Figure 5.44

The decision table can now be altered to take account of these redundant rules, and the slightly trimmed down version is shown in Figure 5.45.

Rules

	1	2	3	4	5	6	7
Age ≥ 25 years	Y	Y	Y	N	N	N	Y
Clean licence?	Y	Y	N	Y	N	–	N
Blame-free accident record for at least 2 years?	Y	N	Y	Y	Y	N	N
Normal premium	X						
Loaded premium			X	X			
Insurance refused		X			X	X	X

Figure 5.45

When eliminating redundancy you look for action entries that are the same where the condition entries differ only in one respect. Examination of this decision table reveals that rules 2 and 7 can be combined, and when this is done the following simplification is obtained (Figure 5.46).

Rules

	1	2	3	4	5	6
Age ≥ 25 years	Y	Y	Y	N	N	N
Clean licence?	Y	–	N	Y	N	–
Blame-free accident record for at least 2 years?	Y	N	Y	Y	Y	N
Normal premium	X					
Loaded premium			X	X		
Insurance refused		X			X	X

Figure 5.46

Redundancy can be taken a stage further in this example. If you look at the decision table in Figure 5.46 you can see that the new rules 2 and 6 differ only in their first lines and their action entries are the same, so these too can be combined. The final version of the decision table, eliminating both impossible and redundant rules, is shown in Figure 5.47.

	Rules				
	1	2	3	4	5
Age ≥ 25 years	Y	–	Y	N	N
Clean licence?	Y	–	N	Y	N
Blame-free accident record for at least 2 years?	Y	N	Y	Y	Y
Normal premium	X				
Loaded premium			X	X	
Insurance refused		X			X

Figure 5.47

Activity 16

a A finance company will grant a loan to applicants who have a bank account or are married house owners. A partly completed decision table applying these rules is shown in Figure 5.48.

- Complete the decision table by filling in the action entries.
- Are any of the rules impossible?
- Some of the rules are redundant. Identify these rules and draw a revised decision table eliminating these rules.

b A CD club offers members the following discounts as part of a pre-Christmas special promotion:

- an order for 3–5 CDs qualifies for a discount of 5%
- for an order of 6 CDs and more, the discount is 10%
- this special offer applies only to members who have been members for 12 months or more.

  i Write down the conditions and actions.
  ii Calculate the number of rules and use this to help you draw the decision table.
  iii Remove any rules that are impossible from your decision table.
  iv Look carefully at the decision table you have just drawn and locate any redundant rules. Produce a final decision table with the redundant rules eliminated.

c To pass an Access course in computing, students must satisfy the following conditions:

- they must pass in a minimum of 21 computing modules
- they must pass all the modules in English and maths
- they must have at least 80% attendance throughout the course (unless they do they will fail even though they have satisfied the other conditions).

▶▶

The actions that can happen are:

- pass the course
- repeat some or all computing modules
- repeat English, maths or both
- fail the course.

Produce a decision table to show these rules and actions. Make sure that you have removed impossible rules and eliminated any redundancy.

	Rules							
	I	2	3	4	5	6	7	8
Bank account?	Y	Y	Y	N	Y	N	N	N
Married?	Y	Y	N	Y	N	Y	N	N
Home owner?	Y	N	Y	Y	N	N	Y	N
Loan granted								
Loan refused								

Figure 5.48

5.5 System specification

A system specification consists of the data and information resulting from the investigation and structured analysis of a system along with statements about the input and output needs of the system. The system specification is used by both systems analysts and programmers to enable them to design and produce a system that meets the user requirements.

The specification will usually include the following:

- a high-level DFD
- low-level DFDs
- an entity relationship diagram (with the many-to-many relationships removed)
- a data dictionary
- process specifications
- input specifications
- output specifications
- details of resource implications.

Output specification

The following items of documentation are included in the output specification:

- data required for output
- screen report layouts (if the output appears on the screen)
- methods of data output (what output devices are used to produce the output data)
- printed report layouts (any bills, invoices, picking lists, management reports, etc. are designed and included).

The production of an output specification is the first stage in the design process because it determines how the rest of the system must operate. To produce a certain type of output it is necessary to have inputs, then to perform a variety of processes on them. If the system does not produce the outputs required by the user, it is of no use.

You need to say what sort of output is required and whether it is hard copy (i.e. a printout) or on a screen. In many systems it will be both. For instance, a customer might ask a travel agent for the availability and price of a holiday. If it comes up on screen as being too expensive they may not be interested, but they might want the details printed out so they can take them away to consider. The output needs to be determined in advance because it is possible to output information only if the required data has been input and the correct series of processes applied to it.

Reports

A report is defined as the production of output from software (such as a database) for a specific purpose, such as a telephone list, a list of orders, an invoice or a statement of account. Reports can be divided into two types: screen reports and printed reports.

What is a report and what are the names of the two types of report?

This is the first stage in the design process and determines how the rest of the system must operate. To produce a certain type of output it is necessary to start with the inputs and then to perform a variety of processes on them.

The user requirements will have been investigated before this stage so the output can be designed using the requirements specification.

Data required for output

A consideration of the data to be output either on screen or in the form of a printed report is important because the data has to be either input or calculated from the input data. By determining the data required for output you can then think about how it is to be input or processed from the inputted data.

It is useful to present the data required for output in a grid chart like the one shown in Figure 5.49. Notice that the output reports often contain the same data so this is a convenient way of summarising it.

It is possible to work back from an output document through the processing to the input required. To do this you should take the following steps:

1 Determine which of the items on the output document are produced by calculation or logical deduction. All the other items will have to be input into the system.
2 Break down the input items into those requiring input every time and those which can be obtained from a stored file.

The above steps have been taken to produce the diagram in Figure 5.50, which shows the input and output for a sales invoicing system.

The output, which is the invoice sent to the customer, is worked out first with the items of data on the invoice being listed. You could use the existing invoice as a guide or collect invoices from other similar organisations to see what items of data they include.

You can then go through the list of items of data in the output box and decide whether they are obtained from previously stored files or need to be input. There

Output reports / Items of data required	Customer order	Packing list	Invoice	Stock record	Customer record
Customer no.	✓	✓	✓		✓
Customer name	✓	✓	✓		✓
Delivery code	✓	✓	✓		✓
Product no.		✓	✓	✓	
Product name	✓	✓	✓	✓	
Quantity ordered	✓	✓	✓		
Order date	✓	✓	✓		
Customer order no.	✓	✓	✓		
Our order no.		✓	✓		

Figure 5.49 Example of a grid chart

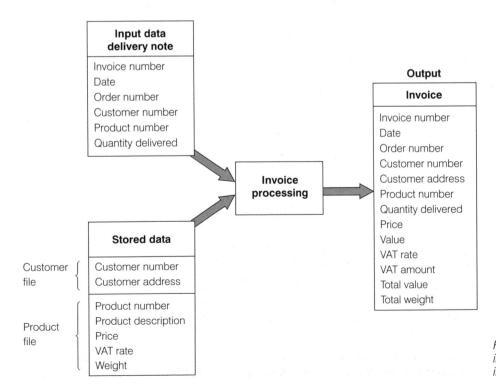

Figure 5.50 Designing the input and output for a sales invoicing system

335

will also be some items of data in the output, such as Total Value or Total Weight, which are not input or stored. This data is calculated from the input and stored data.

Notice that the data in the stored data is put into two groups according to the file it would be stored in.

Activity 17

a Obtain a utility bill (gas, electricity, telephone, water) and classify the data items on it according to whether they are calculated or input. For those items that are input, decide whether they need to be input each time or obtained from data already stored.
b Draw a diagram similar to the one in Figure 5.50 to summarise the above.

Methods of data output

There are many different methods of output, but by far the most common are output on paper (called hard copy) and output on screen. There are various questions that need to be answered when deciding about output.

- Are paper documents needed? If so, what quality is required? A document that is being sent out to customers may need to be of a much higher quality than one being used internally. Also, is multi-part stationery being used (e.g. where several copies such as white, green and yellow are produced)?
- What volume of output is required and how often? Each different form of output needs to be investigated to gain some idea of the capacity and speed of the printer needed. For instance, for a CD club, each member will be given an account at the end of each month, so the volume of this output would be the same as the number of members.
- Is the output subsequently used as an input document to be read using OCR (optical character recognition)? This is the case with the tear-off sections of bills, which are returned with the payment and are used as input media for the system. Such a document is called a turnaround document.
- What is the content of each form of output? Here we need to specify what information is displayed on the screen and what is printed out for each form of output.
- What is the format of the output? Here we would actually design the layout of any screens and forms used as output.
- What conditions prompt the production of output? Some reports are needed only now and again. For instance, a company may want to produce a list of customers who have overdue accounts at the end of every month.

It is always a good idea to show the user the designs of screens or documents because they will be the best people to spot any omissions and should be able to comment constructively on your proposed designs.

When designing all the different output formats you should bear in mind that all forms of output cost money and try to economise wherever possible.

/ **Activity 18** /

The operation of a video library can be summarised as follows:

- recording new videos ordered from the suppliers
- recording new members and any changes to existing members' details
- recording the borrowing of the videos
- writing letters to members who have not returned their videos.

a List the possible contents of a report for a shop manager on overdue videos.
b List the possible contents of a screen layout for adding new members' details to the members file.

Screen report layout

In many cases the information from a computer can be displayed as a screen report – provided that it is brief. Lengthy reports, containing large amounts of factual information, may be too difficult to absorb from the screen and might need printing out for the contents to be analysed and studied at length. Screen reports are ideally suited to on-line enquiries such as:

- Is a particular book in stock?
- When is the next train from A to B?
- How many articles of item X do we have in stock?
- What is the balance in my account?
- I would like to pay off my car loan. How much do I still owe?

All these basic, short requests can be answered quickly, and a brief on-screen report is usually sufficient. As more and more business is conducted over the telephone, screen reports have increased in importance. Direct insurance and mortgage companies are now very popular and details given over the phone can be confirmed later in writing, if required.

Printed report layout

Many reports are lengthy and need to be taken away and studied; this is easier if the data is on paper. Printed reports include:

- lists of goods on order
- lists of customers with overdue accounts
- lists of members of a video club who have not brought their videos back
- bank statements
- a list of students enrolled on a college course.

We still rely on reports printed as paper documents for external communication so they remain an important part of the output from any computer system.

The printed report will contain the attributes of one or more entities and when designing the report it will be necessary to have their names to hand. The data dictionary contains the attribute names, their size and data type and is a useful source of information when designing reports. The report layout is sketched by the systems analyst and then shown to the user for their comments. Figure 5.51 shows a typical sketch.

Date:						
Employee name	Dept	Hours today	Regular	Overtime	Weekly total	Year-to-date total
Full-time Employees						
Part-time Employees						
Total						

Figure 5.51 A sketch for the report 'Daily payroll register for gross earnings'

Here are some hints to help you when you are designing reports.

- Ensure that similar documents are designed consistently. For example, an invoice and an order form should have a similar look to them.
- Attribute names should appear only once in a document. In an order where several different products are ordered, the product number, product description, quantity, price, etc. should appear once as column headings and then the values placed in rows.
- Key fields are used most often and therefore should be in a prominent place – usually at the top right-hand side of a printed document or at the top left-hand side if the other fields form a list.
- Try not to mix too many fonts and font sizes.
- Use case (i.e. upper and lower) carefully. Avoid the use of capitals except perhaps where a key field is used.
- Abbreviations and codes are acceptable but are better used only on documents that are used internally.
- Use features such as bold, reverse, etc. to highlight certain important items.

Give three examples of information best presented as a screen report and three examples that are best presented as a printed report.

Activity 19

a Collect three invoices or photocopies of invoices and compare their design features. Is there one which works better than the others? What is it about it which makes it so? Produce a report describing the features used in the invoices.

b Here are the fields used in a sales invoice.

- Invoice Number
- Date
- Order Number
- Customer Number
- Customer Address
- Product Number
- Product Description
- Quantity Delivered
- Price
- Value (NB this is obtained by multiplying the Quantity Delivered by the Price)
- VAT Rate
- VAT Amount
- Total Value
- Total Weight

i Why is Total Weight included in this invoice?

ii Produce a suitable design for this invoice on paper. Do not just do the one – produce several draft versions and then choose the best.

Input specification

An **input specification** details how the data is to be input and the measures taken to ensure that only valid data is accepted for processing. The following documentation would normally be included in the input specification:

- sources of the data used for input
- methods of data capture
- validation methods used
- data input forms or screen layouts
- verification methods used.

Sources of the data used for input

This gives information about the origin of the data and the transcription method used. Transcription is the process of transferring information either from a document (such as an order form) or from a conversation directly onto disk by keying in the details. Such a method is tedious, but more and more transactions are now being done over the phone and the keyboard operator types in the details directly while talking to the client. The rise in direct insurance and mortgage companies shows there are savings to be made in cutting out as much paperwork as possible, and some of these savings are passed on to the customer. Internet-based businesses rely on the customer supplying the data straight into their computer system. Getting the customer to type in their own details (name, address, items ordered, credit card details, etc.) reduces the administration costs, and sometimes these savings are passed on to the customer.

Data sources may be internal or external. For example, a new customer will supply the external customer data but once it has been entered into the system it becomes an internal source of data.

Methods of data capture

Data capture involves getting the data into the internal system. Usually this means entering information contained on documents or given verbally into the system using the keyboard. The trouble with this system is that it is slow and very prone to errors. In other methods of data capture the data is read automatically from special documents, which eliminates the need for keying in.

For direct methods of data capture such as OCR or OMR data tends to be input in large batches. However, there are disadvantages with these methods, the main one being that the rejection rate can be quite high – so there has to be another system in place to cope with these rejections.

One of the main purposes of input design is to decide how to get the inputs into a system as accurately and quickly as possible. Traditionally, most people use keyboard entry as the method of data capture, but for large amounts of data this can be very costly so alternative methods such as OCR, OMR and barcoding should be considered. Sometimes it may be possible to use a combination of methods. For instance, a student enrolment form could have the course code, fee and number of hours coded in a barcode but other information, supplied by the student, would need to be typed on to the form.

Many Internet-based companies get the customer to do their data capture for them – after all, the customer will take their time entering the details and there will be fewer errors since these are easily spotted. The biggest benefit is not having to pay anyone to enter the data into the system!

List as many methods of data capture as you can think of.

Once the method of data capture has been chosen, you can start designing the input media. The data needed depends on the output specified above, so you must make sure that all the data needed to produce the output has been entered. We can get this information from the data flow diagrams, entity charts or any of the other diagrams we used while designing the system. We could use the context diagram to get information about the functional areas and then use the level 1 DFD to describe the input in more detail.

Validation methods used

Although it is impossible to produce checks that will pick up every mistake, it is possible to reduce errors to a minimum. If keyboard entry is used as the data capture method transcription errors are certain to occur, so alternative methods should be used wherever possible. If keyboard entry is the only feasible method, validation checks should be made on all input data.

There are various validation checks that can be performed on data once it has been entered, depending on the database software used. If you are going to write the program code using a programming language, you will need to ensure that rigorous validation checks are used to prevent the processing of inaccurate data. Such validation checks would, for instance, pick up a character being entered rather than a number.

/ *Activity 20* /

Research the range of validation checks available to a programmer or database designer and produce a brief report outlining your findings.

You should state the types of data on which it is possible to perform a validation check and what types of data cannot be checked.

Data input forms or screen layout

A screen with fields for the user to type into directly is used in many applications. Data is then obtained either from source documents or by direct communication with the customers – in person or over the phone.

Verification methods used

A system is needed to ensure that no errors are introduced during data entry. In most cases this means that the keyboard operator will proofread the data entered against the source document. Sometimes, as with business conducted over the telephone, the keyboard operator will read the data back to the customer to confirm its accuracy.

Data can also be verified by having two operators key information from the same source documents. In this case only if both sets of information are the same will it be accepted for processing. This used to be the only method of verification but the cost implications of using double the number of staff for input tasks means that it is now seldom used.

Producing a process specification

The specification must include all the clerical and computer procedures necessary to produce the output. It is important not to neglect the manual (not computer-based) systems that need to be put in place at the same time as the computer system.

Once the data has been entered it will be subject to various processes, and the DFD can be used to identify the processes that need to be described in more detail. Processes should be described in enough detail that no further analysis is needed and a programmer could, if necessary, write suitable program code to process the data.

The **process specification** consists of a series of techniques that are used to define each process in the DFD. Even processes that are not clearly indicated on the diagram must still be defined with a process definition. Any decisions that need to be made during the process should be identified and a decision table drawn up to summarise them. An overall structure diagram is usually produced, together with structured English descriptions of each subsection.

The process specification for your project should be accompanied by the following documentation:

* structured English description
* structure diagrams
* decision tables
* flow charts.

5.6 Creating a conclusion

This covers the creation of the two main reports:

- the feasibility report
- a system specification.

As with all reports, there should be a conclusion, where the writer gives their opinion on the content of the reports.

Resources

Resources are the tools, equipment and suitably qualified people needed to do a job. If a job is to be completed quickly and efficiently, then all the resources must be available. The types of resource can be divided into three areas.

- *Hardware resources* – these include computers and their peripherals such as printers, scanners or modems, and any other additional equipment such as communication lines.
- *Software resources* – programming languages, operating systems, applications packages and any tailor-made software.
- *Personnel resources* – these are important because, although there might be enough people to perform a job, they may not have the necessary expertise. For instance, if orders are to be processed quickly using keyboard entry, then it will be essential that the people performing this task can type quickly and accurately.

The resources needed to analyse, design and implement a system are hard to estimate at first and it is not until the project is well under way that the resource implications can be assessed accurately. Once the process specification has been produced it is possible to gain some idea of the amount of programming code and the number of hours needed. The use of applications packages with applications generator features will reduce both these considerably, but the final system will be inferior to bespoke software.

Specification

In the conclusion, the following will need to be specified clearly:

- possible alternatives
- resource requirements
- possible risks of the change
- potential benefits of the change
- cost–benefit analysis
- recommendations.

Possible alternatives

Possible alternatives will look at the available options. Usually there are several different ways in which a particular problem can be solved and these ways should be mentioned. Usually they will differ in complexity – and hence cost.

Resource requirements

Resource requirements will discuss what resources will be needed to complete the project and when these resources will be needed. For example, many systems are developed on 'test' computers that are not used operationally for the day-to-day running of the organisation. The project team would need to make sure that this test computer was not being used for other system development at the same time.

Possible risks of the change

Possible risks of the change will look at what could go wrong and what the likely consequences would be. In many systems things have gone wrong during the development and implementation, which has cost the companies involved a lot of money. One such large company developed a new order-processing system which caused so many problems that the shops it supplied had to wait 10 weeks for deliveries of stock. Many of the customers who ordered items from the shops got fed up with waiting and cancelled the orders. This caused huge losses for the manufacturer and nearly resulted in them going bankrupt.

Risks can include the costs spiralling out of control and a project costing many times its initial estimate. Some projects have to be abandoned even though a lot of work has been put into them and no benefit is gained from all the work and money put in.

Another risk is that, with the rapid pace of developments in IT, you could end up replacing old equipment with equally out-of-date equipment.

Potential benefits of the change

Possible benefits of the change should be summarised from the section that was part of the feasibility report.

Cost–benefit analysis

A cost–benefit analysis may also be included here as a summary. Take it from the same section in the feasibility report.

Recommendations

Recommendations need to made by the system developer. Obviously, if you have put in a lot of hard work you will want to recommend your system and go ahead with the development. In this section you need to give your opinions in favour of development of the system.

Possible constraints

If a particular resource is unavailable, it is called a constraint on the system.

Case study

Tutorial Services
John, who was the Head of Mathematics at a large comprehensive school, decided to take early retirement and start a tutorial agency matching private tutors to students. His business has been a great success and he has had to develop an administrative system to cope with the increase in paperwork.

At the moment he advertises in all the local papers around Merseyside, and these adverts produce about 50 enquiries per week, although there is a

You can use this opportunity to provide evidence for the following key skills: Communication C2.3 or C3.3.

peak in the number of enquiries between Christmas and the exams in June. The enquiries are answered by a team of advisors (usually people who are at home all day). These advisors collect information about the student – name, address, telephone number, age, subject, etc. – and fill in a form which they send to head office. At head office another form is filled in, photocopied and sent to any tutors able to teach the student (usually the ones chosen are in the student's area). The tutors then contact the student with information on their rates and availability. If the student and tutor are able to make suitable arrangements another form is sent back to the agency, informing them that the student has been taken on.

As well as advertising for students, the agency advertises for suitably qualified and experienced tutors. This is done in the same advert and the advisors fill in a form and record brief details for applicant tutors. This form is also passed to head office, which sends a more detailed application form for the potential tutor to fill in, along with a covering letter. John looks at each tutor's application form and decides whether to place them on the file of available tutors. If an application is successful, the new tutor's form is transferred to a card file in a filing cabinet called the 'main tutor file'. Each tutor is given a unique tutor number and tutor details are stored in number order. In addition to this file, there are sections in another filing cabinet which contain the numbers of tutors who are able to teach a certain level and subject. The main headings for these files are:

- infants
- junior
- lower secondary level
- GCSE
- A-level, BTEC, Advanced GNVQ
- degree and professional qualifications.

These files are further divided by subject, into which the tutor numbers are written. When a tutor is needed for a particular level such as GCSE, the file is located, the subject is found and a list of tutor numbers retrieved, which can be looked up in the tutor file to obtain personal details.

This approach is giving rise to a variety of problems. If someone from a certain area rings up and wants a GCSE maths tutor, head office needs to look though the GCSE file and locate the list under the maths section. Staff must then look through all the addresses and, with the aid of a map, produce a list of suitable tutors, with their addresses and phone numbers, to send to the enquirer. The whole process is time-consuming, especially now that the files have become so large.

Once payment has been agreed between tutor and student (or the student's parents), a form is filled in and sent back to the agency informing it of the arrangements made. Each month the tutor fills in a form which gives the personal number and name of the student and the number of hours of tuition provided. The agency charges 10% of the fee negotiated with the tutor, and this is returned as a cheque to the agency. This does rely on the honesty of the tutors but the system is fairly simple and it is easy to check by asking the students or their parents about the tuition completed during the month.

This is the system used at present, as described by the owner of the business. The owner is prepared to change the existing system if it will help run the business more efficiently. He has a limited knowledge of computing but has agreed to take on a full-time member of staff with some experience of setting up databases, etc. In addition to this he has set aside £5000 for the purchase of equipment, specialist training and any other goods or services needed.

1 *The case study might not reveal all the facts about the business. Are there any other questions you would like to ask about the business before continuing? Write a list and discuss them with your tutor.*

2 *Draw a diagram to illustrate how the system works. Make up your own. It need not conform to any standards so you are free to use your artistic talents.*

3 *Produce a feasibility report for this system, making sure that you have included all of the following in this report:*

- *a statement of the purpose of the system*
- *a definition of the system scope*
- *a list of deficiencies of the current system*
- *a statement of the user requirements*
- *an outline of the costs and benefits of development*
- *a conclusion and recommendations.*

4 *To encourage established and experienced tutors to stay with the agency rather than advertising for students on their own, the following arrangement has been proposed:*

- *If the tutors have been with the agency for three years or more or they have achieved a total of 400 or more tutorial hours since they started, they will pay the agency only a 5% fee.*
- *If the tutor teaches professional courses only, the fee will drop down to 5% after 300 hours. The time period proviso is, however, the same.*

Draw a decision table to illustrate these rules, making sure that both impossible rules and redundancy are eliminated.

5 *Produce a refined data model in first normal form (i.e. with any many-to-many relationships removed) for the proposed new system.*

6 *Produce a system specification for the system outlined in the case study. Normally the user will let you know what is required, but in this case he has left it entirely up to you. Make sure that all parts of the system specification are included. Here is a check list to help you.*

- *input specification*
- *output specification*
- *process specification*
- *resources*
- *constraints.*

Activity 21

The owner of a small shop, Harry Jones, decides to automate some of the processes that are at present performed manually and has asked you for your advice. Harry has taken an introductory course in the Basic programming language. He suggests that you could give him a hand to write a program in Basic which could sort out a number of problems.

a You tell Harry that Basic is just one of a range of languages that are available and each one has its advantages and disadvantages. Research the range of programming languages available and produce a brief description of each one.

b After further discussion with you, Harry comes round to the idea that designing and writing software from scratch is both complicated and time consuming. He has heard about applications packages and how they can be adapted for different types of businesses. Write brief comments on this.

c Harry has agreed to look at computerising the present manual system, especially for recording purchases as they are made and for stock control, although he is willing to look at other areas of the business as well.

To be able to help him do this you will need some information from him about the purpose of the system:

- a description of the present system (inputs, processes and outputs)
- expectations for the new system
- constraints (timescale, costs, etc.).

You send Harry this list, expecting him to supply you with a list of answers, but instead he sends the following note:

> *Dear Student*
>
> *You're going to have to help me with this. I don't understand what the above list means. Maybe I'm being stupid but could you please spell out in language that an idiot like me can understand what each of the items in your list means.*
>
> *If you could give me some ideas of the answers, it would help but as things are I'm completely lost.*
>
> *Regards*
>
> *Harry Jones*

Produce a more detailed explanation of the information you supplied which will enable Harry to supply you with the required information.

d Harry is enthusiastic about the whole project and suggests that you both go to a computer superstore and buy a computer and any software he might need. You are against this idea because you have no idea of the type of system that will be needed without further careful analysis and design. You tell him that this is the wrong way to go about computerisation of the business and suggest that you tell him how to go about it in the right way. Someone suggested to Harry that he gets a book out of the library on systems analysis but he says that he finds it difficult to understand all the computer jargon and would prefer to be told about it by someone who knows what they are talking about.

Produce a word-processed document which explains the stages involved in systems analysis, making it as easy to understand as possible. Any technical words, abbreviations, etc. should be fully explained – and remember to make your explanation relevant to Harry's type of business (i.e. the running of a small shop).

Activity 22

The following tasks are to be performed in groups/pairs. Your tutor will let you choose who you want to team up with and you will need to tell them the names in each group.

Each group should contact a business or organisation in your area to arrange an interview to obtain information about the type of organisation, what it does and what data-processing requirements it has. The organisation chosen should be fairly small, or a department of a larger organisation, and should either be using computers already or be capable of using them in the future if they had the expertise. In some ways it may be easier to find a business which uses a manual system at the moment but may not need a computerised one. If you have no luck in finding such a business, your tutor may be able to create a fictitious company – in which case your tutor will be your contact for the business.

When contacting businesses you might have more success if you make use of friends or relatives to arrange an introduction. You could, as an alternative, look at a business where you have worked in the past or where you now work part time. If the business is part of a large organisation you will need to investigate just a small part of it, perhaps just one department.

The whole group should perhaps attend the interview; one can take notes while the other/others ask the questions.

a Design a questionnaire to be used to find out about the type of business (or the area of business) you are looking at. Remember that you need to determine not just the way computers are, or could be, used but also something of the nature of the organisation: what it does, how many people it employs, etc.

b From the information you gleaned identify either a new system or an alteration to the existing system, which can be made to form part of an ongoing project that will be assessed in the form of assignments. Your system should have a relational database at its centre, so identify a problem (or problems) that can be solved using a database. Make sure that you keep your tutor informed about your choice of project: if it is too ambitious, your tutor may recommend that you investigate and produce a system specification for only part of the system. You should be realistic in your choice of project and what you will be able to achieve in the time available.

For this task you will need to:

- investigate the current or proposed system and collate the materials collected during the fact finding
- produce high-level and low-level DFDs to describe the events in the existing or proposed system
- produce a feasibility report.

c   Now that you have identified a particular problem which can be solved using a relational database solution, you need to extend your project to the production of a system specification. Prepare a system specification for the new system which includes the following:

- process specification
- an input specification
- an output specification
- the resources implications of the new system.

5.7 Standard ways of working

As part of Unit 5 you will have to show that you have used standard ways of working. The whole point of having a standard way of working is to help you to manage your work more effectively. You can save a lot of time and effort if you think logically about how you intend working. All of the standard ways of working are looked at in detail in Unit 1. Here is a list of the ways of working you should apply to this unit.

- Plan your work to produce what is required to given deadlines.
- Edit and save work regularly, using appropriate names that remind you of their contents.
- Store your work where others can find it easily in the directory/folder structure.
- Keep dated backup copies of your work on another disk and in another location.
- Keep a log of the ICT problems and how they were resolved.
- Make sure that confidentiality (i.e. Data Protection Act 1998) and copyright laws are obeyed.
- Avoid bad posture, physical stress, eye strain and hazards from workplace layout.

Key terms

After reading this unit you should be able to understand the following words and phrases. If you do not, go back through the unit and find out, or look them up in the glossary.

Attributes
Constraint
Context diagram
Data capture
Data dictionary
Data flow diagrams (DFDs)
Decision table
Entities
Entity relationship diagram (ERD)
Entity relationship model (ERM)
Fact finding
Feasibility report
Feasibility study
Foreign key
Input specification
Iteration
Key
Logical design
Many-to-many relationship

One-to-many relationship
One-to-one relationship
Physical design
Primary key
Printed report layout
Process specification
Relationships
Screen report layout
Selection
Sequence
Structure diagram
Structured English
System design
Systems development life cycle
System specification
Systems analysis
Validation
Verification

Review questions

1 It is not always possible to develop the ideal system because there are always constraints imposed on a system.
   a Give the names of three constraints on a system.
   b Briefly describe how each of the constraints named in part (a) might affect the design of a new system.

2 Optical mark reading (OMR) was chosen as the method of data capture for the National Lottery. This is a direct method of data capture.
   a Explain what is meant by the term 'a direct method of data capture'.
   b What are the main advantages in choosing OMR as the method of data capture for the National Lottery?

3 As part of systems analysis and design, a feasibility study is often undertaken. Briefly explain the purpose of a feasibility study and what steps such a study should consist of.

4 Your lecturer has said to you that 'many-to-many relationships' between two entities should be avoided. In a college enrolment system you have identified two entities: STUDENT and COURSE. It is possible for the one student to be enrolled on more than one course and the one course can contain many students. The relationship between the two entities is a many-to-many relationship.
   Explain how to resolve the above problem and draw an entity diagram showing your new diagram. Also explain how you have solved this problem.

5 A data dictionary is an important piece of systems documentation and should always be included. Explain what is likely to be included in the data dictionary.

6   A systems analyst has been brought in to investigate the current manual system used by a video library. The systems analyst has no detailed knowledge of how a video library works. Describe the methods they could use to find out about the existing system.

7   Data modelling is a useful technique used during systems analysis. What is data modelling, and in what ways does it help the analyst?

8   What are the main differences between data verification and data validation? Explain three different types of validation check that may be performed on data?

9   Data flow models must be made valid by the removal of many-to-many relationships. What is meant by this statement?

10 Explain the difference between a high-level data flow diagram and a low-level data flow diagram.

Assignment 5.1

This assignment will be developed further in Unit 6, where the analysis will be used for the design and eventual implementation of a database.

For this assignment you will need to identify a problem that is to be solved using a database. Your lecturer/teacher will either let you choose your own problem or identify one for you. Unit 5 precedes the unit on database design (Unit 6) and the problem used for this unit can also be used for Unit 6.

For the assessment evidence for this unit, you must produce:

- a feasibility report
- a system specification to meet the requirements.

As part of the system specification, you will need to show evidence of data modelling with an entity-relationship diagram (ERD) that has at least three related entities.

1   Produce clear statements of the purpose of the system and the user requirements for the system. The user requirements will include a definition of the scope of the system and high level and low level data flow diagrams with views at appropriate levels to help explain the system. You must also produce an entity relationship diagram (ERD) and a data dictionary that clearly lists the entities, their attributes and relationships. Accurate input and output specifications need to be produced along with details of the resource implications.

2   Produce suitable process specifications using one of the appropriate methods.

3   Provide a conclusion that makes recommendations for development of the new system.

4   Show a good understanding and effective use of structured analysis tools in the development of your data flow diagrams, the identification of events and the production of process specifications.

5. Show a good understanding and method in the development of your entity relationship diagram and data dictionary to resolve problems and ensure that first normal form is reached.

6. Show that you can work independently to produce your work to agreed deadlines.

7. Show a systematic approach to your analysis of the existing system, investigation of potential improvements and selection of priorities for development.

8. You must define clearly in your input specification the following:

   • appropriate sources of data
   • methods of data capture
   • layout of screen input forms
   • validation and verification techniques.

9. You must define clearly in your output specification the information to be output in screen or printed reports and appropriate ways of organising and presenting it.

10. You should specify clearly in your conclusion the possible alternatives, constraints, risks and potential benefits, and include a cost-benefit analysis to support your recommendations.

Get the grade

To achieve a grade E you must complete tasks 1–3.

To achieve a grade C you must complete tasks 1–6.

To achieve a grade A you must complete tasks 1–10.

Key Skills) *Opportunity*

	You can use this opportunity to provide evidence for the Key Skills listed opposite.
Communication C2.3	Write two different types of document about straightforward subjects. One piece of writing should be an extended document and include at least one image.
C3.3	Write two different types of document about complex subjects. One piece of writing should be an extended document and include at least one image.

Database design

What is covered in this unit:

6.1 Database concepts
6.2 Logical data modelling
6.3 Normalisation
6.4 Relational database structures
6.5 Relational database construction
6.6 Documentation
6.7 Standard ways of working

In this unit we will look at the concept of an organised collection of data and how a structured collection of data can be stored and subsequently processed by a computerised database. We also look at how record-structured databases are used in organisations. This unit will expand on logical data modelling, which was developed in the previous unit. You will then learn how to refine the logical data model until it reflects the data in reality and then use this model to create the framework for a relational database.

Materials you will need to complete this unit:

- computer hardware including a printer
- Microsoft Access (preferably Access 2000 or Office 2000). There is a difference between Access 97 or Access 95 and the new version Access 2000, and many of the screens look different. However, the packages still work in a similar way.

In this unit we will be looking at how a system can be designed and built using a database. Unit 6 follows on from the systems analysis covered in Unit 5, as it is necessary to perform systems analysis before starting on the design of a database.

6.1 Database concepts

In everyday use, we use the words data and information to mean the same thing but in computer usage these words have subtly different meanings.

Data is the raw facts and figures at the collecting stage before they are processed. In many cases, data is meaningless until it is processed in some way (having a calculation performed on it, sorting it or even grouping it into a certain order). The task of performing such operations on the data is referred to as **data processing**. Data processing results in **information**, and in some cases this information may be processed further in an information-processing system to yield higher quality and often more valuable information.

Figure 6.1 How data is
turned into information
by processing

Unlike raw data, information has a context, its meaning is derived from that context. In the sentence 'Mr A. Hughes has a credit limit of £800 on his credit card' 'Mr A. Hughes' and '£800' are both items of data but on their own they have little meaning – it is only when put with other words in a sentence that the data items are given a context to provide meaningful information. Putting data into a structure such as a sentence turns it into information.

When data is entered into a database, the structure imposed by the database turns it into information. For example, if we enter the data 'Mr A. Hughes' into the customer **field** of a credit limit database, we now know that Mr A. Hughes is a customer who is given a credit limit.

Later in this unit we will look at the creation of database structures. It is always tempting to try to create databases without giving enough thought to their development and expansion. Although a simple database (such as one for names and addresses) could be developed in this way, a larger application created without careful planning and design may not be able to do what is eventually required, and much time and money can be wasted in trying to get it to work.

quick fire

Explain the subtle
difference between the
terms 'data' and
'information'.

Activity 1

To be able to understand data models and databases it is important to understand the information needs of organisations. Here are a few simple systems that you can investigate. The general areas are in capitals.

*   HEALTH – General Practitioner (Doctors, Patients, Appointments)
*   EMPLOYMENT – Personnel Department (Personal details, Pay, Departments)
*   AGENCIES – Travel (Clients, Services, Reservations)
*   SALE OF GOODS – (Orders, Goods, Invoices)
*   LIBRARIES – (Books, Members, Loans)
*   POLICE – (Offenders, Crimes, Officers, Victims)

The AGENCIES system need not be a travel agency. It could be theatre booking, model, cleaning, house sitting, baby sitting, nursing, nanny, private tutoring, talent, etc.

The LIBRARIES systems need not be for books: they could be for videos, DVDs, computer games, music CDs, etc.

a   Find out for each one in the above list what data would be held in each of the entities in brackets.
b   Produce a list of attributes for each of the entities listed for each system.

This unit begins looking at database design by building a **data model**. Later we look at how this data model can be converted into a valid data model on which a database structure can be based.

The following sections will demonstrate how to produce a data model to meet a given specification, refine the data model using **normalisation** procedures and produce **relational database** structures to suit the data model.

Creating a data model to meet the specification

Before a database can be designed and built, a clear idea is required of what it needs to do. This understanding is obtained by careful systems analysis and culminates in the system specification, consisting of the data and information needed to allow that system to be analysed and for the design and development of a database system to meet user requirements. The specification can include some or all of the following: a data-definition model, a process specification, an input specification, an output specification and resource requirements.

6.2 Logical data modelling

A **logical data model** is used to give an insight into the way data is used within an organisation and, like all models, it can be progressively refined in the light of further investigation until it represents as nearly as possible the real system.

The purpose of a data model is to identify and present user requirements in a way easily understood by users and computer professionals. It is also important that the model can be easily converted to a technical implementation and provide rules and criteria for implementing the system. A data model may consist of some or all of the following: high-level information flow diagrams, data flow diagrams, **data dictionary** and **entity relationship diagrams**.

Entity relationship diagrams

We looked at entity relationship diagrams in Unit 5, and it would be advisable for you to re-read this section (pages 311–318) to refresh your memory.

Entity relationship modelling is a technique for defining the information needs of an organisation to provide a firm foundation on which an appropriate system can be built. Putting it simply, entity relationship modelling identifies the most important factors in the organisation being looked at (the entities). The properties which these factors possess (called attributes) and how they are related to one another (their relationships) are also studied. Entity models are logical, which means that they do not depend on the method of implementation. For example, if two departments in an organisation perform identical tasks but in different ways their entity models would be identical because they would be using the same entities and relationships – but their data flow diagrams would probably be different because their information flows may well differ.

There are two main objectives when producing an entity relationship model. The first is to provide an accurate representation of the information needs of the organisation; the second is to provide a model that looks at the data independently of the method of storage or access to the data.

Entity relationship modelling, although important, is only an intermediate step towards the successful implementation of a system. However, a carefully thought-out system can be developed faster and will work better, so the extra time is well spent.

At the core of entity relationship modelling is the entity relationship diagram. We looked at these in Unit 5.

Activity 2

In your own words, define the following terms:

- entity
- attribute
- relationship
- primary key
- foreign key.

If you get stuck you will find some definitions in Unit 5.

Example: A database for the police

A database is to be set up to hold details about the crimes, offenders and officers to be used by a police force. Before setting up the database, an initial investigation was conducted and some analysis performed.

The initial list of entities was:

- OFFICER
- CRIME
- OFFENDER
- VICTIM

The attributes for these entities were as follows:

OFFICER	*OFFENDER*	*CRIME*	*VICTIM*
<u>Officer number</u>	<u>Offender number</u>	<u>Crime number</u>	<u>Victim number</u>
Title	Title	Crime code	Title
Surname	Surname	Officer number	Surname
Forename	Street	Date/time	Forename
Rank	Town	Location	Street
Department	Date of birth	Postcode	Town
Sex		Offender number	Postcode

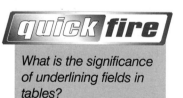

What is the significance of underlining fields in tables?

The **primary key** attributes are underlined.

We can now start to develop this logical data model.

Looking at the relationship between VICTIM and CRIME we see that one VICTIM can have more than one CRIME done against them and that one CRIME could have more than one VICTIM. This means that the relationship between the entities VICTIM and CRIME is many-to-many (Figure 6.2).

Figure 6.2 The many-to-many relationship between the entities VICTIM and CRIME

Many-to-many relations need to be removed by creating a new entity containing some of the attributes in the existing entities.

A new entity called VICTIM HISTORY containing the attributes Crime number and Victim number is created. A particular crime can appear more than once in the VICTIM HISTORY entity because there can be more than one victim of the same crime but a Crime number in the VICTIM HISTORY entity can appear only once in the CRIME entity. Hence the relationship between CRIME and VICTIM HISTORY is one-to-many. As the same victim could appear several times in VICTIM HISTORY entity if they were the victim on more than one crime, the relationship between VICTIM and VICTIM HISTORY is one-to-many. The new entity relationship diagram can now be drawn (Figure 6.3).

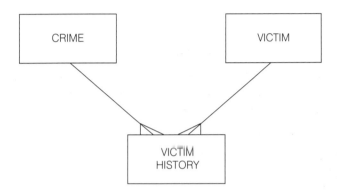

Figure 6.3 The creation of an intersection entity called VICTIM HISTORY breaks the many-to-many relationship with the creation of two one-to-many relationships

It was found during the initial investigation that only the senior investigating officer has their officer number recorded in the CRIME entity. This means that a particular officer investigates more than one crime but a particular crime is dealt with by only one senior officer. Hence the relationship between the entities OFFICER and CRIME is one-to-many.

On investigating the relationship between OFFENDER and CRIME is was found that an OFFENDER can commit more than one crime but (to make this simple) a CRIME can be committed by only one OFFENDER.

On investigating the other entities, the entity relationship model shown in Figure 6.4 was constructed.

The VICTIM HISTORY entity will contain the following attributes:

- Victim number
- Crime number

Now suppose there are two crimes, crime numbers 134 and 428, that have been committed against the same victim (victim number 298). This data can be summarised in the table to the right, which also shows some other data.

Victim number	Crime number
298	134
300	200
298	428
299	134

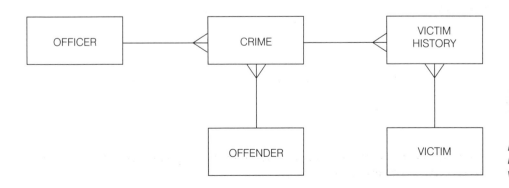

Figure 6.4 The entity relationship model for the whole system

Notice that you can have the same crime number with different victims and the same victim number with different crime numbers. This means that neither the victim number nor the crime number are unique on their own but when used together as a pair they *are* unique. Hence we have to use two attributes together as a composite primary key. We underline the two primary keys: <u>Victim number</u>, <u>Crime number</u>.

Like all models, entity relationship models can be refined before a final version is obtained that reflects the true position.

Activity 3

There are still a number of problems with the model we have here. For example, we have an offender for a crime, but this assumes that the person is guilty. We really need to distinguish between those people who have been found guilty in a court of law of the crime and those people who are merely suspects. We therefore need another entity which we shall call SUSPECT.

a Develop this model further by including the new entity SUSPECT and any other entities you feel are necessary to give a more realistic picture of the data.

b Briefly explain what you have done and also list the entities along with their associated attributes. Underline any primary keys.

c Finally, produce a final entity relationship diagram showing the entities and their relationships to each other. Use the primary and foreign keys to determine which of the tables should have relationships between them.

Relational databases

Once a logical data model has been built and refined the next step is to think about the physical design of the system. If the problem is to be solved using a database then the next step is to turn the entities into tables and the attributes of the entities into fields in each table.

The basic component of any relational database is the table, which is a collection of information arranged in rows and columns. Relational databases can consist of many tables containing related data. The field names are shown at the top of the columns and a record, which is more correctly called a **tuple**, is shown as a row. Figure 6.5 shows a table used to store the data about the vehicles in a car hire firm.

What is a tuple?

If you look at the table, the first column contains the primary key, which must be unique. The registration number can be used as the primary key in this table, but in other tables it might be necessary to create a unique reference number if there isn't one already. It is important to note that each row in the table must be unique so that there cannot be any duplicate attributes for a given value of the primary key. To ensure that this does not happen, the database designer/analyst will need to be careful.

In a video library, for instance, we might assign each video a unique number and list its details (title, category and hire price) in one table. Another table could list the library members, who are also each given a unique reference number along with information such as name, address and telephone number. A further table, for

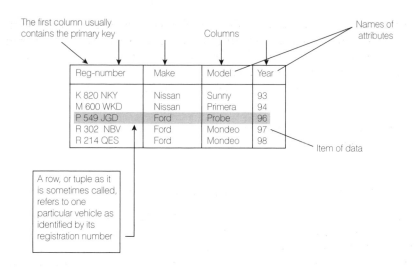

Figure 6.5   A table used to store data about the vehicles owned by a car rental firm

rentals, can be used to record which member has which video, using the appropriate customer and video numbers. The rental table would also record the date when a video is borrowed so that the library knows when its return is overdue.

It might be thought that all the information can be most conveniently kept in a single table, but there are problems with this. First, only a few members may have a video at any one time, so much of the part of the table devoted to video details would be blank. A second problem is that the video library has no limits to the number of videos a particular member can borrow at once, so many sections would have to be provided for each member, although most of the time this space would not be needed. There is, in addition, one very serious problem. When a new video arrives from the supplier it must be given a unique code number and its details recorded on the database but, because members, videos and rentals are recorded in the same file, we cannot add a video to the table without a member borrowing it. For these (and similar) reasons relational databases use several separate tables for information storage.

Although tables are the basic building blocks of relational databases other features – such as **queries**, **reports** and **forms** – are also stored as part of the database. These tools allow the user to view and manipulate the data in a variety of different ways.

Once the data model has been created and refined using the normalisation process (explained later in this unit) we have a series of entities, each with its associated attributes. We can now turn our attention to the design and structure of the relational database. When talking about database structures we use the term 'table' in place of 'entity' and 'field name' instead of 'attribute'.

With a relational database management system, data from different subject areas is first placed in tables and then relationships which link the tables are constructed. Creating separate tables rather than combining all the information in a single table avoids duplication of data – saving storage space and increasing the speed of the machine when accessing data.

Database relationships

In the same way as attributes can have relationships between them, the tables (which now represent the entities) can also have relationships between them. Most databases allow the user to create either a one-to-many relationship or a one-to-one relationship (these were defined in the last unit).

To link tables, a relationship must be created between them using a matching field: a field that appears in both tables. To create the link the primary key field in the primary table is related to the same field in the related table. If such a field does not appear in the related table, it will need to be added.

In most databases, key fields (ones through which the tables are related) cannot be altered or deleted without first deleting the relationship between the tables. It is also necessary to make sure that the data in linked key fields is of the same type in both tables – that is, character fields should link only with character fields, etc.

Definitions

A **key** is any **attribute** of an **entity** with which an **index** has been created or a relation has been set. There are three main types of key: primary, secondary and foreign.

Primary key

The primary key is the set of mandatory columns that make the rows in the table unique. Rows in the table are normally accessed using the primary key. If, for example, a surname were chosen as the primary key, the database would not allow details to be entered for two people with the same surname. To avoid this problem it would be necessary to create a unique reference, such as an employee number, member number or customer number. Database software allows users to create their own primary keys or will supply a sequential number, called the record counter, which is automatically used as the primary key.

Secondary keys

A secondary key is a key that can identify more than one record. If, for example, an employee number is the primary key then a surname could act as the secondary key. In this case the surname might not be unique as several employees could have the same surname.

Foreign keys

A **foreign key** is a column in one table containing data that corresponds to a key in another table (the key can be either primary or secondary).

Consider the two tables shown below, which we will come across later when we study the development of a database for a video library.

In the MEMBER table, Membership_number uniquely defines a row (which can also be called a record) and is chosen as the primary key for this table. In the VIDEO RENTAL table, Video_number is the primary key because it uniquely defines a particular video being borrowed. A member can borrow more than one video, for instance Membership_number 0001 has borrowed two videos, 0004 and 0005. With respect to the VIDEO RENTAL table, Membership_number is a foreign key.

MEMBER table

Membership _number	Surname	Forename	Title	Street	Town	Postcode	Date_of _birth	Tel_no
0001	Bell	John	Mr	12 Queens Rd	Crosby	L23 6NN	12/12/56	924-8882
0002	Smith	Jenny	Ms	1 Firs Close	Crosby	L23 5TT	01/08/79	924-9090

VIDEO RENTAL table

Video _number	Membership _number	Date _borrowed
0004	0001	12/01/98
0001	0002	12/01/98
0003	0005	12/01/98
0005	0001	12/01/98
0002	0006	12/01/98

Candidate keys and alternate keys

All columns that contain unique values for every row are called **candidate keys**. A primary key will therefore always be a candidate key. The candidate keys left after the primary key has been selected are called alternate keys.

Activity 4

a There are lots of terms with 'key ' in their title. You have to know what they all mean. Can you say what each term means?
- primary key
- foreign key
- secondary key
- candidate key
- alternate key.

b On page 362 is a table containing some of the fields used in a personnel system along with some of the data used. Employee No is to be set with the data type AutoNumber.
  i Write down the field names of those fields that would be unique.
  ii Look at the data in the table and using the data types AutoNumber, Number, Date of birth and Logical fill in the next table with the most appropriate data type for the field.

Employee No	Surname	Date of birth	Car Reg No	National Insurance number
1020	Bell	12/12/80	L200HGT	YY643838T
1021	Jones	03/09/67	W430YWS	RG782312U
1022	Freeman	23/02/78	V109JJG	YY349867R

Field Name	Data Type
Employee No	AutoNumber
Surname	
Date of birth	
Car Reg No	
National Insurance Number	

▶▶

iii Which one of the fields is best chosen as the primary field? Why?

iv List the fields that are

- candidate keys
- alternate keys.

c A primary key is the unique field in a group of fields. Here are some fields in a table. The name of the table is shown along with a list of the fields that the table will hold. Also shown is a sample of the data which will be entered into the field.

For each of the tables shown, state with a reason the name of the field that is best chosen as the primary field. Remember to think about the purpose of the table when deciding on the primary key.

Student table

Surname	Jones
Forename	Peter
Title	Mr
Gender	Male
Student ID	980006

Car Park Spaces

Parking space number	190
Car registration number	W215AKD
Employee number	180041
Department code	104

Employee table

Tax code	416L
Surname	Jones
Date of birth	15/06/78
National insurance no	AB100136Y
Department code	029

Product table

Product description	A4 printer paper
Price per unit	£2.50
Number in stock	127
Product code	1381
Supplier number	2015

Patient table

Patient surname	Graham
Patient forename	Julie
Title	Miss
NHS number	09-09809-8
Date of birth	16/12/87

6.3 Normalisation

Normalisation is a mathematical technique for analysing data.

Reasons for normalisation

- Normalisation minimises the duplication of data. You might think that each table will need to have all its data input, so there will still be duplication of data. However, although there *is* some duplication, we need type the data in only once. Where the same data is needed in several tables, the computer will be able to read the data into the different tables automatically.

- The normalised data enables the data model to be mapped onto a wide variety of different database designs.
- The final tables in third normal form provide the flexibility to extract data efficiently.

Example 1

A simple database containing the records of students in a college would contain the data in an un-normalised form. As each different course a student takes has a different number, this part of the student record would need to be repeated for any student enrolled on more than one course. If, for instance, a student is taking five GCSEs each one will have a different course number and course title. The un-normalised data would be as follows:

STUDENT
Student number
Surname
Forenames
Address
Tel. no.
Date of birth
Course number
Course title
Course cost
Date enrolled
Lecturer name
Lecturer number

A shorthand method can be used to present this data, and would be written as follows:

STUDENT (Student number, Surname, Forenames, Address, Tel. no., Date of birth, Course number, Course title, Course cost, Date enrolled, Lecturer name, Lecturer number)

Notice the following in the shorthand method:

- The entity is written in capital letters and in singular form.
- The attributes of the entity are placed in brackets, with the attribute which uniquely defines the entity (the primary key) underlined.
- The items in the above list are in their un-normalised form.

Going from un-normalised form to first normal form

The first stage of normalisation is the removal of repeating items and showing them grouped together by the creation of a new entity. The attribute used as the key from the original entity will still need to be included as this is used to provide the link between the two entities. So in this case we have the original key field as Student number and this must be included under the courses record entity ENROLMENT.

Data in first normal form:

STUDENT (Student number, Surname, Forenames, Address, Tel. no., Date of birth)
ENROLMENT (Student number, Course number, Course title, Course cost, Date enrolled, Lecturer name, Lecturer number)

Going from first normal form to second normal form

To go from first normal form to second normal form the entities containing more than one key must be examined (in this case the two underlined in the ENROLMENT entity) to check if each attribute relates to only part of the key. In our example, Course title refers to part of the key attribute Course number. When this happens the attribute is removed with its key attribute and transferred to form a new entity, which in this case is called COURSE. In a similar way the Course cost depends only on the course and should be put into the COURSE entity.

Data in second normal form:

> STUDENT (<u>Student number</u>, Surname, Forenames, Address, Tel. no., Date of birth)
> ENROLMENT (<u>Student number</u>, <u>Course number</u>, Date enrolled, Lecturer name, Lecturer number)
> COURSE (<u>Course number</u>, Course title, Course cost)

Going from second normal form to third normal form

To reduce the data to third normal form, entities must be examined to see if any of the data is mutually dependent. Mutually dependent items are moved to a separate entity, leaving behind one of the items in the original entity to use as the key for the newly created entity. Lecturer number and Lecturer name are mutually dependent because each lecturer has a number and for each lecturer number there is a corresponding name.

Data in third normal form:

> STUDENT (<u>Student number</u>, Surname, Forenames, Address, Tel. no., Date of birth)
> ENROLMENT (<u>Student number</u>, <u>Course number</u>, <u>Lecturer number</u>, Date enrolled)
> COURSE (<u>Course number</u>, Course title, Course cost)
> LECTURER (<u>Lecturer number</u>, Lecturer name)

Notice that there are now four entities where previously there was only one.

The beauty of the normalisation process is that all the entities and attributes can now be used when designing the relational database. It is clear that these entities could be used as the names of the tables and that attributes can become the field names. This defines four tables in which to store all the data with a minimum of duplication.

Summary of the normalisation steps

1 Conversion into first normal form removes all repeating data elements.
2 Conversion into second normal form ensures that the data items are all dependent on the primary key.
3 Conversion to third normal form removes any mutual dependence between non-key attributes.

Example 2

A book publisher decides to keep records on each of its authors using a relational database such as Microsoft Access. After conversations with users about the attributes for the entity AUTHOR the following list was obtained:

> AUTHOR (this is the name of the entity, which is the name of the proposed table)

Author number
Surname
Forename
Address
Phone number
ISBN
Book title
Book category
Royalty rate
Agent number
Agent name
Agent address

A shorthand way of writing this list is:

AUTHOR (<u>Author number</u>, Surname, Forename, Address, Phone number, ISBN, Book title, Book category, Royalty rate, Agent number, Agent name, Agent address)

AUTHOR is the name of the entity and in the brackets is a list of the attributes (i.e. further information about the entity). The unique identifying attribute, the primary key, is underlined.

We now need to go through the process of normalisation. Before attempting to normalise, however, it is important to note the following:

- An author may have written more than one book (this is important when converting into first normal form).
- The ISBN is different for all book titles and is used by bookshops to identify a certain book. To convert data elements into second normal form the non-key attributes are examined to make sure that they have full functional dependency.
- When the data is in second normal form each of the entities is checked to see if there are any functional dependencies between pairs of non-key attributes. In this case, Book category determines the royalty rate – for instance a paperback might have a royalty rate of 15%, while hardbacks have a rate of 20% and textbooks 10%.

We will now work through the normalisation process.

Un-normalised form (UNF)

AUTHOR (<u>Author number</u>, Surname, Forename, Address, Phone number, ISBN, Book title, Book category, Royalty rate, Agent number, Agent name, Agent address)

This is not in first normalised form because the book details (ISBN, Book title, Book category, Royalty rate, Agent number, Agent name, Agent address) are a repeating group – one author could have written more than one book.

First normal form (1NF)

Each of the books an author has written will have a different ISBN, Book title, etc., so this is the repeating group. We need to create a new entity called PUBLICATION but still include the key of the original entity. We now obtain the following:

AUTHOR (<u>Author number</u>, Surname, Forename, Address, Phone number, Agent number, Agent name, Agent address)
PUBLICATION (<u>Author number</u>, <u>ISBN</u>, Book title, Book category, Royalty rate)

What does UNF mean?

Second normal form (2NF)

To move the data elements from 1NF to 2NF any entities containing more than one key need to be examined. PUBLICATION contains two keys, Author number and ISBN, so all its attributes must be checked to make sure they all depend on both keys – and, if not, that they are taken out with their key to form a new entity.

In this case we do not need to be concerned with the entity AUTHOR, because it has only one key.

The attributes Book title, Book category and Royalty rate are dependent only on the ISBN, so these need to be removed with their key and a new entity BOOK created. We now have the following:

AUTHOR (<u>Author number</u>, Surname, Forename, Address, Phone number, Agent number, Agent name, Agent address)
PUBLICATION (<u>Author number</u>, <u>ISBN</u>)
BOOK (<u>ISBN</u>, Book title, Book category, Royalty rate)

The list is now in 2NF.

Third normal form (3NF)

To go from 2NF to 3NF all the attributes are checked in case any of them are mutually dependent. Any that are mutually dependent need to be moved with their keys to a newly created entity. For instance, in the entity BOOK the royalty rate is determined by the Book category, so we move these to a new entity called COMMISSION, leaving Book category in the original entity and using it as the key for the newly created attribute.

The attributes Book category and Royalty rate are mutually dependent, because for a particular book type there will be a certain royalty rate. Also, in the AUTHOR entity we find that Agent number determines both the Agent name and Agent address, so these can be moved to a new entity called AGENT. All these features should have come to light during analysis.

So, in 3NF we have:

AUTHOR (<u>Author number</u>, Surname, Forename, Address, Phone number, Agent number)
PUBLICATION (<u>Author number</u>, <u>ISBN</u>)
BOOK (<u>ISBN</u>, Book title, Book category)
ROYALTY (<u>Book type</u>, Royalty rate)
AGENT (<u>Agent number</u>, Agent name, Agent address)

Activity 5

The following data items are in un-normalised form. They need to be fully normalised (converted to 3NF) so that tables can be created which minimise data duplication across them, thereby solving many of the problems associated with data redundancy.

CUSTOMER ORDER (<u>Customer order</u>, Customer number, Customer name, Customer address, Customer tel. no., Depot number, Depot name, Product number, Product name, Product quantity, Product price)

Go through the process of normalisation showing the various stages (1NF, 2NF and 3NF). To help you through the processes, here are a few reminders:

- primary keys are underlined;
- 1NF: a table is in 1NF if it contains no repeating groups;
- 2NF: the table must first be in 1NF and then have non-key attributes removed which are dependent on only part of the primary key;
- 3NF: the table must be in 2NF and in addition have no non-key attributes which depend on other non-key attributes.

Defining normalisation

Now that we know what the normalisation process entails, it is possible to define what normalisation is. Here is one definition:

Normalisation is the process of converting an invalid data model into a valid data model, ensuring consistency and integrity of the data model. Normalisation reduces the entity to an atomic structure (i.e. the entity cannot be broken down any further) and removes any repeating groups of attributes.

It could be argued that the results of carrying out the process of normalisation on a set of attributes could have been arrived at equally well by a good systems analyst. However, by normalising data you ensure that the solution is one that can be implemented successfully.

What is meant by an atomic structure of an entity?

Activity 6

A data model is to be constructed for a new in-patient system in a hospital. Each ward in the hospital has its own name and a unique reference number. The number of beds in each ward is also recorded, along with its name and reference number. Each ward has a complement of nurses who are given unique staff numbers which are recorded along with their names. Each nurse works in only one ward.

In-patients are given a patient number when they arrive, and this is recorded with the patient's name, address, telephone number and date of birth. When admitted to one of the wards, a patient is assigned to one consultant who is responsible for his or her medical care. Consultants have their own unique staff numbers, recorded with their names and specialism.

a Draw an entity relationship diagram for this system.
b Write down the names of all the attributes and put them in one entity, called PATIENT.
c Go through all the stages of normalisation from UNF to 3NF. You should show the names of the entities and their corresponding attributes in each form and also explain what you are doing at each stage.

6.4 Relational database structures

A considerable amount of analysis is needed to construct a relational database successfully, and only when this has been done will the developer know what fields are required within the tables. When devising the fields, the developer needs to ask 'what output is needed from the database?' Remember that nothing can be output that has not been input, or cannot be obtained from the input. The following steps should then be taken:

- Identify the entities required to form the structure. Each entity will correspond to a table in the database structure.
- Decide on the attributes for each of the entities identified above. Each of these attributes can be considered to be a database field.
- List all the fields likely to be used. Do not include any fields that are unlikely to be used and use sensible field names.
- Try to put the fields into some sort of order, grouping those that go together logically.
- Normalise the data, converting it from UNF to 3NF. The normalisation process ensures that data is in a form that can be successfully and efficiently implemented, and places data in groups that can be used as tables in a relational database.
- Decide on the data type for each of the fields and the length of each field.
- Identify the fields that are to be used as the primary key fields. These will be the fields underlined when 3NF has been reached.
- Identify the foreign keys.
- Determine the relationships between the tables and create them using the database software.
- Refine your design. It is quite hard to get everything right first time, so you may have to make changes to the structure as you go along. Do not worry about this too much as most databases are flexible and it is easy to make changes to the structure even once data has been entered in tables.

The more time you spend thinking about the design of the tables (e.g. how many tables there are, what they are called, what fields should be put into each table, what the primary keys are, etc.) the less time you will need to spend on making changes. Although it is quite easy to make trivial changes to a database even if data has been entered into it, it becomes increasingly difficult as the database grows and becomes more complex. It is important to complete the design of the database before any data is entered as the data can cause complications if the structure needs to be changed.

All relational database management systems have a framework (or structure) that has to be created before it is possible to enter data. The basic structures are as follows:

- **tables** (records, fields)
- **indexes**
- **relationships**
- **keys** – primary, secondary and foreign.

Tables

Entity types in a relationship model are represented by a table of values, where the columns represent attributes of the entity and each row of the table corresponds to an entity occurrence. The table below is used to hold details about members of a video library.

MEMBER table

Membership _number	Surname	Forename	Title	Street	Town	Postcode	Date_of _birth	Tel_no
0001	Bell	John	Mr	12 Queens Rd	Crosby	L23 6BB	12/12/56	924-8882
0002	Smith	Jenny	Ms	1 Firs Close	Crosby	L23 5TT	01/08/79	924-9090
0003	Cannon	Paul	Mr	12 Bells Rake	Crosby	L23 5FD	09/03/65	924-0098
0004	Charles	Steve	Mr	8 Moor Grove	Crosby	L23 7YY	02/07/69	924-1121
0005	James	Karen	Miss	3 Meols Rd	Crosby	L23 4RR	01/09/45	924-8111
0006	Brady	June	Mrs	9 Fox Close	Crosby	L23 5EE	20/01/59	924-0232

The columns represent attributes of the entity MEMBER. When we come to implement the database the entity name MEMBER can be used as the table name and the attributes (the column headings) used as field names. Each table needs to have a primary key which uniquely defines each row or record in the table, and in this case it will be Membership_number.

We have already seen how easy it is to convert the fully normalised data model into a table structure, using entities and attributes for table names and field names. Within a relational database, the data is held in tables consisting of columns that represent the fields of information, and rows that represent the individual records. Each record consists of a number of fields that are determined by the columns in the table.

Establishing relationships between tables enables data from more than one table to be combined for queries, forms and reports. In many cases a relationship involves tables where the primary key in one table matches a field (called a foreign key) in another.

Indexes

Suppose you want to find a particular topic in a book. One way to do so would be to look through the whole book until you find what you want. It would be faster, however, to use the index at the back of the book, and it is for exactly the same reason that we construct and use an index within a database.

Data is usually put into the various tables in chronological order – i.e. it is arranged according to time or date. This makes it harder for a database to locate the record because the date is not necessarily unique, like the primary key field. There is, however, a way round this – if the rows in the table are not in the primary key order, the database will automatically create an index which can be used to locate them quickly in primary key order. An index is always created on the primary key but other fields can also be indexed. The database software can process queries and searches much faster if a field has been indexed.

Any field can be indexed, provided that it is of one of the following types: text, numeric, currency or date. However, the temptation to index all fields should be resisted because any subsequent editing or adding of records will be slowed down. This is because an index table is created at the same time as the original table, so in any subsequent operations two tables are being manipulated. For this reason indexes should be created only on fields that are used repeatedly for searching, sorting, etc.

Let us now have a look at how an index works. Suppose we have a personnel file with the first two columns containing the record number and surname, as shown below.

In database design, what is an index?

Record number	Data
1	Pearson
2	Jones
3	Edwards
4	Jackson
5	Adams

What are the advantages in using an index?

We want to create an index according to surname so that we can print out the employee details in alphabetical order by surname. A new table is created with a new record number and the index referring back to the original record number. Adams will come first: the record number in the index table is 1 and the index will refer to 5, which is the record number in the original table. This process is repeated until all the fields on which the index has been created have been dealt with. The final index table is shown below.

Record number	Data
1	5
2	3
3	4
4	2
5	1

Building a database

A video library operates in a similar way to any other library, with members, videos and rental details to record. There are many more things we might want from such a system but for now let us look at just the system for recording members, videos and borrowings.

The first thing to do is to write down a list of attributes for the whole system. It is tempting to immediately put the attributes into separate tables but, although in this case it is fairly obvious which tables are required, it is not always so simple, so first we write down all the attributes. In choosing attributes, again it is tempting to include every attribute that could possibly be needed, but this should be resisted: specify only those that are definitely needed. It is better to add other attributes as they are needed.

For the video library system we could have the following attributes:

MEMBER (<u>Membership number</u>, Surname, Forename, Title, Street, Town, Postcode, Date of birth, Tel. no., Video number, Video title, Category, Cost price, Rental price, Date borrowed)

We have placed all the attributes in an entity called MEMBER. We must now go through the normalisation process to determine how many tables to use and what attributes should go into each table.

The video details are repeating attributes and therefore need to be removed into their own entity to give the data in the 1NF, thus:

MEMBER (<u>Membership number</u>, Surname, Forename, Title, Street, Town, Postcode, Date of birth, Tel. no.)

VIDEO RENTAL (<u>Membership number</u>, Video number, Video title, Category, Cost price, Rental price, Date borrowed)

We now look at the VIDEO RENTAL entity with the two keys and check to see if the attributes depend on both of the keys. Video number, Video title, Category, Cost price and Rental price depend only on Video number and therefore need to be placed in a new entity which we will call VIDEO. This converts the data to 2NF.

MEMBER (<u>Membership number</u>, Surname, Forename, Title, Street, Town, Postcode, Date of birth, Tel. no.)

VIDEO RENTAL (<u>Membership number</u>, <u>Video number</u>, Date borrowed)

VIDEO (<u>Video number</u>, Video title, Category, Cost price, Rental price)

To convert the data to 3NF we see if any of the non-key attributes are dependent on other non-key attributes. In this case no change is needed and so the 2NF becomes the 3NF and the data is now fully normalised in the form shown above.

We now have three tables: MEMBER, VIDEO RENTAL and VIDEO, and we can use the attribute names as the field names in the tables.

The following tables are set up. Notice the columns contain the attributes or field names, and the rows represent a particular record in each table.

MEMBER table

Membership number	Surname	Forename	Title	Street	Town	Postcode	Date of birth	Tel no
0001	Bell	John	Mr	12 Queens Rd	Crosby	L23 6BB	12/12/56	924-8882
0002	Smith	Jenny	Ms	1 Firs Close	Crosby	L23 5TT	01/08/79	924-9090
0003	Cannon	Paul	Mr	12 Bells Rake	Crosby	L23 5FD	09/03/65	924-0098
0004	Charles	Steve	Mr	8 Moor Grove	Crosby	L23 7YY	02/07/69	924-1121
0005	James	Karen	Miss	3 Meols Rd	Crosby	L23 4RR	01/09/45	924-8111
0006	Brady	June	Mrs	9 Fox Close	Crosby	L23 5EE	20/01/59	924-0232

VIDEO RENTAL table

Video number	Membership number	Date borrowed
0004	0001	12/01/98
0001	0002	12/01/98
0003	0005	12/01/98
0005	0001	12/01/98
0002	0006	12/01/98

VIDEO table

Video number	Video title	Category	Cost price	Rental price
0001	Independence Day	12	13.99	1.00
0002	Eraser	PG	13.99	1.00
0003	Bambi	U	13.99	1.00
0004	Evita	U	35.00	1.50
0005	Brave Heart	U	13.99	1.00

The relationships between the tables must now be devised. Let us look first at the relationship between the MEMBER and VIDEO RENTAL tables. They both contain the Membership number field and this is used to provide a link between them. Although Membership number is the primary key in the MEMBER table it is a foreign key in the VIDEO RENTAL table (the field Video number is the primary key in the VIDEO RENTAL table).

We need to assess the relationship between the MEMBER and VIDEO RENTAL tables. As one member can borrow more than one video at a time, one row in the MEMBER table could correspond to several rows in the VIDEO RENTAL table. This situation arises with member number 0001, who has two records in the VIDEO RENTAL table. The relationship between members and videos is clearly a one-to-many relationship.

Let us now consider the relationship between the VIDEO table and the VIDEO RENTAL table. Each video has a unique Video number, so it can appear only once in the VIDEO table and once in the VIDEO RENTAL table. It follows that the relationship between these tables is one-to-one.

After the tables have been created using the database software it is necessary to establish the relationships between them. In Microsoft Access (which is part of the Microsoft Office Professional suite) we place on screen those tables being used, then form relationships by drawing lines between the fields we wish to connect. In our video rental system, a one-to-many relationship line would be drawn between the MEMBER table and the VIDEO RENTAL table because one member can borrow more than one video. The relationship between the VIDEO and VIDEO RENTAL tables would appear as a one-to-one line.

Figure 6.6 shows the screen used in Microsoft Access to establish the relationships between tables.

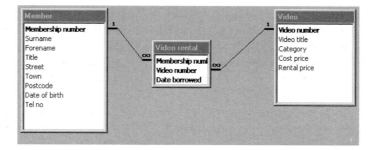

Figure 6.6

Data dictionary

To describe the properties of the fields used in a database, a data dictionary can be used. The data dictionary holds data about data and can be made to appear as a table, either on paper or on the computer screen.

A data dictionary is a file containing descriptions of all the data items in a database. The descriptions might include:

- data type
- element size
- validation range of values
- entity descriptions
- attribute–entity relationships
- purpose.

When setting up the database, various features of the fields have to be defined, such as type (numeric, character, logical, etc.) and this is used to help create the data dictionary.

Here is a part of the data dictionary for a temporary survey staff database.

Attribute	Description	Format	Length	Key	Validation
PAYROLL NO.	A unique number given to an employee e.g. 356991	Number	Double	Primary	N
FIRST NAME	The first name of the person to be employed	Text	30		N
LAST NAME	Family name of the person to be employed	Text	50		N
STREET	Street or house name of the person to be employed	Text	50		N
POSTCODE	Postcode of the person to be employed, e.g. L15 5TR	Text	7		Input mask

Some advice for designing and building a database

- Before listing the fields, think about what you want to get out of the database: remember that you cannot output something that has not been input to the system or derived from such an input.
- When choosing fields, consider how the data in each field will eventually be used. You may, for example, think of having a field for addresses. It is, however, much better to break this into smaller elements, such as street, town and postcode, as the data is more flexible arranged in this way. Having the address stored as one line means it will be printed out this way, which is not the usual format for the top of a letter or an envelope. Include forename, surname and title in preference to just 'name'. You should try, if possible, to break down larger chunks of data into several fields.
- Any calculations that need to be performed on the data (such as adding VAT or totalling an invoice) should not be put in a table; they should be added at the report stage.
- Make sure that any field you choose as the primary key is unique. It is usually best that a number rather than text field is used and that this number is not too large. You should consider using the record counter as a primary key for some tables. In tables where records are not uniquely defined by one field, you need to incorporate more than one field as a primary key.

From now on, we move away from the theory towards the practical side of databases. It is difficult to discuss databases without referring to a particular package, and Microsoft Access has been chosen to illustrate this chapter. All the examples, screenshots, etc. are from this package. Microsoft Access was chosen for its widespread use (it is the most popular database package for PCs) and availability.

The version of Access used is Microsoft Access 2000, which is provided with the package Microsoft Office 2000 but is also available on its own. If you do not have Access 2000, you will still be able to follow the instructions for creating the database system which follows. Some of the screen shots may look slightly different but that is all.

Creating tables

What is a table?

Relational databases store all data in tables, so once the names of the tables (i.e. the names of the entities in the ERM) are decided, we can design the tables using the database software. We have first to plan out the structure of each table and save each table structure separately.

We can refer to entities as table names and, instead of attributes, talk of field names.

Activity 7

Find out how to load your database software and produce a brief user guide, making use of screen dumps. Screen dumps can be produced using the following procedure:

- If you are using Access (or any other Windows-based database), press Alt + Print Screen and the screen design will be pasted into the clipboard.
- Exit the database package and load your word-processing software.
- Open a new document and select 'paste' from the edit menu (if you are using Microsoft Word) to place the screen into your document.

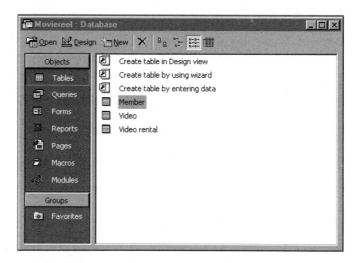

Figure 6.7 A database consists of many tables

Next, the structure of the tables used to hold the data is created; this is done in Microsoft Access using the table definition window shown in Figure 6.8.

Using this window, table structures are created for all the tables in the database. As you can see from Figure 6.9, the window is divided into two parts. The upper part is used to define the field name and the type of data allowed in that field, and to give a brief description (with examples) of the data that the field can hold. The bottom part of the screen is used to define the field properties for each field in turn and ensure that only allowable data is entered. Only the default values are shown – these will need to be changed depending on the type of data to be entered.

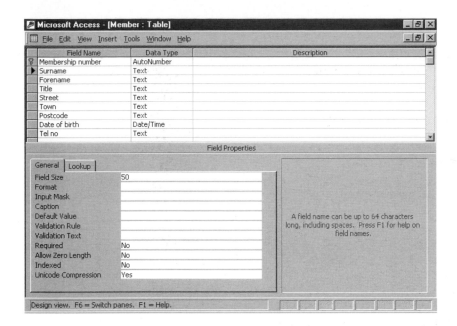

Figure 6.8 The table definition window in Microsoft Access

Choosing field names

Field names should always be descriptive to identify the field. Try to avoid using spaces in fields; instead use dashes or a combination of upper- and lower-case letters. Don't make the field names too long either – these will be the column headings in the tables; if they are too long the columns will be wide and not as much data can be displayed on the screen at a time.

Choosing data types

Once a field has been named, the type of data that can be entered has to be specified. In most databases, the data can be of the following types:

- text (alphanumeric characters)
- memo (used for data in note form; a window will usually open where messages which do not conform to any particular format can be typed in)
- number (for numeric values such as integers, real numbers, etc.)
- date/time
- currency (monetary values to two decimal places)
- AutoNumber (a numeric integer value which the computer automatically increments for each record added)
- yes/no (Boolean values)
- OLE (OLE objects, graphics and other binary data).

Figure 6.9 shows the data type being selected from the pull-down menu for the field Membership number. Data type 'AutoNumber' has been chosen because the database designer wants each member to be given a sequential membership number when they join the video library. A brief description of the field name is also given.

Important note
There are many situations in which we use a number that is not numeric. A proper number, i.e. one that *is* numeric, contains no spaces, no leading zeros, no letters or non-numeric characters, and can have calculations performed on it. The simple test is to ask yourself 'am I likely to perform a calculation using the field?' If the answer is 'yes', you need to store it as a proper number, if 'no', it is best stored as text.

375

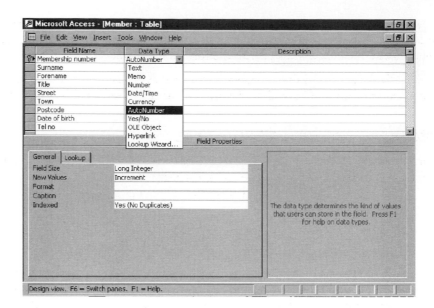

Microsoft Access - [Member : Table]

File   Edit   View   Insert   Tools   Window   Help

Field Name	Data Type	Description
Membership number	AutoNumber	
Surname	Text	
Forename	Memo	
Title	Number	
Street	Date/Time	
Town	Currency	
Postcode	AutoNumber	
Date of birth	Yes/No	
Tel no	OLE Object	
	Hyperlink	
	Lookup Wizard...	Field Properties

General   Lookup

Field Size	Long Integer
New Values	Increment
Format	
Caption	
Indexed	Yes (No Duplicates)

The data type determines the kind of values that users can store in the field. Press F1 for help on data types.

Design view. F6 = Switch panes. F1 = Help.

Figure 6.9 The table definition window in Microsoft Access

Activity 8

Here are some items of data which are to be stored in tables in a database. Identify whether they should be stored as text or numbers. Examples of the data to be entered are shown in brackets.

- the number of units in stock (234)
- an invoice number (0001 to 9999)
- a telephone number (0161 876 2302)
- a tax code (488H)
- an employee number (00234442)
- the rate of VAT (17.5%)
- the rate of pay (£10 per hour)
- a National Insurance number (TT232965A).

Description of fields

In Microsoft Access there is a section where a description of the field can be added to give more information about the field name. For instance, you might have a field called 'street' with the description 'this is the first line of the address'. Adding field descriptions in this way makes the tables you create easier to understand. With some database packages such as Microsoft Access, you can use these field definitions to automatically produce a data dictionary.

Choosing the field size

The sizes of some fields, such as date and currency, are preset, but others (such as numeric and text fields) need to be specified unless the default value is to be used.

When you are choosing the field size for text, the number of characters likely to be required should be considered. If the field size is not set, a default value is entered, which for Microsoft Access is 50. If a field size is too small, you *can* modify the structure of the table once data has been entered. Any data entered will remain intact when the structure is changed, provided that a field is not reduced in size so as to 'chop off' any data already entered.

A student always uses the default setting for the field size. What does this mean and what are the disadvantages in doing this?

Some fields must always contain a fixed number of characters and their field size should be set to this value. This ensures that only a certain maximum number of characters can be entered, thus providing a form of validation.

Numeric field sizes determine the range of values that can be stored in a numeric field; whether or not the value should contain decimal points can also be specified.

Data types and validation checks

If data is being entered into a field, either in a table or a form, before you can move on to the next field the computer checks to make sure that the data entered is allowable. If it is not, the computer will alert you with a message that you can specify in the validation text.

There are various reasons why data might not be allowed.

- It could be of the wrong type for the particular field. For instance, a field that is to hold telephone numbers may have been set up as a numeric field. If a user then tries to enter the telephone number 0151-929-6758, which is not numeric (it contains a leading zero and dashes), it will be refused. If this happens the type of field will have to be altered; telephone numbers should be stored as text.
- It might break a rule for the field that was set up in the field properties section.
- The field properties section may or may not demand that a particular field be filled. For instance, a telephone number field can be left blank if the person does not have a phone, but an employee number field in a personnel system should require the employee number to be entered. An error will be reported if a field that always requires data to be entered is inadvertently left blank.
- The user may have tried to enter the same data more than once into a field set as a primary key. Because the primary key must be unique, the same data is not allowed twice in such a field.

Figure 6.10 shows the structure for a table called MEMBER in the video library database. In the main body (the top half of the screen) the field names, data types and descriptions of the data are input.

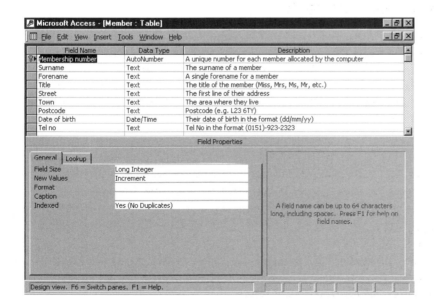

*Figure 6.10
The completed structure for the MEMBER table*

Notice the small key at the side of the Membership number field, which indicates that this field is set as a primary key. The data type for the field is set as AutoNumber and is automatically numbered sequentially by the computer. Note also that the description provides some brief explanation of the field.

You will already have created a data dictionary, but in setting up the database you may find you need to go back to the structure to make changes. If you do, make sure that the structure screen is the same as your data dictionary; if it is not the dictionary will need to be altered.

As well as limiting the type of data that can be entered into a particular field, validation checks can be performed on the data by making use of validation rules. Validation rules are specified as validation expressions. For instance, the following validation expressions will produce the text if the rule is not obeyed (NB: these expressions can be used only with Microsoft Access).

Validation expression	Validation text (appears if the data being entered is not allowable)
>100 And <200	The value entered must be between 100 and 200 but not including 100 or 200
>= 100 And <= 200	The value entered must be between 100 and 200 including 100 and 200
>0	The value entered must be positive
<>0	A non-zero value must be entered
<#12/1/98#	The date entered must be before 12/1/98
Like "A????"	The data entered must be 5 characters long beginning with the letter A
='2.5' Or '5.0'	Value must be 2.5 or 5.0
>=#1/1/97# And <=#31/12/97#	The date entered must be in the year 1997

Screen input forms

Screen input forms are effectively electronic forms which enable data to be entered directly into a computer via the keyboard. They are similar to paper forms, with boxes to be filled in by the user. The screen design will usually incorporate a title for the form and have prompts (usually the field names) for the data which the user has to add.

Many database packages allow the user to enter data directly into tables, but it is usually better to create special forms to allow records to be entered one at a time.

Figure 6.11 shows data being entered into a table via a datasheet.

Most database software allows the use of text, data, pictures, lines and colour to improve the appearance of input forms, and this is particularly important if the screen faces the customer. As well as being used for data entry, input forms can be used to view data on screen. Data for a form can come from a table or a **query**, while information such as company logos, titles, etc. are stored in the form design itself.

Forms can be used to view a single record or all the records for that form. Figure 6.12 shows a form being used to input data into a MEMBER table, one record at a time.

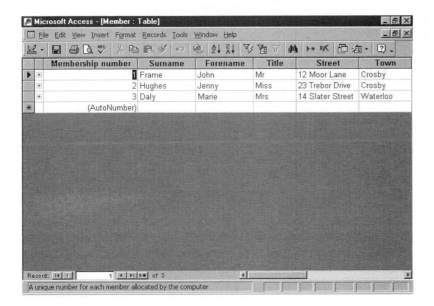

Figure 6.11 Entering the data straight into a table called the datasheet

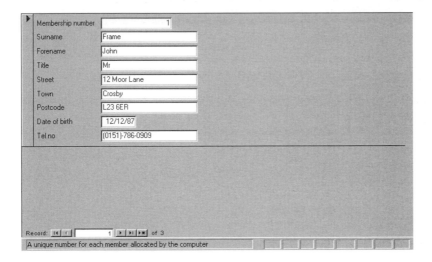

Figure 6.12 A form being used to input data into a MEMBER table

Single data entry forms

In single data entry the data entered into a screen input form goes to a single table, where it is stored and can be accessed as required.

Multiple data entry forms

These are input forms that enable data to be entered into more than one table at a time. Suppose two tables need to contain the customer number. Rather than using separate forms to enter this data into each table, a single form is employed which automatically passes the data to both tables. This saves time and reduces the chances of error. Multiple entry forms can be used only with relational databases; for flat file databases single data entry forms have to be used.

Modifying form fields to enable accurate input of data

Form fields are the blank spaces included in the form on the computer screen, into which variable data can be added. The added data is then passed to the records contained in one or more tables, where it is stored for further use. Like a

What are data entry forms?

379

paper-based form, form fields have headings and prompts to let the user know what sort of data should be keyed into the boxes.

Figure 6.13 shows a form designed to input the hours worked by a temporary worker over a week. Notice the last two fields – Total (the total number of hours worked that week) and Total Pay – are not entered but calculated from other fields. The total hours are worked out by adding the numbers in the Number of hours worked fields on each of the five days, while the total pay is the total number of hours worked multiplied by the rate of pay.

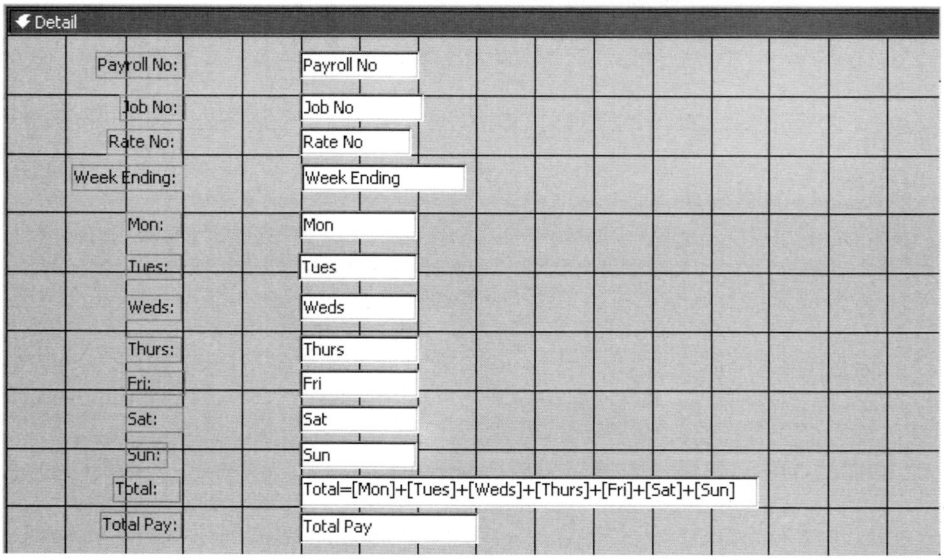

Figure 6.13 Form used to input hours worked in a week

Field length

Field length (also known as field size) is usually taken to be the maximum number of characters an entry is likely to occupy within a field. This is usually specified when the database is set up, but most databases allow the field length specified in the structure to be adjusted without damaging the data already entered.

Data entry fields

Some fields, such as today's date or a sequential order number, can be filled in automatically by the computer, but in many fields the data must be keyed in by the user.

Instruction fields

All screen forms should include instructions to let the user know what each field requires. Sometimes the field name alone is sufficient. 'Surname' is self-explanatory but 'date' can be problematical if an inexperienced user is not familiar with the 'dd/mm/yy' format. Any fields with field names that are not obvious need an example to give the user an idea of the type of format required for input.

Field titles

Field titles are usually the same as the field names, but they can be more detailed. It is more convenient if field names are kept short, as they have to be typed in when performing searches, but the actual field titles can be longer and offer further explanation.

Validation checks

Validation checks are used to prevent inaccurate data from being entered into tables. Although staff who enter data should ensure that it *is* accurate by checking the completed form against the source document, each field can also have certain validation rules attached to it, allowing only data that conforms to the rules to be entered. If the data entered does not conform to the validation rule for a particular field, a message appears informing the operator why the data is incorrect.

Various validation checks can be performed on data – for example a credit limit field may have a range check to make sure that no credit limits are entered that exceed a predetermined value.

Activity 9

There are four types of validation check. Using manuals, books or on-line help, find out how each of these checks is performed. Outline your findings in a brief report.

You can use Activity 9 to provide evidence for Key Skills for Communication C2.3 or C3.3.

Input masks

In some fields, data might all have the same format, so to simplify data entry an **input mask** can be used. The input mask supplies the invariable characters in a field so that the user need enter only the data that differs from one field to another. This is best illustrated by taking an example of entering telephone numbers. If all the telephone numbers to be entered have the same format, (0151) 876 2341 for example, we can use the input mask to supply the brackets and spaces between blocks of numbers.

As well as saving some input time, input masks help to ensure that the data entered adheres to a format. If an order number field contains customer numbers which each start with a single letter of the alphabet followed by four numbers, this can be specified in the input mask.

Microsoft Access uses the following mask characters. Note the difference between 'may be entered' and 'must be entered'.

Mask character	Can be used in the input mask to mean
0	A digit must be entered here
9	A digit may be entered here
#	A positive (+) or negative (–) sign may be entered here
L	A letter must be entered here
?	A letter may be entered here
A	A letter or a digit must be entered here
a	A letter or a number may be entered here
&	Any character or space must be entered here
C	Any character or space may be entered here
<	All characters to the right are converted to lower case
>	All characters to the right are converted to upper case
. , : ; - /	Decimal point, thousands, date and time separators
!	Mask fills from right to left (useful if the optional characters are on the right-hand side of the mask)
\	Character to the right is interpreted as an ordinary character and not part of the mask

Let us now look at constructing some input masks using these characters. For a telephone number such as (0151) 876 2341 we could use the following input mask:

   (0000) 000 0000

But if we want dashes between the groups of numbers we should use this mask instead:

   (0000)-000-0000

Activity 10

a To store a postcode the first task is to find the formats for all postcodes. Our input mask would need to take account of the following possibilities:

   • L23 8VY
   • GL50 1YW
   • L9 0BQ.
   • EC1X 4BD

Design a single input mask suitable for the entry of the above postcodes.

b Using your database software, find out which validation checks can be performed on the data entered into the following fields:

   • date of order (dd/mm/yy)
   • order reference (a letter followed by four numbers e.g. A1729)
   • postcode (e.g. L23 5GT).

Ensuring data entry fields comply with the data dictionary

We have already come across the description of a data dictionary, but what is its purpose? Data dictionaries provide a valuable reference because, even within an organisation, IT users might define an element of data quite differently. (An element of data is a unit of data that cannot be decomposed any further.)

Take, for instance, a college principal who needs to know how many students the college has. They could do a simple count of students on the register, but some of the students may well appear on more than one class register, so there is a danger of counting them more than once. Another approach would be to obtain a total of the hours attended by all students in the college over a one-week period, and divide this by the number of hours per week a full-time student would normally attend. This method would not be accurate, however, if some courses classed as full-time require 16 hours attendance while others are based on 24 hours or more per week. As you can see, data that at first sight appears relatively simple can easily be misunderstood and misused. Two people often think they are talking about the same thing when in fact each has a different definition of the item under discussion. Taking again the example of full-time students, all members of staff need to agree a definition if they are not to risk talking at cross-purposes. A data dictionary would help by defining a student, a full-time course and all the other data elements in the system.

Creating and using special fields

Special fields called form fields are designed to contain special data such as date, time, page number, or calculations like column total.

Calculated fields

Calculated fields are those whose contents are worked out from data in other fields using calculations that the database software performs. It is important to note that calculated fields are not stored in the table. Figure 6.14 shows a form containing two calculated fields of this type.

Date

Sometimes it is necessary to enter a specific date such as a date of birth, but in many situations a current date ('today's date') is all that is required and this can be done by default, using the system date. It should be noted that much of the software used is from the USA where the MM/DD/YY format is used rather than the DD/MM/YY format used in Europe. If you have problems with this, you will need to check that the system clock in the operating system of your computer is set to DD/MM/YY.

Time

When many transactions are taking place on one date it may be useful to record the time of each transaction so that different transactions by the same person can be identified. In most database systems the time field is tagged onto the date field; the time on its own is not very useful without the date to which it refers. When creating the database structure you normally include a date/time field. In Microsoft Access you can display the time in a variety of ways, such as 2:30:00 PM, 14:30 or 02:30 PM. You can use the date and time together to group records producing, for instance, a list of transactions in complete chronological order.

Queries

Queries are requests, written in a special language, for specific information from a database. For instance you may want to obtain a list of all the entries with the surname 'Jones' or details of accounts where outstanding credit exceeds the credit limit. Database software allows users to design specific queries. The following points should be considered when designing such queries.

What is a query?

Choosing fields

You can choose any combination of fields stored in one or more tables. Provided that the data has been stored, it can be retrieved and put into a query. Figure 6.14 shows how a query is set up in the query design screen and Figure 6.15 shows the results when this query is used to interrogate the database.

Notice the tables in the upper part of the screen along with the relationships between the tables. In the lower part of the screen we can choose which fields we wish to display and order the data in one or more fields. In Microsoft Access the fields you want are marked by clicking on 'Show' – a cross will appear to indicate which field will be shown.

Choosing records

You can choose data within a field to form a group of records. For example, the items sold over a certain week can be selected and listed.

Using Microsoft Access you can enter criteria for the search into the query design. For instance, in Figure 6.16, for those employees who work fewer than 20 hours the payroll number, first name, last name, job number, title and the total hours worked are picked out in order of last name. Figure 6.17 shows the results obtained using this query.

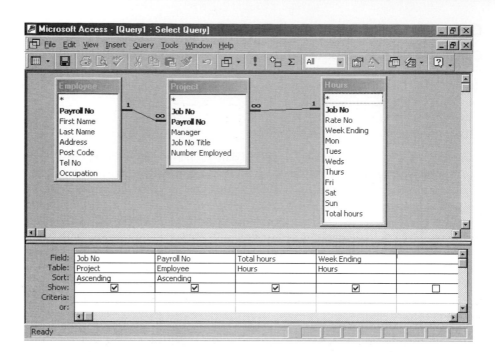

Figure 6.14 Setting up a query using the query design screen

	Job No	Payroll No	Total hours	Week Ending
▶	S232	132241	14	02/05/00
	S232	132785	21	02/05/00
	S344	122122	21	02/05/00
	S344	132456	24	02/05/00
	S344	223242	14	02/05/00
	S344	233478	27	02/05/00
	S345	122243	15	02/05/00
	S345	131234	19	02/05/00
	S345	132121	15	02/05/00
	S349	232247	25	02/05/00
*			0	

Record: 1 of 10

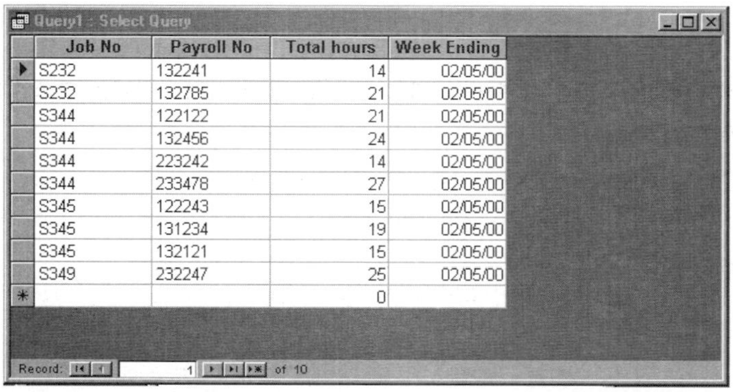

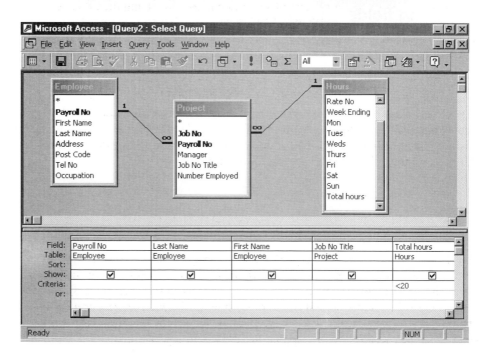

Figure 6.15 Results obtained when a query is used to interrogate a database

Figure 6.16 Entering criteria for the search

	Payroll No	Last Name	First Name	Job No Title	Total Hours
►	122243	Owens	Jane	Skills Survey	15
	131234	Harris	Peter	Skills Survey	19
	132121	Hughes	Dawn	Skills Survey	15
	132241	Jackson	Jane	Crosby Herald Survey	14
	223242	Johnston	Pamela	Crosby Traffic Survey	14
*					0

Figure 6.17 The results from a query

Sorting the data

Records are usually most useful if they are sorted into an order related to the main field or fields used for the query. For example, if we have obtained details of customer accounts to determine who owes the most, it would be helpful to place the records in order of the balance, with those customers owing the largest amounts placed first. If the list were to be used to answer customer enquiries regarding their balance, alphabetical order according to surname would be more useful. Sorted data can be displayed in a form by creating a query to sort the records, then using the query as a source of data for the form.

Notice in Figure 6.15 that two orders are being used. The first is in order of job number, and within this (because several jobs have the same number) the order is ascending payroll number.

Asking questions about the data and deciding which tables to use

Questions can be asked about the data contained in more than one table and the results viewed either on a form or in a report.

Using calculated fields
New fields can be created containing the results of calculations – these are called 'calculated fields'. For example, you might want to multiply the price of an individual article by a quantity to give the total price. To do this a new 'total' field would be created to contain the results of a calculation involving the other two fields. In order to show the calculated field, you need first to construct a query which contains the calculated field and then base a form or report on the results of the query.

Deciding whether to present the search as a form or a report
As we have already seen, queries are used to extract certain data from tables. When the query is run, the results are presented in a table with columns for each field and with records for the tables in the rows underneath. You can also print out the results from the queries, but it is usually better to load the query into a report because this allows more flexibility of arrangement of the data on the page.

Other uses for queries

One very useful feature of queries is that they can be used to update, delete or add to (append) the group or record specified by the query.

Queries are of two types: select queries and action queries. Select queries are used to find and extract certain information, whereas action queries perform actions such as deleting or updating a record. Action queries enable you to make changes to many records in just one operation. For example, if all the employees of a certain organisation were given a 5% pay rise, all the salaries contained in the personnel records can be increased in one operation rather than separately.

Activity 11

Classify the following queries according to whether they are select or action queries.

- Increase the salary of all employees working in the sales department by 5%.
- Identify all the records of employees who have attended a first-aid training course.
- Identify all sales staff whose sales are below target.
- Delete all details in a video library of members who have not borrowed a video in the last two years.
- Alter all the telephone numbers for a certain area from 051 to 0151.

Structured query language

Structured query language (**SQL**) consists of a small number of commands that the user can combine to extract particular details from a database. It is now the industry standard language for the extraction of information from databases.

In SQL, the SELECT command is used to query the database, and is constructed like this:

    SELECT attribute list
    FROM table list
    WHERE condition.

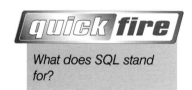

What does SQL stand for?

The attribute list is used to list the attributes (i.e. the fields) we want to retrieve, while the table list indicates in which table the listed attributes can be found. The condition is a Boolean expression used to identify the records to be retrieved.

Using relational logic in queries

SQL also allows the logical operators AND, OR and NOT to be used to combine relational expressions.

AND

Two expressions are combined with AND when both must be true for a record to qualify. For example, we might want a surname Jones AND the town Crosby.

OR

Two expressions are combined with OR when either one needs to be true for the record to qualify. For example, we might want the surname Doyle OR Prescott.

NOT

NOT can be used to display all the records that do *not* meet a criterion. For example, we may want details of all people who live in a town which is NOT Crosby.

Suppose we want to extract the name and address of a video library member whose membership number, stored as text, is 0001. If the table containing the member's details is MEMBER we can use the following SQL statement:

    SELECT Surname, Street, Town, Postcode
    FROM MEMBER
    WHERE Membership number = '0001'

The following examples show how Boolean expressions can be used to narrow down a search.

Example 1

We want to extract the names of all employees in the production department of a company earning over £30 000 per year. The employees' details are stored in a table called PERSONNEL. The SQL instruction to do this is:

```
SELECT EMPLOYEE_NAME
FROM PERSONNEL
WHERE DEPARTMENT = 'PRODUCTION' AND SALARY >30000
```

Example 2

Suppose we want to extract a list of the names and addresses of employees who work in the production or marketing departments from the PERSONNEL table. We could use the following SQL statement:

```
SELECT SURNAME, STREET, TOWN, POSTCODE
FROM PERSONNEL
WHERE DEPARTMENT = 'PRODUCTION' OR 'MARKETING'
```

quick fire

Why is SQL useful when information needs to be extracted from a database?

Activity 12

The table shown below is used to store the details of the members of a video library.

Write down the SQL statements that could be used to query the data contained in the table.

a Extract all the members' membership numbers and surnames, sorted into alphabetical order according to surname.

b Produce a list of the membership numbers, surnames and dates of birth of all the members born before 12/12/70.

c Extract the names and addresses of all the female members of the video club.

MEMBER table

Membership number	Surname	Forename	Title	Street	Town	Postcode	Date of birth	Tel no
0001	Bell	John	Mr	12 Queens Rd	Crosby	L23 6BB	12/12/56	924-8882
0002	Smith	Jenny	Ms	1 Firs Close	Crosby	L23 5TT	01/08/79	924-9090
0003	Cannon	Paul	Mr	12 Bells Rake	Crosby	L23 5FD	09/03/65	924-0098
0004	Charles	Steve	Mr	8 Moor Grove	Crosby	L23 7YY	02/07/69	924-1121
0005	James	Karen	Miss	3 Meols Rd	Crosby	L23 4RR	01/09/45	924-8111
0006	Brady	June	Mrs	9 Fox Close	Crosby	L23 5EE	20/01/59	924-0232

Query by example

The queries we have constructed so far are all of a type called **query by example** (**QBE**), a simplified method of entering SQL queries that enables users to enter a query using a menu or keystroke sequence that is automatically converted into an

SQL command. With Microsoft Access it is possible to use either SQL or QBE. Because it is so easy to use, most users prefer QBE, although some experienced users accustomed to programming and building large applications prefer SQL.

When you create a QBE query in Microsoft Access, the database constructs an SQL statement and carries it out. The SQL statements are not shown on screen as this might confuse the user. However, users familiar with SQL statements may want to use them rather than QBE, and the system allows you to do so by opening an existing query then selecting the SQL option from the view menu.

What is the difference between SQL and QBE?

Designing database report layouts

Reports are used to extract information from a database and produce a printout on paper or display on the screen as a single table or a number of tables. Reports can also be used to present the results of a query. Other information can be included on a report, such as headings on the top of the report and graphs showing numerical data.

Most people consider a report to imply a printed copy, but it may be sufficient to display a report temporarily on screen rather than produce a hard copy.

Screen reports

Screen reports are really only suited to small amounts of information that can be easily assimilated and do not need to be taken away. Screen reports are ideal for summarised data and answers to simple queries such as 'how many of item X do we have in stock?'

Printed reports

Many reports are lengthy and may need to be taken away and studied or used for future reference; such reports are always printed out. Some reports are the result of complex file interrogations and manipulations, and the grouped or sorted data needs to be printed out as neatly formatted hard copy.

Always remember that a table is a way of storing data with a minimum of duplication. It is easy to be confused by relational databases, and it is common to think that you can print out data from only a single table at a time but this is not the case. The table structure allows any data, in any of the tables, to be printed out, provided that relationships exist between them.

How is a report laid out?

There are certain design features of reports.

Pagination
Pagination involves deciding which parts of the report should be on each page. You should try to keep related data on one page if at all possible.

Footers
There are three types of footer: group footer, page footer and report footer. The group footer appears at the end of each group of records and usually gives the total of a particular group. The page footer appears at the bottom of each page in a report, and normally shows the page number in a multi-page report. Report footers appear only once, at the end of the whole report. They are usually placed above the page footer on the last page of a report and should be reserved for items such as report totals.

Headers

There are three types of header: the report header, the page header and the group header. The report header appears at the beginning of a report and usually contains a logo, the report's title and its date. The page header appears at the top of every page in the report and includes the column headings in the case of a tabular report. A group header appears at the start of a group of records. Here you would normally put the name of the group.

Totals

Most reports contain some numerical data and group totals are often useful. Such group totals can be specified in the report design. Figure 6.18 shows a report of the pay earned for the week ending 11/02/00. Notice that records are ordered according to last name and that totals are calculated, along with a grand total at the end.

Single Pay for Week Ending 11/02/00
23-Apr-00

First Name	Last Name	Payroll No	Job No	Total Pay
Gareth	Bean	233478	S727	£127.50
Peter	Belling	711234	S729	£133.28
Jan	Chilcott	823456	S727	£142.50
Anna	Coleman	286752	S729	£145.04
Rolph	Harris	987654	S727	£146.52
Miriam	Higgins	356991	S728	£122.85
Amanda	Jones	982345	S729	£152.88
Stephen	Morris	243167	S727	£105.00
Harry	Secombe	762345	S728	£119.34
Mike	White	234567	S727	£123.75
Dave	Williams	923452	S728	£98.28
Dave	Wood	678213	S729	£156.80
			Grand Total =	£1,573.74

Figure 6.18

Calculations

Rather than perform calculations on the data in the database itself, it is easier to do any necessary arithmetic at the report stage. To produce an invoice or bill the quantity is multiplied by the cost per item, each line in the order totalled and the VAT calculated and added to give the final amount. As well as performing simple calculations, the reporting process can be used to carry out more complex statistical procedures.

Data groupings

The ability to group records in some way in a report can be very useful. For example, you might want to list customers who are over their allowed credit limit in order of the amount they owe. The person using the report can immediately see the important ones at the top of the list. Customers could also be grouped logically according to their department or job title.

Sometimes reports which show overdue accounts are useful if the accounts are ordered according to how overdue they are, and this is done using the date field.

Figure 6.19 shows a report with data groupings according to job title. The report shows how much the staff working on each job earned in the week ending 11/02/00.

Project Group and Individual Pay by Week Ending 11/02/00
23-Apr-00

Job No Title	Week Ending	Payroll No	First Name	Last Name	Total Pay
Bootle Town Centre Traffic Counts	11/02/00	233478	Gareth	Bean	£127.50
		823456	Jan	Chilcott	£142.50
		987654	Rolph	Harris	£146.52
		243167	Stephen	Morris	£105.00
		234567	Mike	White	£123.75
					£645.27
Netherton Residents Skills Survey	11/02/00	711234	Peter	Belling	£133.28
		286752	Anna	Coleman	£145.04
		982345	Amanda	Jones	£152.88
		678213	Dave	Wood	£156.80
					£588.00
Southport Visitors Survey	11/02/00	356991	Miriam	Higgins	£122.85
		762345	Harry	Secombe	£119.34
		923452	Dave	Williams	£98.28
					£340.47
				Grand Total:	£1,573.74

Figure 6.19

Totals are also included for the cost in wages for each job, for that particular week. A grand total comprises the wages for all the jobs carried out that week.

The report in Figure 6.19 demonstrates the wide variety of ways that data held in tables can be extracted and presented and illustrates the power of a relational database.

Print preview

Print preview is a feature included in most relational database management systems and allows the user to view a database report screen before printing. Adjustments can then be made if needed, so using print preview can save both time and paper.

Specification of a report

Whether a report is being produced for your own use or someone else's, you need to consider the purpose of the report and who it is aimed at, and adjust the arrangement of the fields and other information to suit these requirements. The specification of the report needs to cover several areas.

Order

Reports can be arranged in any order so it is best to consider the purpose of the report and order the information accordingly. For instance, customers who have not paid their end-of-month account could be placed in order of how overdue the account has become or in order of the amount owed. Either way, it will help the person who has asked for the report. If the purpose of the report is known, the process can be taken a step further and an exception report produced which prints only items of interest.

Data in reports may be grouped. What is meant by grouping?

The order in which the fields should be printed is also important. If, for instance, there are columns of fields, the more important and unique fields should be positioned to the right-hand side of the report.

Fields

Before creating a report you need to decide which fields should be included in it. There is a tendency to try to cram in as many fields as will fit into the report 'just in case they are needed'. This should be resisted. Reports should contain only information that is essential to fulfil their purpose; adding non-essential information only obscures the main points.

Field positions

Rather than have field names at the tops of columns with their respective data in rows below (much as data is arranged in tables), it is often better to specify where the data is to appear in a report. Many reports contain data in various places. Consider an invoice, for example: details such as customer number, name and address appear only once but the order line for the various items on the order appears many times, usually across the page in the following order:

   product number, product description, unit price, quantity, cost.

Testing the database solution

The finished database should not be handed over to the user without thorough testing. If you were developing a database for a real commercial company and there were problems with the database that resulted in the organisation losing money, then the company could be entitled to compensation from you (you should have insurance against this).

What ways are there of thoroughly testing a database?

As part of the systems analysis documentation, there will be the specification for the project as agreed by the developer and the user. This document can be used to check whether the developed solution meets the specification.

The next stage is to check whether allowable data can be entered. For example, you could check to see if the field widths will accommodate the actual data.

6.5 Relational database construction

A college enrolment system: an example of a relational database system

We saw earlier in the unit how the database to keep details about students and the courses they were enrolled on, along with details of the lecturers who were the course tutors, could be modelled using four entities. When the fields were fully normalised to 3NF, the following tables and their fields were obtained:

   STUDENT (<u>Student number</u>, Surname, Forenames, Address, Tel. no., Date of birth)
   ENROLMENT (<u>Student number</u>, <u>Course number</u>, <u>Lecturer number</u>, Date enrolled)
   COURSE (<u>Course number</u>, Course title, Course cost)
   LECTURER (<u>Lecturer number</u>, Lecturer name)

The next stage is to set up the structure for each of these tables.

As part of setting up the structure of the tables we need to think about the field properties.

When you design a table, each field you set up can have an associated group of properties and these are shown in a box at the bottom of the screen. The choices you can make in this box are determined by the data type of the field selected.

To see the table and how the properties change, load the Student enrolment system database and, when at the database window, select the STUDENTS table.

Here is a list of the field properties and what they mean.

- *Field size* – This is set at 50 for a text field, which means that you can type in a maximum of 50 characters into this field. You *can* alter this, so if a code only ever contained up to four characters you could change the field size to 4. If the field selected is a number there is a variety of different number settings possible. If you require more knowledge about this, use the on-line help.
- *Format* – The format of a field determines how the data is displayed once it is entered into a field. Many formats are already set up for you to choose from – and you can create your own.
- *Input mask* – The input mask is used to present the data in a particular format or pattern. For example, if someone entered the text JONES into a field then, using an input mask, we can keep the first letter of the surname in upper case and turn all the other letters to lower case. This is done to keep the data consistent in the database, thereby ensuring its integrity.
- *Caption* – A caption is a name that appears on either a form or a report that describes the field. If no caption is entered, by default the field name will appear.
- *Default value* – This is what is entered into a field before the data is put in. For example, in the COURSES table for the Course cost field (which is a currency field) £0.00 will appear by default.
- *Validation rule* – This restricts the data that can be entered into a field. Its purpose is to reduce the mistakes during data entry. If the data being entered breaches the validation rule then a pre-determined message is displayed (you will have typed this in the validation text property section).
- *Validation text* – A message can be typed in here that will be displayed if the validation rule is broken. This message should be helpful to the user so it is important, as well as saying what they are doing wrong, to also give them some idea of what is acceptable.
- *Required* - Here you specify whether the field needs to be filled in with data or not. For example, all students have addresses so the Required property could be set to Yes for the fields Street, Town and Postcode. The Tel. no. field should have the Required property set to zero as not all students will have a phone.
- *Allow zero length* – This allows you to enter what is called a null string (" "); this indicates that the data exists but is unknown.
- *Indexed* – This allows an index to be created on the field to speed up searches and sorts.

Worked example:
creating the structure for the STUDENT table

1 Load Microsoft Access. The menu shown in Figure 6.20 appears. Make sure that Blank Access is checked. Click OK.

Figure 6.20

2 The menu shown in Figure 6.21 appears. There are two things you need to specify here.

Figure 6.21

   a A name for the database. Call it 'College Enrolment System'.
   b A place to store the database file. In this case we will use the floppy disk drive. Your teacher will inform you where to save your files.

   It is very important that these steps are performed *now*, as there will be no opportunity to alter the name of the file or its location later on. Click Create. The next screen (Figure 6.22) appears.

3 Double-click on Create table in Design View.

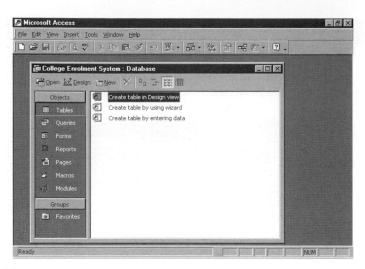

Figure 6.22

4 Using the screen in Figure 6.23, create the structure of the STUDENT table by specifying the field name, data type and description. This is called the design screen.

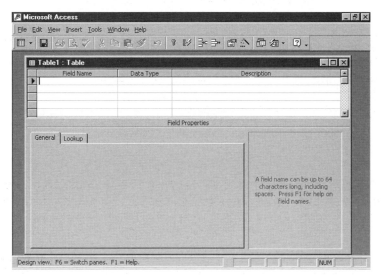

Figure 6.23

5 Enter 'Student number' for the first field name. Press Enter. The cursor moves onto the Data Type. Double click on this and a list of data types will appear in a drop-down menu (Figure 6.24).

Select AutoNumber from this list. This gets the computer to allocate a unique number to each student when his or her details are entered (the first student will be given the number 1 and so on). The computer automatically allocates this number, so we do not need to remember the next number ourselves.

In the description column, type in the description for the field – 'A unique number allocated by the computer'. The description here allows a person who did not develop the database to understand ▶▶

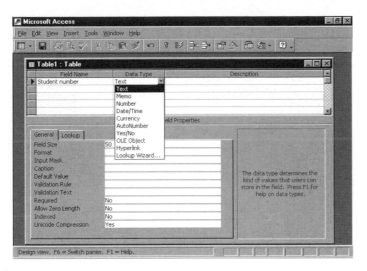

Figure 6.24

what each field name represents. Press Enter and the cursor moves to the next line. Most of the field names used in the structure of this table are easy to understand.

6 Complete the structure, as shown in Figure 6.25.

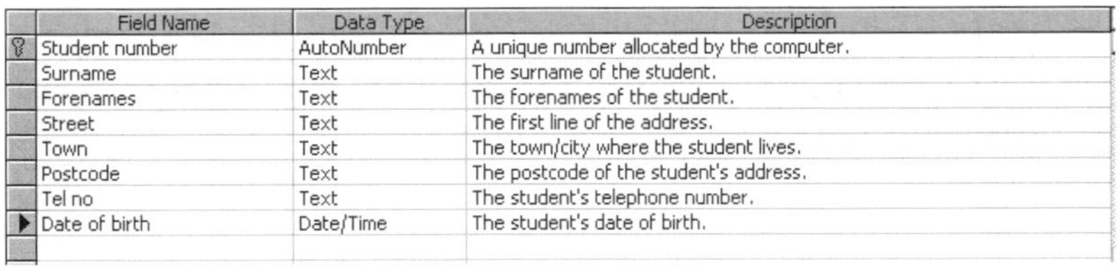

Field Name	Data Type	Description
Student number	AutoNumber	A unique number allocated by the computer.
Surname	Text	The surname of the student.
Forenames	Text	The forenames of the student.
Street	Text	The first line of the address.
Town	Text	The town/city where the student lives.
Postcode	Text	The postcode of the student's address.
Tel no	Text	The student's telephone number.
Date of birth	Date/Time	The student's date of birth.

Figure 6.25

7 The primary key for this table, Student number, must now be set by moving the cursor to anywhere on the Student number field and clicking on the key icon in the toolbar (Figure 6.26).

Figure 6.26

Make sure that a picture of a key has been inserted next to the Student number field.

8 You have now completed the structure, so you need to save it. Click on the close window button (the cross below the top right-hand cross). The box shown in Figure 6.27 appears for you to enter the name for the table.

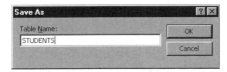

Figure 6.27

▶▶

395

Enter the table name 'STUDENTS' and click OK.

9   The table is now saved and appears in the list of database tables (Figure 6.28).

Figure 6.28

10 Click Open. A blank datasheet will appear, into which you can now enter the data for the table. Enter the student data into the datasheet as shown in Figure 6.29.

	Student number	Surname	Forenames	Street	Town	Postcode	Tel no	Date of birth
+	1	Hughes	John	34 Moor Lane	Crosby	L23 4RT	0151-978-2221	12/03/84
+	2	Prescott	Jayne	12 Fir Close	Formby	L23 7YH	01704-789020	01/06/85
+	3	Edwards	Amy	23 Pit Street	Waterloo	L22 5ED	0151-345-0989	30/12/80
+	4	Johnson	Paul	14 Main Street	Waterloo	L22 7YH	0151-345-0234	17/10/69
+	5	Roberts	Hayley	23 Morton Road	Crosby	L23 5ER	0151-978-2838	20/05/70
+	6	Scott	Jenny	56 Empress Drive	Waterloo	L22 7YH	0151-345-8766	29/04/80
+	7	Bulmer	Grant	12 Seel Street	Crosby	L23 6HT	0151-978-0011	01/12/83
+	8	Butler	Colin	111 Moor Avenue	Crosby	L23 7YR	0151-978-0034	27/11/82
+	9	Hale	Paul	12 Hall Lane	Crosby	L23 5EF	0151-978-0192	09/07/85
+	10	Grant	Richard	67 Elm Drive	Waterloo	L22 7BV	0151-978-6234	02/04/81
+	11	Whitley	Frank	45 Moor Avenue	Crosby	L23 6TH	0151-978-0654	26/05/79
+	12	Doyle	Shirley	102 Copps Drive	Waterloo	L23 7DE	0151-345-4297	13/12/79
+	13	Edwards	Paul	23 Pit Street	Waterloo	L22 5ED	0151-345-0989	30/06/81
+	14	Hanks	Anne	87 Seel Street	Crosby	L23 4CD	0151-978-0021	06/03/81
+	15	Hall	Tanya	34 College Road	Waterloo	L22 8YG	0151-978-4282	07/09/86
▶	(AutoNumber)							

Figure 6.29

11 Close the table window (by clicking on the cross). The data in this table will be saved automatically.

Creating the structures for the other three tables

We now need to create the structures for the other three tables: COURSES, LECTURERS and ENROLMENTS.

Set up these tables with the structures shown in Figure 6.30.

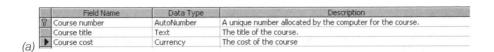

Field Name	Data Type	Description
Course number	AutoNumber	A unique number allocated by the computer for the course.
Course title	Text	The title of the course.
Course cost	Currency	The cost of the course

(a)

Field Name	Data Type	Description
Lecturer number	AutoNumber	A unique number allocated by the computer to each lecturer.
Lecturer name	Text	The name of the lecturer

(b)

Field Name	Data Type	Description
Student number	Number	The unique number given to the student.
Course number	Number	The unique number for the course the student is to be enrolled on.
Lecturer number	Number	The unique number for the lecturer who is responsible for the course.
Date enrolled	Date/Time	The date the enrolment for the course was made.

(c)

Figure 6.30
(a) The COURSES table
(b) the LECTURERS table
(c) the ENROLMENTS table

There are three primary keys in the ENROLMENTS table. To turn the three fields into primary keys, highlight them using the field select arrow, then click on the primary key icon in the toolbar (Figure 6.31).

Field Name	Data Type	Description
Student number	Number	The unique number given to the student.
Course number	Number	The unique number for the course the student is to be enrolled on.
Lecturer number	Number	The unique number for the lecturer who is responsible for the course.
Date enrolled	Date/Time	The date the enrolment for the course was made.

Figure 6.31

Creating the relationships between the tables

Before creating the relationships between the tables it is necessary to have an accurate representation of the data in a entity relationship model. Once the fields in the tables have been normalised to 3NF it is best to make sure that the original entity relationship model reflects the true position of the system. During the normalisation process, new entities may have been produced or the entities may no longer have the original attributes. There are now four entities (one for each table). The entity relationship model for the college enrolment system is shown in Figure 6.32.

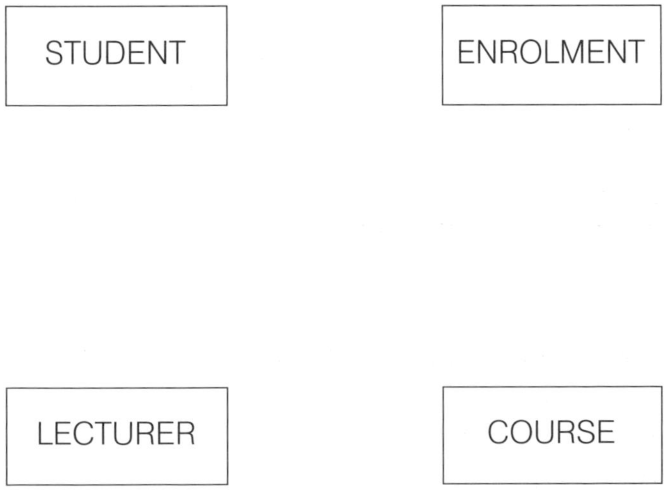

STUDENT   ENROLMENT

LECTURER   COURSE

Figure 6.32 The entities in the college enrolment system

We can use the data in our tables to help us draw this model. We only need to have relationships between those entities that have attributes used as primary or foreign keys. For example Student number appears in both the STUDENT and ENROLMENT entities/tables. This means that there must be a relationship between them.

Looking in the direction STUDENT to ENROLMENT we can see that a particular Student number will appear in many enrolments if the student is on several courses. In the direction ENROLMENT to STUDENT we see that an enrolment (i.e. a particular student on a certain course with a particular lecturer in charge) is only ever for one student. The relationship between STUDENT and ENROLMENT is one-to-many.

/ **Activity 13** /

a   Write down a list of the entities that have a relationship between them and establish in the way shown above the types of relationships involved.

b   Use the above details about the system to produce a final entity relationship diagram.

In a similar way, all the other relationships are established until the following entity relationship model shown in Figure 6.33 is obtained.

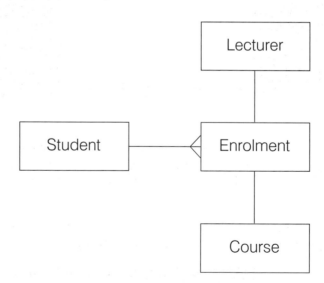

Figure 6.33 The correct entity relationship diagram

This entity relationship diagram can now be used to determine the types of relationships between the tables for our database.

A word on forming relationships in Microsoft Access

In order to form a link (called a relationship) between two tables you must have the same field in both tables. They do not both need to be primary key fields – one can be a primary field and the other a foreign field.

It is important that these common fields have exactly the same name, and the data type must be the same. If AutoNumber has been used as a field in one table then you can use a number field in another table and still be able to link them because they are still both numbers, but you cannot link a text field in one table with a number field in another. Make sure that the name of the field is spelt the same in both fields and check that the use of upper case and lower case letters is consistent. You can create a link between two fields that have identical field names.

It is usually better to create the relationships before data is put into the tables. This is because some data may cause conflicts with existing data.

Worked example:
creating the relationships between the tables

To be able to combine the data from each table the tables must be linked using relationships.

1 Load Microsoft Access and load the college enrolment system. A screen like the one in Figure 6.34 will appear.

Figure 6.34

2 Click on the relationships button on the toolbar (Figure 6.35).

Figure 6.35

3 A screen appears to allow you to tell the computer which tables you will be using to create the relationships (Figure 6.36).

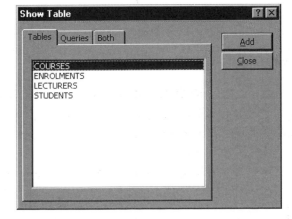

Figure 6.36

▶▶

Highlight each table in turn and click on the Add button.

Then click Close.

4   The four tables are now shown (Figure 6.37). Notice the names of the tables and the highlighted primary key fields.

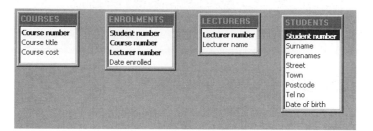

Figure 6.37

Click on Course number in the COURSES table and, keeping the left mouse button pressed down, drag the small rectangle that appears to the Course number field in the ENROLMENTS table. Now release the mouse button. The screen in Figure 6.38 appears.

Figure 6.38

This relationship is one-to-many. Click on Enforce Referential Integrity (i.e. make sure that there is a tick in the box). Now click Create. A line will be drawn between the two tables. Notice how the one-to-many relationship is indicated on the line.

5   Now make a similar relationship between STUDENTS and ENROLMENTS. It is important to click on the field that is the one part of the relationship and drag to the many part. This means that you should click on Student number in STUDENTS first and drag across to Student number in the ENROLMENTS table. Again ensure that referential integrity is enforced. Your screen showing the relationships will now look like Figure 6.39.

6   The final relationship – between the LECTURERS and ENROLMENTS tables – is a one-to-many relationship. Create this relationship. Your screen should now look like Figure 6.40.

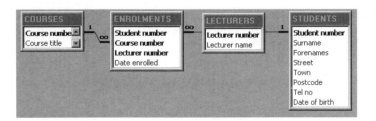

Figure 6.39

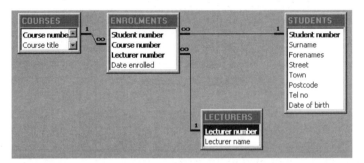

Figure 6.40

Some of the tables have been moved because some of the relationship lines crossed each other, which was confusing. To re-arrange the boxes click on the blue part of the table and, with the left mouse button pressed down, drag it into position.

7 Click on the cross to close the window. You will be asked if you want to save the relationships. Click on Yes.

Worked example: entering data into the COURSES, LECTURERS and ENROLMENTS tables

What does the term 'enforcing referential integrity' mean?

1 Open the COURSES table in datasheet view (the view with the rows and columns) and enter the data shown in Figure 6.41. To make the column for the Course title wider click on the line between the two columns and drag it until the required width is reached.

	Course number	Course title	Course cost
⊞	1	BND Computer Studies Year 1	£3,500.00
⊞	2	BND Computer Studies Year 2	£3,500.00
⊞	3	HND Business IT Year 1	£4,500.00
⊞	4	HND Business IT Year 2	£4,500.00
⊞	5	Access To Business IT	£3,750.00
⊞	6	BNC Computer Studies Year 1	£2,100.00
⊞	7	BNC Computer Studies Year 2	£2,100.00
⊞	8	AAT Final Daytime	£1,750.00
⊞	9	AAT Final Evening	£1,750.00
⊞	10	ILEX Year 1	£2,000.00
⊞	11	ILEX Year 2	£2,200.00

Figure 6.41

2 Close the window and click Yes when you are asked if you want to save.

▶▶

401

3 Now place the data into the LECTURERS table (Figure 6.42). You will need to widen the Lecturer name column in order to accommodate the data. Save the data in this table.

	Lecturer number	Lecturer name
⊞	1	Peter Hughes
⊞	2	Liz Patterson
⊞	3	James Inghams
⊞	4	Pat Fitzroy
⊞	5	Jack Trainor
⊞	6	Paul Wods
⊞	7	Suzanne Sanderson
⊞	8	Frank Morley
⊞	9	Jane Dickson
⊞	10	Anne Dickson
▶	(AutoNumber)	

Figure 6.42

4 Now enter the data for the ENROLMENTS table, but before you do so it is important to note the following:

   a Do not try to enrol a student on a course we do not have. As referential integrity has been enforced the computer will check to make sure that any Course number we enter appears in the COURSES table.

   b Do not allow anyone who is not a student to enrol on a course. Again, by enforcing referential integrity the computer checks any Student number with the STUDENTS table to make sure that it exists.

   We have only 15 students so the only valid numbers that can be entered into the Student number field are from 1 to 15. There are also only 11 courses to enrol on, so only numbers from 1 to 11 can be entered for the Course number.

5 Enter the data as shown in Figure 6.43 into the ENROLMENT table.

Student number	Course number	Lecturer number	Date enrolled
1	2	8	10/09/00
2	1	7	11/09/00
3	2	8	10/09/00
4	5	1	12/10/00
5	5	1	05/10/00
6	4	3	07/09/00
6	11	5	10/09/00
7	9	7	10/09/00
8	11	5	12/10/00
9	5	1	08/12/00
9	9	6	03/10/00
10	1	7	04/10/00
11	2	8	10/09/00
12	4	3	10/10/00
13	4	3	10/09/00
14	5	1	05/09/00
15	2	8	05/09/00

Figure 6.43

6 Close the table and save the data. You have now completed the relational database.

Asking questions of the database

Because there are relationships between the four tables, we can combine and then extract the data from more than one table by using a query. Queries are used to ask questions of the database.

Creating queries to extract information using the Query Wizard

Suppose we want details of all the students and the courses they are taking along with the names of the lecturers responsible for the course. To do this we will need to use the data from more than one table.

1 Load Microsoft Access and select the College Enrolment System database. The screen shown in Figure 6.34 appears.

2 Click on Queries in the Objects section of the screen. The queries screen appears (Figure 6.44). Click on 'Create query by using wizard'.

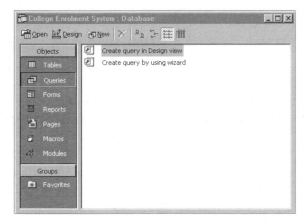

Figure 6.44

3 The Query Wizard will guide you through the steps in producing a query. The first step is to select the table from which the first fields come. Select the table STUDENT and then click on Surname in the 'Available Fields' box. Click on the Add fields button indicated by the single arrow. 'Surname' will now appear in the Selected Fields box. Your screen will now look like the one in Figure 6.45.

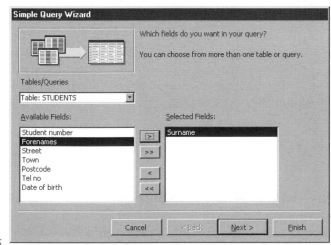

Figure 6.45

▶▶

403

4 In a similar way add the following fields to the 'Selected Fields' box.

- Forenames
- Course title
- Lecturer name

(NB: you will need to change the table in the Tables/Queries box to find these fields).

After doing this, your screen will look like Figure 6.46.

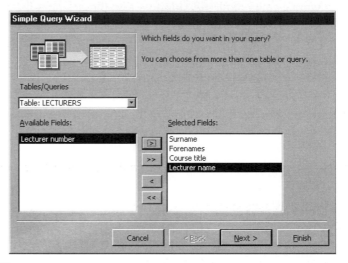

Figure 6.46

5 Click Next>. The dialogue box shown in Figure 6.47 appears.

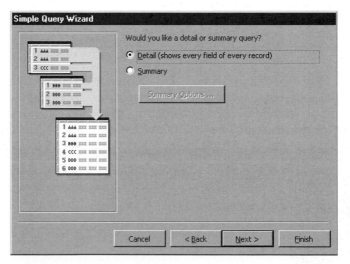

Figure 6.47

Make sure that 'Detail' has been selected and then click Next>.

6 Change the name of the query to 'Enrolment details' (Figure 6.48).

Figure 6.48

Click on Finish.

The query will now run automatically and the results will be shown (Figure 6.49).

Surname	Forenames	Course title	Lecturer name
Hughes	John	BND Computer Studies Year 2	Frank Morley
Edwards	Amy	BND Computer Studies Year 2	Frank Morley
Prescott	Jayne	BND Computer Studies Year 1	Suzanne Sanderson
Grant	Richard	BND Computer Studies Year 1	Suzanne Sanderson
Whitley	Frank	BND Computer Studies Year 2	Frank Morley
Hall	Tanya	BND Computer Studies Year 2	Frank Morley
Johnson	Paul	Access To Business IT	Peter Hughes
Roberts	Hayley	Access To Business IT	Peter Hughes
Hanks	Anne	Access To Business IT	Peter Hughes
Bulmer	Grant	AAT Final Evening	Suzanne Sanderson
Hale	Paul	AAT Final Evening	Paul Woods
Hale	Paul	Access To Business IT	Peter Hughes
Scott	Jenny	HND Business IT Year 2	James Inghams
Doyle	Shirley	HND Business IT Year 2	James Inghams
Edwards	Paul	HND Business IT Year 2	James Inghams
Scott	Jenny	ILEX Year 2	Jack Trainor
Butler	Colin	ILEX Year 2	Jack Trainor

Figure 6.49

7   Now close the window. The query will now be listed (Figure 6.50).

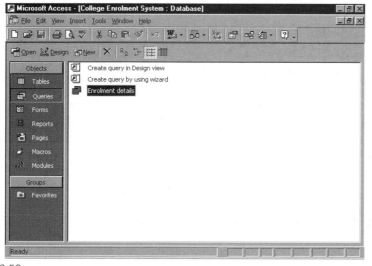

Figure 6.50

▶▶

405

8 Click on the design icon and you will see the query design (Figure 6.51).

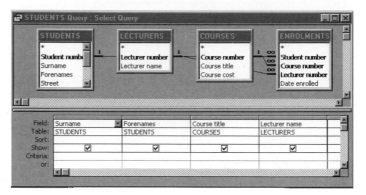

Figure 6.51

Notice the list of the fields and the tables they come from. Using the query design you can sort the query into different orders as well as specify search criteria.

Click on the Sort box under the Surname field. A drop-down list appears (Figure 6.52), from which you should select Ascending.

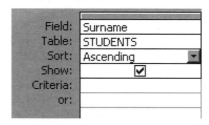

Figure 6.52

9 Close the window. You will now be asked if you want to save the query. Click Yes. The screen shown in Figure 6.50 appears.

10 Make sure that the name of the query is highlighted and click on the Open button. The results of the query will be sorted into ascending order, as shown in Figure 6.53.

Surname	Forenames	Course title	Lecturer name
Bulmer	Grant	AAT Final Evening	Suzanne Sanderson
Butler	Colin	ILEX Year 2	Jack Trainor
Doyle	Shirley	HND Business IT Year 2	James Inghams
Edwards	Amy	BND Computer Studies Year 2	Frank Morley
Edwards	Paul	HND Business IT Year 2	James Inghams
Grant	Richard	BND Computer Studies Year 1	Suzanne Sanderson
Hale	Paul	Access To Business IT	Peter Hughes
Hale	Paul	AAT Final Evening	Paul Woods
Hall	Tanya	BND Computer Studies Year 2	Frank Morley
Hanks	Anne	Access To Business IT	Peter Hughes
Hughes	John	BND Computer Studies Year 2	Frank Morley
Johnson	Paul	Access To Business IT	Peter Hughes
Prescott	Jayne	BND Computer Studies Year 1	Suzanne Sanderson
Roberts	Hayley	Access To Business IT	Peter Hughes
Scott	Jenny	HND Business IT Year 2	James Inghams
Scott	Jenny	ILEX Year 2	Jack Trainor
Whitley	Frank	BND Computer Studies Year 2	Frank Morley

Figure 6.53

Putting search criteria into queries

All of the following can be used for the search criteria in a query.

=	Equal to
>	Greater than
@	Less than
>=	Greater than or equal to
@=	Less than or equal to
@>	Is not equal to

1   Open the previous query in design view.

    In the criteria section of the Lecturer name field enter the criteria =Peter Hughes.

    The computer will automatically put the text in inverted commas as soon as you enter the criteria (Figure 6.54).

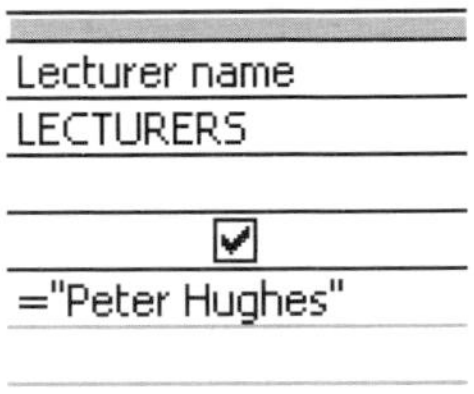

2   Click on the Open button. The results of the query will be displayed (Figure 6.55).

Lecturer name
LECTURERS

☑
="Peter Hughes"

Figure 6.54

Surname	Forenames	Course title	Lecturer name
Hale	Paul	Access To Business IT	Peter Hughes
Hanks	Anne	Access To Business IT	Peter Hughes
Johnson	Paul	Access To Business IT	Peter Hughes
Roberts	Hayley	Access To Business IT	Peter Hughes

Figure 6.55

3   Close the window and, when asked if you want to save the query, click Yes.

Worked example: performing an advanced query

The principal of the college has announced that some funds will be made available for the more mature students in the college (21 years or older). He would like you to produce a list of the students eligible for the award showing their surname, forenames, street, postcode, date of birth and age.

1   Load the College Enrolment System database and create a new query using the wizard.

2   Add the fields to the Selected Fields list, as shown in Figure 6.56.

3   Click on Next>. Notice that we cannot add the field Age – this has not been stored in the table as it can be calculated by subtracting the student's date of birth from today's date.

    Enter the name for the query (Figure 6.57).

    Click on Finish.

4   The query results are displayed (Figure 6.58). All the records are shown because we have not entered any search criteria. We also need to calculate the age and display it.   ▶▶

407

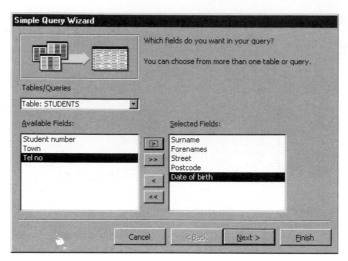

Figure 6.56

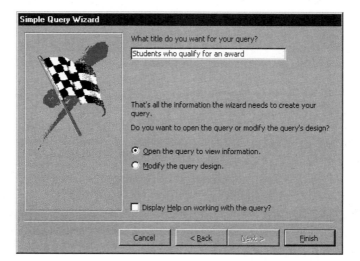

Figure 6.57

Figure 6.58

5 Click on the query design icon in the toolbar. In the next blank column, enter the formula Age: Year(Now())-Year([Date of birth]) to calculate age. Your screen should look like Figure 6.59.

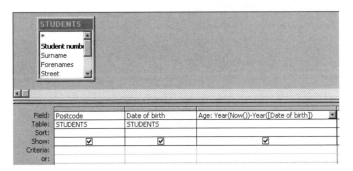

Figure 6.59

6 Close the window and save the changed design for the query.

7 From the database screen double-click on the query 'Students who qualify for an award' to see the results (Figure 6.60).

Surname	Forenames	Street	Postcode	Date of birth	Age
Hughes	John	34 Moor Lane	L23 4RT	12/03/84	16
Prescott	Jayne	12 Fir Close	L23 7YH	01/06/85	15
Edwards	Amy	23 Pit Street	L22 5ED	30/12/80	20
Johnson	Paul	14 Main Street	L22 7YH	17/10/69	31
Roberts	Hayley	23 Morton Road	L23 5ER	20/05/70	30
Scott	Jenny	56 Empress Drive	L22 7YH	29/04/80	20
Bulmer	Grant	12 Seel Street	L23 6HT	01/12/83	17
Butler	Colin	111 Moor Avenue	L23 7YR	27/11/82	18
Hale	Paul	12 Hall Lane	L23 5EF	09/07/85	15
Grant	Richard	67 Elm Drive	L22 7BV	02/04/81	19
Whitley	Frank	45 Moor Avenue	L23 6TH	26/05/79	21
Doyle	Shirley	102 Copps Drive	L23 7DE	13/12/79	21
Edwards	Paul	23 Pit Street	L22 5ED	30/06/81	19
Hanks	Anne	87 Seel Street	L23 4CD	06/03/81	19
Hall	Tanya	34 College Road	L22 8YG	07/09/86	14

Figure 6.60

8 The ages are displayed using a calculated field in the query. However, we still need to pick out those students who are 21 years or older, so we need to add this search criterion for the calculated Age field: >=21. Enter this as shown in Figure 6.61.

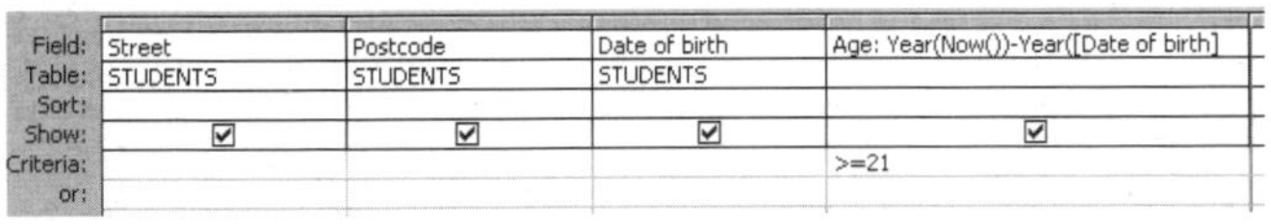

Field:	Street	Postcode	Date of birth	Age: Year(Now())-Year([Date of birth])
Table:	STUDENTS	STUDENTS	STUDENTS	
Sort:				
Show:	☑	☑	☑	☑
Criteria:				>=21
or:				

Figure 6.61

9 Close the window, save the new design and then run the query. The results shown in Figure 6.62 will be obtained.

	Surname	Forenames	Street	Postcode	Date of birth	Age
▶	Johnson	Paul	14 Main Street	L22 7YH	17/10/69	31
	Roberts	Hayley	23 Morton Road	L23 5ER	20/05/70	30
	Whitley	Frank	45 Moor Avenue	L23 6TH	26/05/79	21
	Doyle	Shirley	102 Copps Drive	L23 7DE	13/12/79	21
*						

Figure 6.62

You have now completed this query.

You can use Activity 14 to provide evidence for key skills for Information Technology IT2.1 or IT3.1.

Activity 14

a Here are some search criteria used in queries. Explain in simple terms what each one does:

- =2001 (in a date field)
- <=12/02/98 (in a date field)
- ='Jones' (in a text field for a surname)
- >=45 (in a number field for quantity)
- <>0 (in a number field for quantity)
- <0 (in a currency field for account balance)
- <31/12/00 AND <01/01/02 (in a date field).

b Create the search criteria from the description given:

- The date equal to 1999
- A date before the date 03/05/91
- A date equal to or after the date 1999
- A date between the dates 07/09/41 and 06/09/45
- A number greater than 20
- A number less than or equal to 45
- A number that is not equal to zero
- A currency amount that is non zero.

c There is to be a student satisfaction survey, and the students will be questioned about the quality of the course they are on. The survey only applies to students who are aged 18 or under. The survey is to be conducted by phone and the staff who are doing the survey have asked for a list containing the following fields: Surname, Forenames, Tel. no. and Course Title. Construct the search criteria.

d The head of the business studies department at the college needs a list of the lecturers who are responsible for the course along with the names of all the students on the course. The list needs to be organised so that the lecturer name and course title are shown and then the student number and their surname and forenames are shown.

e Some students (or their employers if they are working), have to pay their own fees. Where the course fees are over £3000 there is the possibility of an interest-free loan. The office needs a list of the course details (Course number, Course title and Course cost) for all the courses where the fees are over £3000. Produce a query to supply this information.

f Students frequently forget their student number. The office has suggested that lists of students need to be produced containing their surname in alphabetical order along with their forenames, date of birth, student number and the course numbers of the course/courses they are doing. Construct the query to give you this list.

Printing a query

To avoid wasting time and paper it is always best to preview a query before printing to check it. To preview the query you use the print preview icon on the toolbar (Figure 6.63).

Print preview allows you to see how the report appears on the page before printing (Figure 6.64).

Once you are happy with the results, click on the print icon for a printout.

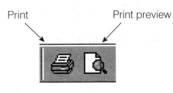

Print   Print preview

Figure 6.63

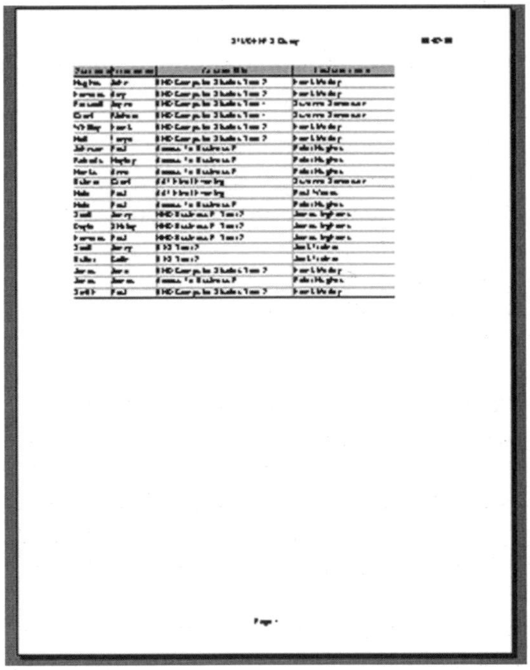

Why should print preview always be used before printing?

Figure 6.64

Deleting a query

Some queries are only needed once so, to avoid cluttering up the query selection screen with lots of old ones, it is best to get into the habit of deleting them. To delete a query, select the query object in the database screen and then select the query you want deleting. Click on the delete icon in the toolbar (Figure 6.65).

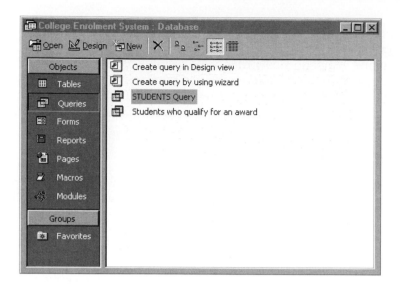

Figure 6.65

Worked example: producing a report

Reports are used to present the output from a database. You can of course print the results of queries or contents of tables and forms but you get more control over the appearance when you use reports. Reports are ideal when the data needs to be presented on paper.

1   Load Microsoft Access and then the Student Enrolment Database.

2   Click on Reports in the list of objects. The screen shown in Figure 6.66 should appear.

Figure 6.66

3   Click on 'Create report by using wizard'. The screen shown in Figure 6.67 will appear. As with the queries, you can select the tables and the fields you want to include in the report. You can also select previously stored queries.

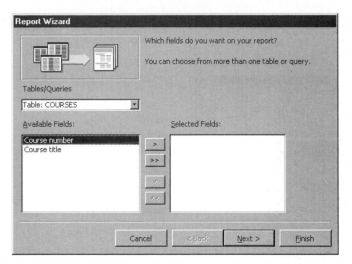

Figure 6.67

4 Select the fields Course number, Course title and Lecturer name (you will need to change the table to get this field). Your screen will now look like Figure 6.68.

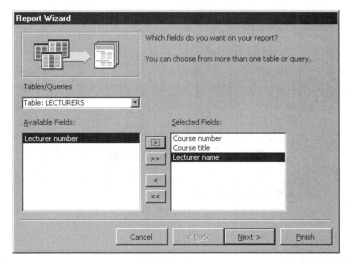

Figure 6.68

All the fields we want to include in the report have now been included so click Next>. The next screen of the report wizard appears.

5 We want to see the course numbers and the names of the courses each lecturer is involved with, so we need to change How do you want to view your data? to by COURSES (Figure 6.69) – you can see the arrangement of the report on the left-hand side of the screen.

6 Click Next>.The next step appears (Figure 6.70). Here we can group the data, but as the Course number and Course title fields are both unique, we do not need to bother about any groupings. Just click Next>.

7 You can now sort the report according to certain fields (Figure 6.71). Click on the drop-down arrow and select Course number from the list.

▶▶

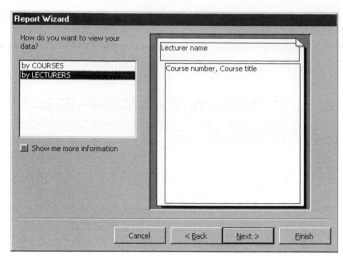

Figure 6.69

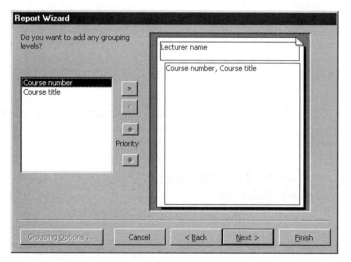

Figure 6.70

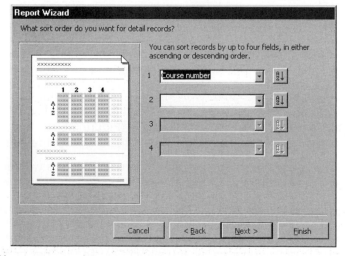

Figure 6.71

8 Click Next>. The screen in Figure 6.72 appears, to let you select the way the report is laid out. You can see the results of making alterations to the layout on the preview screen, and you can alter the orientation of the paper. Keep the settings as shown in Figure 6.72.

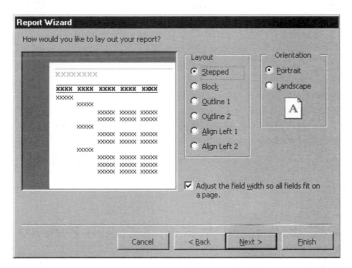

Figure 6.72

9 Click Next>. You can now select the style of the report (Figure 6.73). Select Corporate and then click Next>.

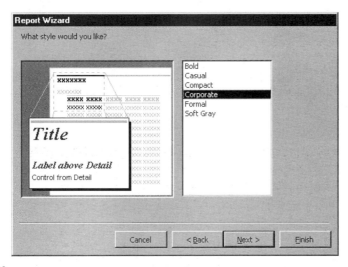

Figure 6.73

10 Use the screen shown in Figure 6.74 to give the report a title. This is the name it will be saved under. Type in 'Course list according to lecturer'.

11 Click on Finish. The report will be saved and displayed on the screen (Figure 6.75).

▶▶

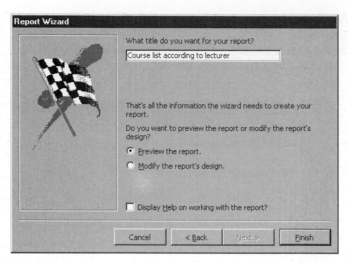

Figure 6.74

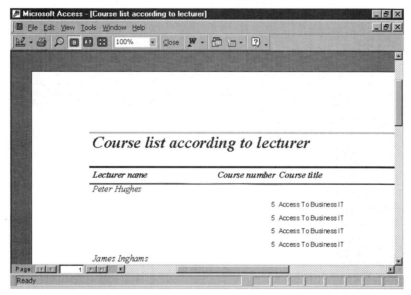

Figure 6.75

12 Click on the left mouse button and the report will appear as a single page (Figure 6.76). The magnifying glass can be positioned on any part of the report you want to see in more detail.

13 Close the screen and the database screen appears with the report listed (Figure 6.77).

14 To open the report, just double-click on it. To produce a printout of the report, check first that the printer is switched on, on-line and has paper, then click on the print button in the toolbar.

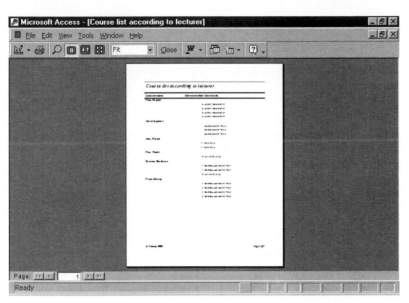

Figure 6.76

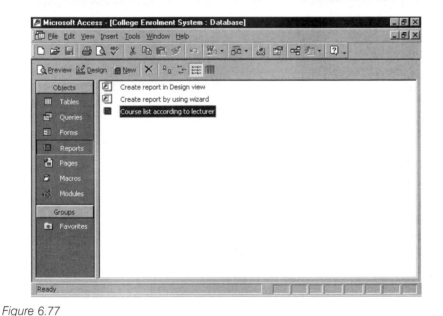

Figure 6.77

Printing the results of a query in report format

1   Create the query.

2   At the database screen click on Report by using wizard.

3   In the Tables/Query box, click on the pull-down arrow and click on the query 'Students who qualify for an award'.

4   The available fields are listed. You can choose to select just some of the fields or all of them. In this case select all of them by clicking on the button with the two right-pointing arrows on it.   ▶▶

5 Click Next>. The screen shown in Figure 6.78 will appear. We do not need any groupings in this report so just click on Next>.

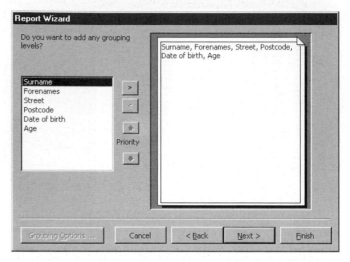

Figure 6.78

6 You can now decide the order you want any of the fields sorted into (Figure 6.79). Choose Surname in ascending order (i.e. A to Z). Your screen will look the same as the one shown here. Click Next>.

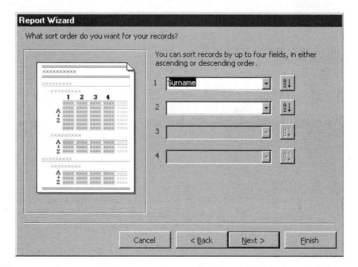

Figure 6.79

7 Select Tabular for the layout (Figure 6.80).

8 Click on Next>. The screen in Figure 6.81 appears.

   Select the Corporate styles and then click Next> (Figure 6.82).

9 Check that 'Students who qualify for an award' is entered in the box.

10 Click Finish. The report will now be shown on the screen (Figure 6.83).

11 Check the results and produce a printout.

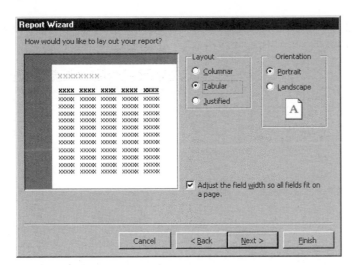

Figure 6.80

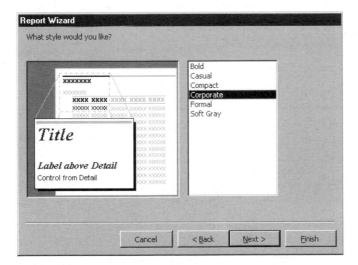

Figure 6.81

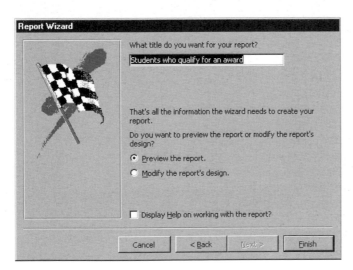

Figure 6.82

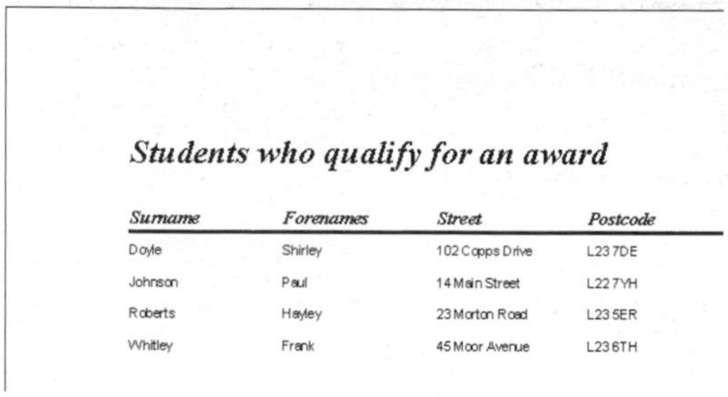

Students who qualify for an award

Surname	Forenames	Street	Postcode
Doyle	Shirley	102 Copps Drive	L23 7DE
Johnson	Paul	14 Main Street	L22 7YH
Roberts	Hayley	23 Morton Road	L23 5ER
Whitley	Frank	45 Moor Avenue	L23 6TH

Figure 6.83

Creating a form to enter data into a single table

Up till now, data has been entered into each table using the datasheet view but there are a number of problems with this. Firstly, because the data in the datasheet is arranged in columns, when there are lots of columns, you cannot see the contents of a single record on the screen at once. Secondly, an inexperienced user can find the entry of the data into the datasheet daunting.

It is much better to enter the data into a form. Here we will create a form that can be used to enter the data into a single table (the Courses table).

1   Load Microsoft Access and open the College Enrolment System database.

2   At the database screen, select the object Forms. You will now see the screen shown in Figure 6.84.

Figure 6.84

3   Double-click on 'Create form by using wizard'. The wizard starts and you will be asked which tables and fields you want to include in the form (Figure 6.85).

▶▶

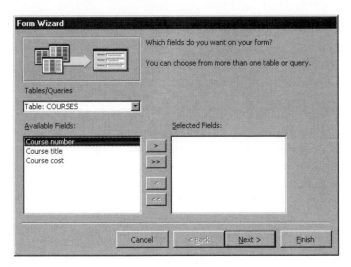

Figure 6.85

4   Click on the button with the two right-pointing arrows. This adds all the fields in Available Fields to Selected Fields. Your screen will now look like Figure 6.86.

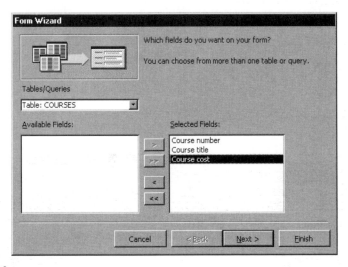

Figure 6.86

5   Click on Next>. You will now be asked what layout you would like for your form (Figure 6.87). Click on each layout to see what they look like. Select Columnar. Click Next>.

6   You will now be asked to choose the style you would like (Figure 6.88). Click on each style in turn to see what each of them looks like. Select Standard. Click on Next>.

7   You will now be asked for a name for the form (Figure 6.89). Enter 'Form for entering data into the COURSES table'.

8   The design of the form is now complete. Ensure that 'Open the form to view or enter information' has been checked and then click on Finish.

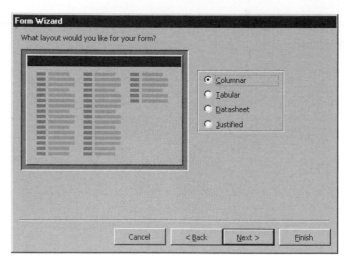

Figure 6.87

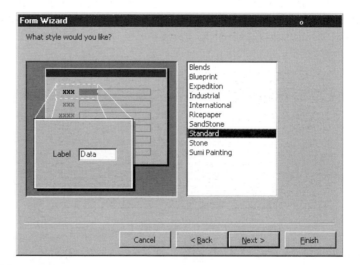

Figure 6.88

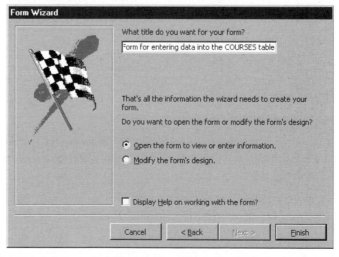

Figure 6.89

9   The form will be saved and opened ready to view or enter data (Figure 6.90).

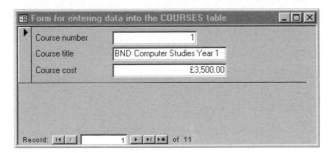

Figure 6.90

10   Using the record indicators shown at the bottom of the form you can move from one record to another or obtain a blank record ready to enter data. Choose New record and enter the new course details, as shown in Figure 6.91. Close the window.

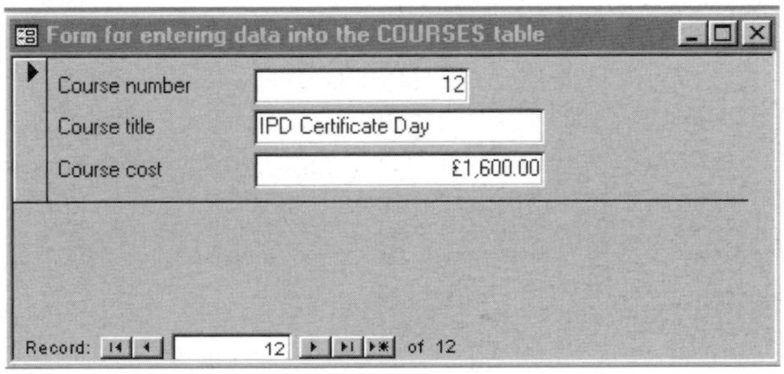

Figure 6.91

11   The form is now listed as a form object (Figure 6.92).

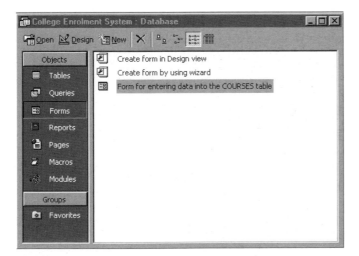

Figure 6.92

/ **Activity 15** /

Using the Form Wizard, create a form to enable entry of the lecturers' details into the lecturers table. Call this form 'Form for entering data into the LECTURERS table'.

Creating forms to enable entry of data into multiple tables

A form can be based on a table and used to enter the details into a single table. However, as well as creating forms from tables, it is useful to be able to create them from queries. Queries can be constructed to use more than one table so a form created from a query can be used to enter the data into more than one table at a time.

Creating a form to enter data into two tables at once

If a student comes along to the college to enrol on a course, the enrolling tutor will take their personal details (name, address, date of birth, etc.) and the details of the course or courses they are enrolling on. Rather than fill in the details into the STUDENT table and then the ENROLMENT table it would be easier if we used the same form to do both. The same student can be enrolled on more than one course so we will need to enter all the courses that they want to enrol on once their personal details have been entered.

To combine two forms we need to create them separately, thinking carefully about the design of the form and in particular the fields that will need to be input. As we will be using this form to enter the data into both the STUDENTS and the ENROLMENTS tables, we will need to include all the fields in these tables. Any field that appears in each table need only be entered once. Not having to enter this duplicate data is one big advantage in using one form to fill in the contents of two tables.

You are not restricted to just filling in two tables. You can use one form to enter the details into lots of tables provided it is sensible to do so.

Two forms need to be created: one to hold the personal details about the students and another to hold the details of the courses they are on.

A main form called 'Student's data entry form' is created to enable the entry into the STUDENTS table. Another form called 'Enrolment details sub form' is created separately for entering the details into the ENROLMENTS table. Look carefully at the design of the form that we are trying to create. It is always best to produce a form like this first (Figure 6.93).

As you can see from the proposed form, it is best to arrange fields on the main (**parent**) **form** in rows and the enrolment fields in the **sub form** in columns.

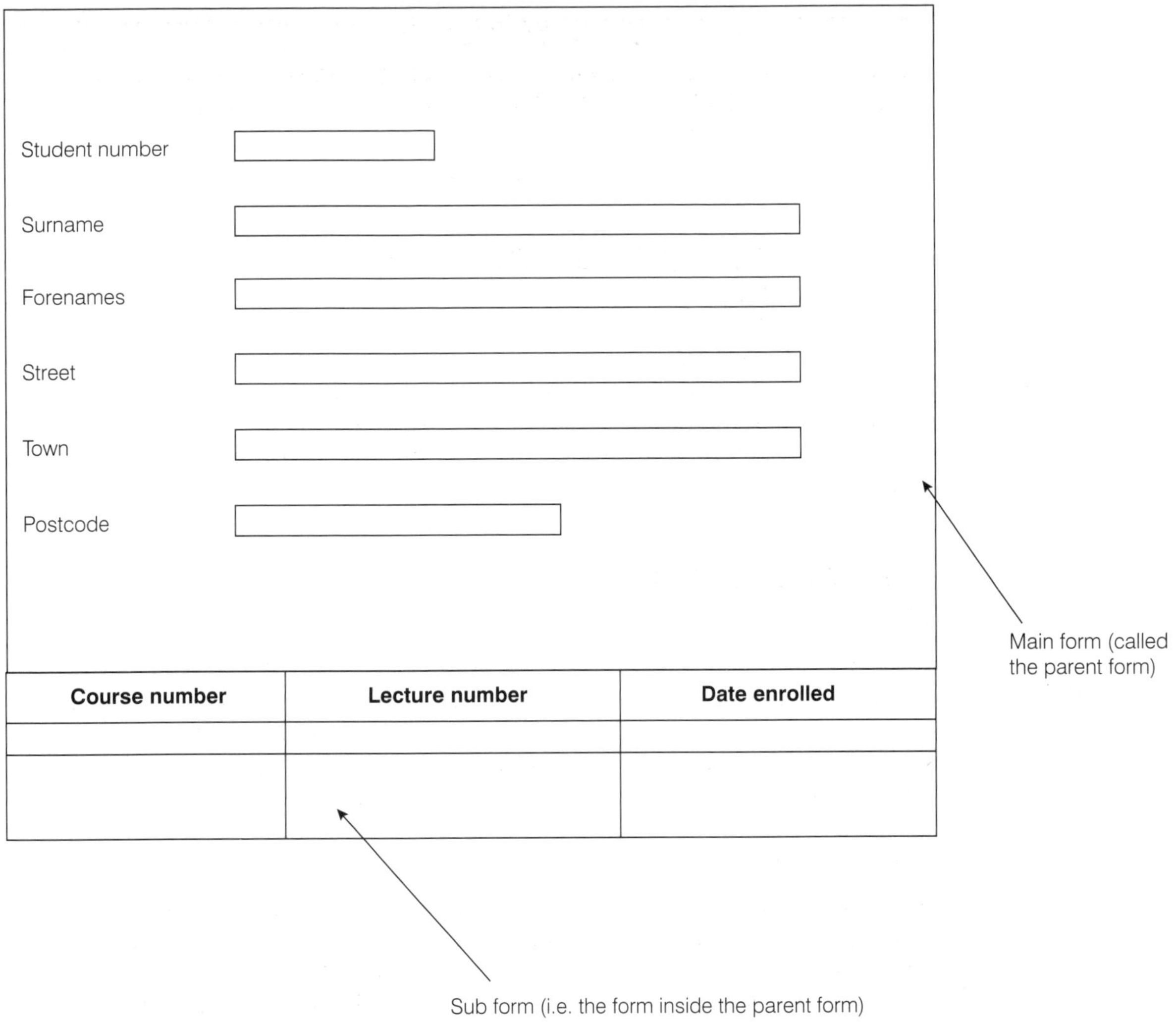

Student number [   ]

Surname [   ]

Forenames [   ]

Street [   ]

Town [   ]

Postcode [   ]

Course number	Lecture number	Date enrolled

Main form (called the parent form)

Sub form (i.e. the form inside the parent form)

Figure 6.93

Creating forms and sub forms

As one student can be on many courses, we can show the many courses part of the relationship as a subform, and the one part of the relationship can be the parent form.

1   Load the College enrolment system database.

2   Create a form called 'Student's data entry form' to enter all the fields into the STUDENTS table (i.e. Student number, Surname, Forenames, Street, Town, Postcode, Tel. no. and Date of birth).

3   Create a new form called 'Enrolment details sub form' and include in this form all the fields included in the ENROLMENTS table – except for Student number because this appears in the main 'Student's data entry form' (which is the parent form).

   The following fields should be included in this form: Course number, Lecturer number and Date enrolled.   ▶▶

4 Go back to the database window (Figure 6.94) and you will see the two forms just created. Open the main/parent form (i.e. the Student's data entry form) in design view. The screen you will see is shown in Figure 6.95.

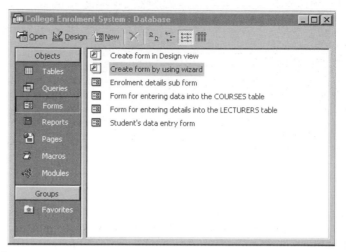

Figure 6.94

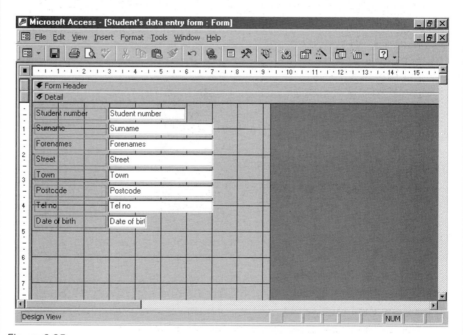

Figure 6.95

5 Click on Window and from the drop-down menu select Tile vertically. The screen now divides in two, with the subform on one side and the database screen with the forms listed on the other (Figure 6.96).

6 Click Form Footer. The cursor changes to a double-headed arrow. Keep your finger on the mouse button and drag the footer further down to allow more room between the Date of birth field and the Form Footer to enable the sub form to be inserted.

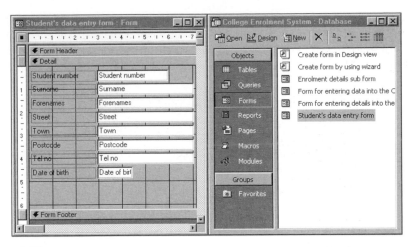

Figure 6.96

7 In the database window click Enrolment details sub form and drag it across to the gap you have just made between the Date of birth field and the Form Footer on the parent form (Figure 6.97).

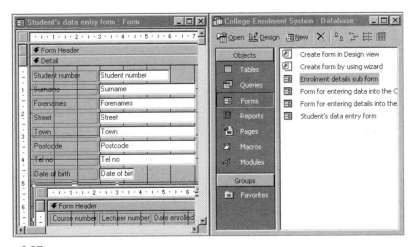

Figure 6.97

8 You can now size and move this window into position using the handles. Close the parent window (i.e. the one containing the forms). When asked to save, do so. Close the other window.

9 At the database window, the new form called 'Student's data entry form' can be opened by clicking on it. The form shown in Figure 6.98 appears.

10 You have now created the parent form and subform and have saved them under the parent name 'Student's data entry form'. You can use this form to view existing records and put data into both the ENROLMENTS and STUDENTS tables simultaneously.

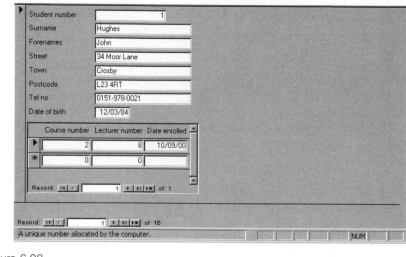

Figure 6.98

Activity 16

a Use the form you have just created to enter the following new students and their enrolment details.

Surname	Jones
Forenames	Jane
Street	4 Court Hey
Town	Waterloo
Postcode	L22 6TT
Tel no	0151-345-0128
Date of birth	12/09/86
Course number	2
Lecturer number	8
Date enrolled	07/09/00

The completed form for the first student is shown in Figure 6.99.

Surname	Jones
Forenames	James
Street	4 Court Hey
Town	Waterloo
Postcode	L22 6TT
Tel no	0151-345-0128
Date of birth	13/03/85
Course number	5
Lecturer number	1
Date enrolled	07/09/00

Surname	Smith
Forenames	Paul
Street	46 Geraints Way
Town	Waterloo
Postcode	L22 7TY
Tel no	0151-345-0092
Date of birth	12/08/84

▶▶

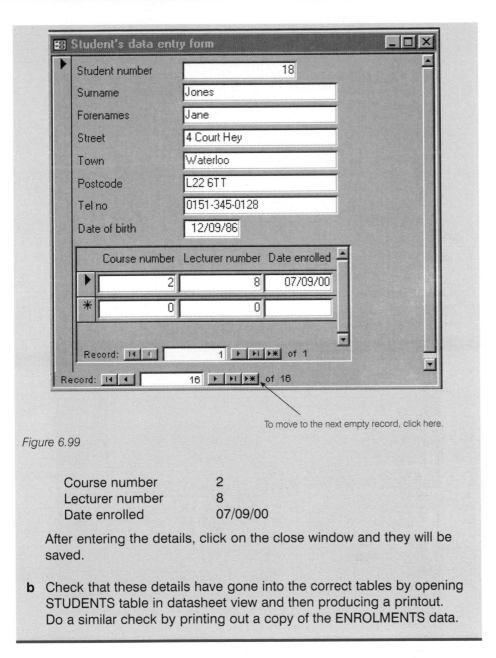

Figure 6.99

Course number	2
Lecturer number	8
Date enrolled	07/09/00

After entering the details, click on the close window and they will be saved.

b Check that these details have gone into the correct tables by opening STUDENTS table in datasheet view and then producing a printout. Do a similar check by printing out a copy of the ENROLMENTS data.

Creating a label

Sometimes we need to produce a stand-alone label in a form, perhaps because we want to produce a message or give a title to the form. We can also produce labels in reports.

Worked example: producing a label in a form

1 Open the form 'Form for entering data into the COURSES table' in design view.

2 Increase the distance between the Form Header and the Detail sections by clicking on the boundary (i.e. the line) above Detail and drag it to make the area above bigger or smaller. In this case we need to enlarge it to a size similar to that shown in Figure 6.100. ▶▶

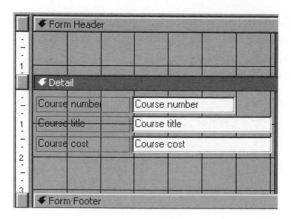

Figure 6.100

3 Open the toolbar by clicking the toolbox icon 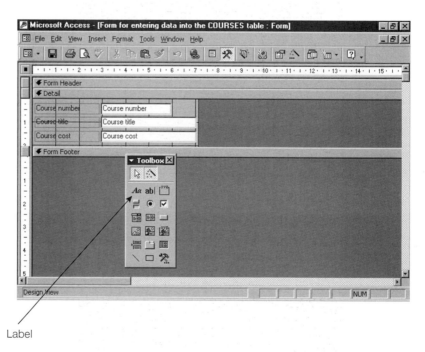 in the toolbar. The toolbox window is opened (Figure 6.101).

Label

Figure 6.101

4 Click on the Label icon in the toolbox as shown in Figure 6.101. Use the cross-wires to position the corner of the text box and then drag the box to the required size, as shown in Figure 6.102.

5 Enter 'Form for entering data about courses' into the label box. Your screen should now look like Figure 6.103.

6 Close the form by closing the window. When asked if you want to save the changes to design, do so.

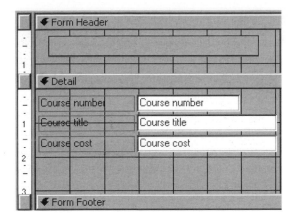

Figure 6.102

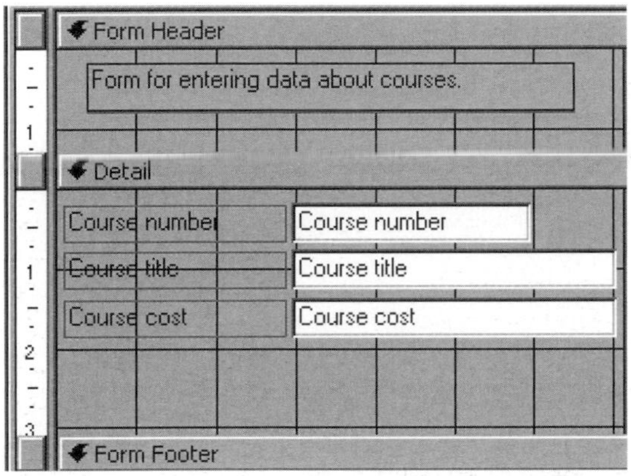

Figure 6.103

7 At the database window, select Form for entering details into the COURSES table by double-clicking it. The form in Figure 6.104 now appears, with the label containing the title at the top.

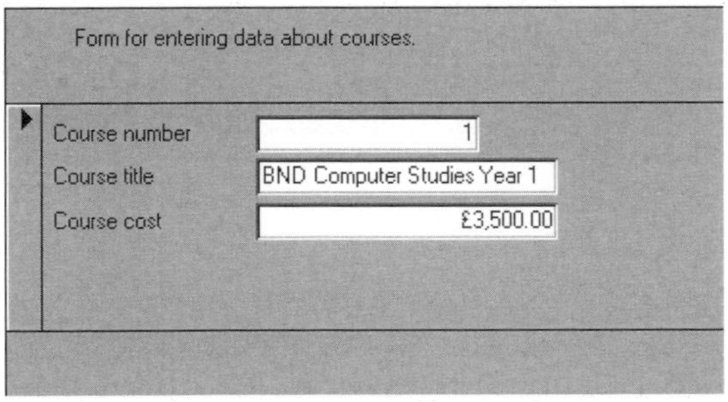

Figure 6.104

Creating tips to help the user know what they are expected to type in

As the user moves the cursor into a data entry point for a field, it is useful for a tip to appear automatically to give them some idea about the type of data that is expected for the field.

1　Open the form called 'Form for entering details into the COURSES table' in design view.

2　Select the control (i.e. the rectangle where the data is entered) for the Course title field. You will see the handles around the field when it has been selected (Figure 6.105).

Figure 6.105

3　Click on the Properties icon  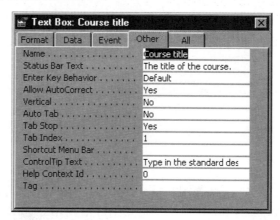 on the toolbar. The box in Figure 6.106 appears. Type in the message 'Type in the standard description of the course'.

Figure 6.106

4　Close the text box window. Open the query and move the mouse so that the pointer is over the field 'Course title'. You will see the message appear (Figure 6.107).

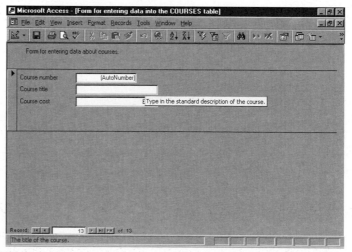

Figure 6.107

Validating data

You can validate data by including validation checks in the tables or in the forms used to enter the data. Some fields are particularly difficult to validate (Surname, Forenames, Street, Town, etc.). Others – such as quantity or date of birth – are easier because the data has to be within a certain range of values.

Validation will pick out:

- data that is entered and is not appropriate for the data type for that field. For example, the user could be trying to enter text into a field that has been set as a Number data type
- data which breaches the validation rule that has been set up for the field
- a field that has been left blank but requires data.

Activity 17

a For each of the tables created, alter the properties in the field properties box for each field so that the user is restricted in their entry of data. The main aim is to narrow down the mistakes that the user can make as much as possible. Make use of as many of the field properties as you can. Save the changes to each of the table designs.

Figure 6.108 shows the field properties for the COURSES table and the Course Title field.

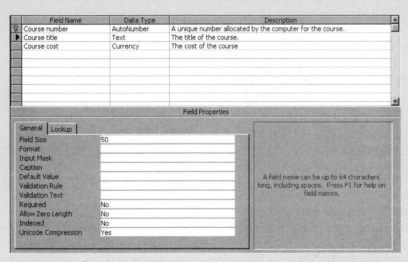

Figure 6.108

b As well as controlling the type of data that can be entered into a table directly, you can validate the data being entered into a table via a form. Find out how this is done and include suitable validation checks on all of the forms that have been created in this section.

Finding out about Microsoft Access for yourself

Microsoft Access is a very large piece of software and in this section we can really only learn a few things about it. Next time you are in a bookshop look at the huge numbers of books devoted to just this one database.

Computer books can be expensive, so build up your knowledge of the software using the on-line help. The on-line help icon has a question mark on it and can be found in the toolbar.

Activity 18

Use the on-line help to find out how to:

- insert a field in a table
- delete a field from a table
- delete a record
- delete a relationship
- create a list box in a form.

6.6 Documentation

It is important to document the development of a database because the original developer might not be around when the structure of the database needs to be changed or developed further. The person who has to do this will need to know how the developer went about the development, and this will be much easier if there is clear documentation to refer to. As well as providing documentation for the specialist (called technical documentation) documentation is needed for the people who will be using the database on a day-to-day basis.

Technical documentation

Technical documentation records the design and development of the database. It would normally include:

- a copy of the specification agreed with the user (this gives an idea of what the initial brief was)
- details of the hardware, software and the other resources used
- a detailed entity relationship diagram
- details of any program code (databases such as Microsoft Access have their limitations and to get them to do certain tasks it may be necessary to write some program code in a computer language called Visual Basic – the language Access is written in)
- details of validation and verification procedures (validation checks should be performed on all fields if possible)
- details of all input and output screens and printed reports.

Remember that, because technical specialists will be reading this documentation, you do not need to explain each technical term used.

User documentation

User documentation helps people to use your database. These people will probably use computers on a day-to-day basis but may have a limited knowledge of IT, and of databases in particular. They are unlikely to understand the technical terminology associated with databases and so any technical terms used will need to be clearly explained.

User documentation would normally include:

- how to start the database (use screen dumps and drawings to help you produce the user documentation)
- how to append, delete and edit records
- examples of screens and data entry forms.

Testing

As with all IT solutions, a database must be thoroughly tested before handing it over to the user. The database must be able to cope with all the data for which it was designed and reject data that is extreme or abnormal. It must also meet the specification agreed with the user, otherwise he or she might reject the database because it does not meet their requirements.

Database applications should be easy to use, especially if most of the users are inexperienced in handling databases. People will not be happy to use a system that is unnecessarily complicated.

Above all, the database solution must be robust and should not crash unexpectedly.

Making your database as easy to use as possible

The user interface
Produce simple and intuitive user interfaces by creating forms that have the same basic features in the same place. You can do this by using the same menus, pick lists, buttons, etc. in the same positions in all of your forms. You can then let users test the form and observe them as they work. By doing this you can find out where they are having difficulty and which steps are taking the most time.

Design of forms
Forms should not be created with too many fields and buttons. It is best if only those items the user is likely to use at the time are shown. Any buttons or fields not needed should be hidden away until they are required.

Design of reports
Reports should be easy to read and understand. For many people, especially managers who will usually act on the contents of the report, the reports are the only part of the database they will see. Include only information on the report that is essential for the purpose of the report. There are many different design options for the production of reports and you should choose one that will include a lot of white space on the page. This ensures that the report is uncluttered because many users will want space to write comments on it themselves.

You might find your database easy to use, but will others?
The best way to find out if other users of the database find it easy to use is to ask them. Usually it is best to use an evaluation questionnaire which asks questions about the ease of use of the system. If the user has been asked for feedback during the development of the database they will have helped in the design of screens, reports, etc. and will therefore be less critical of the new system. It should be borne in mind that many people do not like to change their working practices, so a certain amount of criticism is usual but it is important to find out whether this criticism is justified or not.

Users are often quite critical and will tell you if there is something about the system that makes it hard for them to use.

Activity 19

You need to find out how easy the college student enrolment system is to use. You have selected five staff in the college office who will be using the system, although at least 25 people will use it on a day-to-day basis.

Produce a short questionnaire that these five potential users can fill in to evaluate how easy the system is to use. You should have at least ten questions in your questionnaire.

Checking with abnormal or extreme data

Abnormal data is data that would not normally be entered in the fields. For example, a user might try to enter letters into a numeric field or a product code containing too many or too few letters. He or she may also make mistakes and enter values that are too large or too small. Such data, although it fits the field type, should be trapped by the use of a range check on the field.

When creating the database, you may have developed pick lists, menus and buttons and all these features need to be thoroughly tested to make sure that they do what they are supposed to do and do not cause the system to crash. Crashes are very annoying because they can result in the loss of data and often mean that the computer will need to be re-booted (i.e. have the operating system loaded again).

To ensure that the database is thoroughly tested, a test specification is developed to help define the tests to be applied to the data. This document may be used to show that thorough testing has taken place. A test specification will normally consist of the following:

- acceptable data input values (including maximum and minimum values)
- unacceptable data values that should automatically be rejected
- inputs such as mouse movements that require a specific response
- inputs such as mouse or key depressions to which the system should not respond
- check of facility provided by the database, for example, a check that data entry forms, queries, reports etc. all work properly
- check that all functions and/or formulas work correctly
- check that the system meets the user requirements.

6.7 Standard ways of working

As part of Unit 6 you will have to show that you have used standard ways of working. The whole point of having a standard way of working is to help you to manage your work more effectively. You can save a lot of time and effort if you think logically about how you intend working. All of the standard ways of working are looked at in detail in Unit 1, but here is a list of the ways of working you should apply to Unit 6.

- Edit and save work regularly, using appropriate names for your files (database, tables, queries, reports, etc).
- Store your work where others can find it easily in the directory/folder structure.
- Keep dated backup copies of your work on another disk and in another location.

- Keep a log of the ICT problems and how they were resolved.
- Protect confidentiality and observe copyright laws.
- Avoid bad posture, physical stress, eyestrain and hazards from workplace layout.

Key terms

After reading this unit you should be able to understand the following words and phrases. If you do not, go back through the unit and find out, or look them up in the glossary.

Alternate key	*Key*
Attribute	*Logical data model*
Calculated field	*Normalisation*
Candidate key	*Parent form*
Data	*Primary key*
Data dictionary	*Printed report*
Data model	*QBE (query by example)*
Data processing	*Query*
Data type	*Relational database*
Entity	*Relational logic*
Entity relationship diagram	*Relationship*
Field	*Report*
Field size	*Screen input form*
Foreign key	*Screen report*
Form	*SQL (Structured Query Language)*
Index	*Sub form*
Information	*Table*
Input mask	*Tuple*

Review questions

1 A library in a school uses a relational database management system. This system records the details of members and the loans of books. There is no limit to the number of books a pupil can borrow.
   a Give the names of three tables that this database would need.
   b For each table you name, give the name of the fields it would be likely to hold.
   c State the primary key or keys in each table and give reasons for your choice.
   d Draw an entity relationship diagram to illustrate the types of relationships between the tables.

2 Distinguish between a 'flat file database' and a 'relational database'.

3 A video store manager records the details of videos, members and rentals manually. A friend, who is a systems analyst, suggests that he should computerise the system and use a relationship database. There is no restriction on the number of videos a customer can borrow in one go.
   a The systems analyst has suggested that there should be three tables: Videos, Members and Rentals. Write a list of the fields that each table should hold. Underline any key field or fields in this list.
   b Using three tables means that the system can be maintained with minimum redundancy. What is meant by 'minimum redundancy'?
   c Describe three advantages that this new system will have over the previous, manual system.

4   Outline the steps involved in creating one table in a database.

5   **a**   Explain, using suitable examples, the differences between 'data' and 'information'.

   b   Several types of data are used in a database. Give the names of four data types.

   c   It is important that each line in a database table is uniquely defined. What does this mean?

6   Databases consist of many objects such as tables, forms, queries and reports. Explain each of these and the uses to which they can be put.

7   You are designing a relational database for a company and have come up with three tables containing the following fields.

   CUSTOMERS: Surname, Forename, Street , Town, City, Postcode
   ORDERS: Surname, Forename, Postcode, Date Ordered, Product Code, Product Description, Quantity Bought, Price, Total Cost
   STOCK: Product code, Product Description, Price, Quantity In Stock

   A number of things are wrong with the fields included in each table, and fields need to be fully normalised. Normalise the database – feel free to create and name any new entities you may require. Give your reasons for any changes you make.

8   Outline the ways in which a database may be thoroughly tested before being handed over to the user.

9   To avoid incorrect data being entered into a database, the data must be verified and validated. Explain, with reference to a database you may have created yourself or seen, how these two processes can be done.

10   A data dictionary should be provided as part of the documentation for a database. What is a data dictionary and what are the reasons for its inclusion as part of the documentation?

Assignment 6.1

Develop a relational database

For this assignment you will need to develop a relational database. In many cases you will be able to continue on from the last unit on systems analysis and use the same problem. Your lecturer/teacher will either let you choose your own problem or identify one for you. This assignment follows on from the assignment developed in Unit 5 and the problem used for that unit can also be used here.

For the assessment evidence for this unit, you must produce:

* a relational database to a given specification requiring at least three related tables
* design and analysis notes for the database
* annotated printed copy and test results for the database
* a user guide and technical documentation.

Tasks

1 Present the initial draft design and final model correctly normalised to at least first normal form.

2 Clearly show the entities, attributes, keys, relationships and internally generated or processed data in your design notes.

3 Produce a working relational database that allows users to append, delete and edit data, initiate queries and print reports.

4 Produce suitable and correct data input forms.

5 Produce a user guide that enables novice users to make efficient use of the database.

6 You must define clearly and accurately in the technical documentation:

   • the database structure and data relationships
   • a data dictionary
   • a range of acceptable data (for testing validation rules)
   • example output from queries and reports
   • test procedures.

7 Design and create reports that make correct and effective use of queries, grouping, arithmetic formulae and related tables.

8 Use technical language fluently.

9 Make good use of graphic images and use annotated screen dumps to create effective user instructions and technical documentation.

10 Show that you can work independently and meet agreed deadlines by carrying out your work plans effectively.

11 Provide detailed design and analysis notes that include graphic images to define the data model clearly and demonstrate that it is correctly normalised to third normal form.

12 Make effective use of validation and of automatic counter, date or time fields in the data input forms.

13 Design and implement test procedures that check reliable operation including rejection of data outside the acceptable range.

14 Produce user-friendly, well-laid-out screen data input forms with title labels, field names, set widths, pull-down lists and instructions as appropriate to enable data entry into multiple tables.

Get the grade

To achieve a grade E you must complete tasks 1–6.

To achieve a grade C you must complete tasks 1–10.

To achieve a grade A you must complete tasks 1–14.

Key Skills

Opportunity

You can use this opportunity to provide evidence for the Key Skills listed opposite.

Information Technology IT2.3	Present combined information for two different purposes. Your work must include at least one example of text, one example of images and one example of numbers.
IT3.3	Present information from different sources for two different purposes and audiences. Your work must include at least one example of text, one example of images and one example of numbers.

Glossary

Absolute referencing A formula is told to use a certain cell and when this formula is copied to a new position, it will not alter the address of the cell to which the formula refers.

ADC (analogue-to-digital converter) A device that changes continuously variable quantities (such as temperature) into digital quantities.

Address The location of an item in a computer's memory.

Agenda A programme of business for a meeting.

AI (artificial intelligence) The science of developing computers that 'think' like humans.

Algorithm A set of rules which gives a sequence of operations for solving a problem.

Alternate key Candidate keys left after the primary key has been selected.

ALU (arithmetic and logic unit) Part of the central processing unit. It performs all the arithmetic and logic operations.

Analogue signal Signals that have an infinite number of in-between positions.

Antivirus software Software used to discover and then delete computer viruses.

Appendix An addition or supplement to a document.

Application What a computer can be used to do.

Applications software A program or a set of programs to carry out a particular application such as accounts or payroll; also known as an applications package or package.

Applications software routines Routines that provide user automation facilities such as macros, styles and templates.

Array A structure used to store a collection of data objects of the same type. An array can be one-dimensional or two-dimensional.

ASCII (American standard code for information interchange) A code for representing characters in binary.

Assembler A program that converts assembly language into machine code.

Assembly language Low-level language where one programming instruction corresponds to one machine code instruction.

Asymmetric duplex Transmission of data in two directions at the same time.

ATM (asynchronous transfer mode) A high-speed method of data transmission making use of packet switching which uses fixed-length packets.

Attribute A single data item representing an individual property of an object or entity.

Audit trail The documentary evidence which enables the path of a particular transaction in a system to be followed.

Autofill A feature provided by some software for filling in the rest of a list of numbers, days of the week or months of the year.

Automatic access control Controlling access to a computer room automatically, usually by the use of a code being entered via a keypad.

Auxiliary storage   Storage that is in addition to the normal storage facilities on a file server.

Backing store   Memory storage outside the CPU. It is non-volatile, which means the data does not disappear when the computer is switched off.

Back-up file   A copy of a file which is used in the event of the original file being corrupted (damaged).

BACS (bankers' automated clearing service)   Service set up by the major banks to deal with the payment of wages directly into employees' accounts, the payment of standing orders and direct debits, etc.

Barcode   Lines of varying width used to encode data such as the price of goods.

Barcode reader   An input device used to scan a barcode; also known as a laser scanner.

Basic   A high-level programming language.

Batch file   A file, created using operating systems software in order to automate a processing activity.

Batch processing   Collecting jobs over a period of time and then processing them in one go.

Baud rate   A data transmission rate; the number of bits per second.

Bespoke software   Software specially written for a particular user.

Bibliography   A list of books that have been referred to during the completion of a document.

Binary code   Code made up from a series of binary digits: 0 and 1.

BIOS (basic input output system)   Part of the operating system stored on a chip on the motherboard.

Bit   A binary digit; 0 or 1.

Bit-map image/graphics   An image represented by patterns of tiny dots called pixels.

BNC   Connector used to connect coaxial cables.

Boolean operators   The operators AND, OR and NOT; also known as logical operators.

Booting-up rom-bios   The program in the chip that gives basic instructions to start the computer.

Border   The area around the edge of a page.

Break-even analysis   Technique to determine the number of goods to be sold in order to just offset the costs.

Brief   The instructions to act in a professional capacity.

Buffer   A storage area where data is stored temporarily. Printers have buffers so that the user can get on with something else while data is waiting to be printed.

Bug   A mistake or error in a program.

Bullet points   A paragraph or section of text that has a symbol placed in front to make the section of text stand out.

Bus   A high-speed transportation system for sending data and control signals in a computer.

Bus topology   A type of network arrangement where the terminals are connected off a main wire.

Byte   The amount of memory needed to store one character such as a letter or a number. A kilobyte (kb) is 1024 bytes and a megabyte (mb) is 1 million bytes.

C   A block-structured programming language used mainly for producing applications software for PCs.

C++   A general purpose programming language which is easier to understand than assembly language but runs almost as fast.

Cache   A special part of the memory where a computer can store data temporarily to avoid accessing much slower storage such as hard or floppy disks over and over again.

CAD (computer-aided design)   Using a computer to design a system and produce technical drawings.

Calculated field   A field whose contents are worked out using the contents of other fields.

Candidate key   All those columns in a table that contain unique values for every row.

Case (computer-aided system engineering)   A software tool used by programmers to help in the analysis, planning, design and documentation of computer programs.

CD-R drive   CD-drive where data can be recorded onto the cd once only.

CD-ROM   Abbreviation which stands for compact disk read-only memory.

CD-RW drive   CD-drive where data can be stored onto the disk many times.

Cell   An area on a spreadsheet produced by the intersection of a column and a row in which data can be placed.

Cell format   An option used to increase the cell impression on the person looking at it.

Cell referencing   The way in which a cell is referred to.

Centralised database system   A system where all the data is kept in one place with terminals being able to access it.

Character   Any symbol that can be entered into a computer via a keyboard.

Chart   Another name for a graph (bar chart, pie chart, etc.)

Check digit   Number placed after a string of numbers to check that they have all been input correctly into the computer.

CIP (current instruction register)   A memory location within the CPU which holds the instruction currently being executed and which is also responsible for decoding it.

Client/server   A network where several PCs are connected to one or more servers.

Clipart   Predrawn computer artwork that is available for people to use.

Clock speed   The speed that the processor is able to execute instructions.

CMOS RAM   Abbreviation for Complementary Metal Oxide Semi-conductor Random Access Memory. It is used to store the date and time, for example, because of its very low power consumption.

Cobol A high-level programming language used mainly in business because of its good file-handling facilities.

Command line The place on a computer screen where an instruction is typed in.

Communications software Software which, with the use of modems, allows computers to communicate with each other.

Compiler Software that converts a high-level language program into machine code.

Compuserve An Internet service provider.

Configuration The pieces of hardware that are needed to set up a computer system; also the setting variables that need to be altered to enable software to be used with particular hardware.

Consistency check Check to ensure that an item of data is compatible with all other occurrences of the data item.

Constraint A restriction on the development of a system (e.g. time, money, expertise, etc.).

Contents page A list of the sections in a document.

Context diagram The first dataflow diagram drawn to place the system being investigated into the context of the entire system.

Control device A device connected either within the main processor housing (e.g. disk controller, video card, input/output card, serial and parallel output cards) or to the input/output port of the main processing unit via a buffer.

Control procedure Program created to operate a process control system.

Control system A computer system which automatically controls a process or mechanical device by sensing the need to vary the output.

Control unit Part of the central processing unit used to control the pulses which travel around the CPU.

Controller board The circuit board which contains the circuits responsible for controlling data transfer to and from the disk drives.

Co-processor An additional processor that performs a specific function.

Copyright, Designs and Patents Act 1988 An act which, amongst other things, makes it an offence to copy or steal software.

CPA (critical path analysis) A method used to control and schedule large projects. The project is broken down into a series of smaller activities and a diagram constructed to determine the order in which they need to be carried out. The path along which delay in any one activity will cause a delay in the whole project is known as the critical path.

CPS (characters per second) A measure of the speed of data transfer between hardware devices.

CPU (central processing unit) A computer's 'brain' where data is stored and processed. It has three parts: the ALU, the control unit and the memory.

Customer information Details about the customer and their purchasing habits.

CV (curriculum vitae) A biographical sketch of the course of your life.

DAC (digital-to-analogue converter) A device that changes digital quantities into analogue ones.

Data   Information in a form a computer can understand.

Data bit lengths   The number of bits used to store an item of data.

Data capture   The way a computer obtains its data for processing.

Data compression   Using software to reduce the size of files so that they take up less space on a disk.

Data dictionary   A document or file which contains descriptions of all the data items in a database. The descriptions can include data type, element size, format of data, validation range of values, entity descriptions, attribute–entity relations, purpose and synonyms.

Data entry message   A message that appears when the user moves to a field to tell them what type of data is required.

Data flow diagram   A diagram showing the information flows in an organisation.

Data model   A series of diagrams and text (information flow diagrams, data flow diagrams, data dictionary, entity relationship diagrams, etc.) which are used to describe a system.

Data processing   Doing something with data to produce some form of useful output.

Data Protection Act 1998   An act which restricts the way personal information is stored and processed using a computer.

Data representation   Using a code such as ASCII to represent characters.

Data template   The framework for a document which sets the text, graphics and formatting.

Data transmission media   The media through which data is transmitted from one place to another.

Data type   The type of data, including character, number, graphic, logical and date.

Data type check   A check performed to make sure that data is of the correct type for processing, e.g. that numbers have been entered in numeric fields, etc.

Database   Series of files in which data is stored and which can be accessed in a variety of different ways.

Database report   Output from a database that is to be used for a specific purpose.

Debug   Remove all the errors in a program.

Decision table   A table which shows the relationships between variables and the actions which must be taken under certain conditions.

Default settings   The settings a computer automatically reverts to, unless instructed otherwise.

Design specification   A list of the requirements for a graphic such as type of graphic, the purpose and context of the graphic, etc.

DFD (data flow diagram)   A diagram showing the information flow in an organisation.

Digital computer   Computer that works on data represented by numbers. Most ordinary computers are digital.

Digital signal   A signal consisting of 0s and 1s.

Dip switches Small switches on a circuit board that allow the computer to be configured.

Directory A list of the programs and files stored on a disk. The directory enables the operating system to retrieve files from backing store.

Disk Storage medium used to hold data.

Distributed system Data processing system where the processing, storage, input/output devices, etc. are dispersed geographically.

Distributor A person or company that is given exclusive rights to sell or service goods from a certain manufacturer in a certain area.

Ditto drive backup storage device.

DML (data manipulation language) A programming language, supported by a database management system, for accessing, deleting data, etc.

Document A text file produced by a wordprocessor.

Document imaging system System whereby all paperwork (letters, bills, etc.) are scanned into a computer system and then disposed of.

Documentation The documents that accompany a system explaining how the system works. It can also refer to the manuals that accompany a program.

DOS (disk operating system) A program which tells a computer how to work. Controlling data storage on the disk drives is one of the many tasks it performs.

Dot matrix printer A printer which uses numerous tiny dots to make up each printed character.

DPI (dots per inch) Used to describe the resolution of printers. a 600 dpi printer will produce a clearer image than a 300 dpi printer.

Draft A rough version of a document often provided for comment by others.

Drum plotter Type of plotter in which paper is rolled backwards and forwards over a drum while pens draw lines on the paper.

Dry run Checking the results of each step of a program manually to make sure the program is correct.

DSVD (digital simultaneous voice and data) The ability to transfer both voice and data simultaneously using a single telephone wire.

DTE (data terminal equipment) The equipment which acts as a terminal at the end of a communication line.

DTP (desktop publishing) The use of software to combine text and graphics on a computer screen to produce posters, newsletters, brochures, etc.

Dumb terminal A terminal which has no processing power of its own.

DVD (digital versatile disc) Modern replacement to the CD-ROM that is able to store a lot more data.

Echo Enables transmitted characters to appear on the sender's screen for checking.

e-commerce Business conducted electronically using the Internet.

EDI (electronic data interchange) Network link that allows shops to pay suppliers electronically without the need for invoices and cheques.

Edit Change something stored on the computer.

EFT (electronic funds transfer)   The process of transferring money electronically without the need for paperwork

EFTPOS (electronic funds transfer at point of sale)   Electronic funds transfer that takes place at a point-of-sale terminal.

EIS (executive information system)   A system used to extract strategic information for use by senior executives or directors of a company to aid decision making.

Electronic information   Information stored as a series of signals.

e-mail (electronic mail)   Messages and documents that are created and sent electronically and read on screen, eliminating the need for them to be printed out.

Emulation   Software or hardware which acts in the same way as another system.

Encrypt   Encode. Sensitive files can be encrypted, which means they have to be decoded in order to be read.

Entity   An object of the real world which is of relevance to an information system, e.g. customer, invoice, goods.

EPOS (electronic point of sale)   A computerised till that can be used for stock control.

ERD (entity relationship diagram)   A diagram showing the relationships between entities and their attributes in a system.

Ergonomics   The science of the correct design of working equipment and the working environment.

ERM (entity relationship model)   A model which represents a view of an organisation based on its entity relationships.

Execution error   An error detected during the running of a program.

Existence check   Check to ensure that a certain field in a database contains data.

Expansion card   Circuit board added to improve the sound/graphics capabilities of the computer.

Expert system   A program designed to behave like a human expert in a specialist field.

Face-to-face meeting   A meeting in person.

Fact finding   Investigation of a system prior to performing a feasibility study.

Fax   A machine capable of sending and receiving text and pictures along a telephone line.

Fax cover sheet   A sheet attached to a fax outlining who the fax is from.

Feasibility study   Study carried out before a new system is developed to see what type of system is needed.

Fetch/execute cycle   The cycle of events a computer carries out when executing a program.

Field   A space in a database in which data is entered; for instance, you may have a field for surname, date of birth, etc.

Field check   A check performed by a computer to see if data is of the right type to be put into a field, e.g. that numbers only are entered into a numeric field.

Field size   A setting for the width to be allocated to a field.

File   A collection of related data.

File attachment   a file that is attached to an e-mail.

File protection   A facility offered by some networks to enable only certain users to see particular directories, subdirectories and files.

File server   A network computer used for storing all the users' programs and data.

File transfer   Using communications links to transfer a file from one computer to another.

Flat file database   A database that is only able to use one file at a time, unlike a relational database which is able to use two or more files at a time.

Flat organisation   The structure of an organisation where there a only a few tiers of staff.

Floating point number   Another name for a real number such as 12.0, 112.22, -0.09, 14.008, 0, etc.

Floppy disk   A magnetically coated disk used to store data. The 3.5-inch disk is inside a hard case.

Flow chart   Chart or diagram used to break down a task into smaller parts; also known as a flow diagram.

Flow control   Controlling the flow of data so as to prevent data from being sent too fast or from being lost due to the receiving device not being ready to accept it.

Flyer   A quickly produced document that is handed to passer's by or put through doors.

Folder   A name for a subdirectory on a Macintosh computer.

Font   A style of type.

Footer   Text placed at the bottom of a document.

Foreign key   A key that is the primary key in another table/entity but only an ordinary key in this table.

Form   A screen used to enter data into a database; can also be a document used to collect data.

Form field   A space on a data collection or data entry form into which data can be entered.

Format   The process of allowing the operating system to electronically mark out the surface of a disk ready to store data.

Fortran   A high-level programming language mainly suited to mathematical or scientific programs.

Function   A pre-defined formula for making a certain kind of calculation.

4GL (fourth generation language)   Software tool that allows applications programs to be developed more quickly.

Gantt chart   A type of chart, with horizontal bars, used to plan and schedule jobs.

Gateway   The computer link which connects and translates between two different kinds of computer network.

General purpose software   Software such as spreadsheets, databases and wordprocessing software which can be used for many different applications.

Generation   Every time a file is updated, a new generation of the file is produced.

Gigabyte   Roughly 1000 megabytes.

Gigo (garbage in garbage out)   If you put rubbish into the computer then you get rubbish out.

Grammar checker   A program (usually part of a wordprocessing package) which checks a document for grammatical errors and suggests corrections.

Grandfather-father-son principle   A method of back-up where the last three versions of a master file are kept along with their transaction files, so if one file becomes corrupt it is possible to recreate it from the others.

Graph plotter   A device which draws by moving a pen. It is useful for scale drawings and is used mainly with CAD packages.

Graphic design software   Applications software used to produce and manipulate images.

Graphics   Diagrams, charts and graphs that are either examined on screen or printed out.

Graphics adapter   A circuit board that contains the electronic circuitry needed to supply data in a form that the monitor can display.

Graphics tablet   An input device which makes use of a large tablet containing a number of shapes and commands that can be selected by the user by moving a cursor and clicking. It works by moving the toolbars onto the tablet rather than cluttering up the screen when doing large technical drawings using CAD software.

GUI (graphic user interface)   Interface that allows users to communicate with the computer using icons and pull-down menus. Windows is a GUI, and Macintosh computers use a GUI.

Gutter   In desktop publishing or wordprocessing, the space between two columns.

Hacker   A computer enthusiast who tries to break into a secure computer system.

Half-duplex   Two-way communication, but only one way at a time.

Handshaking   Process whereby signals are sent between devices to indicate that successful data transmission can take place.

Hard copy   Printed output from a computer.

Hard disk   A rigid magnetic disk which provides more storage and faster access than a floppy disk.

Hard drive   Unit containing a hard disk.

Hardware   The physical components of a computer system. These include the VDU, processor, printer and modem.

Hash total   A meaningless total of numbers used to check that all the numbers have been entered into a computer.

Header   Text placed at the top of a document.

Heading and title page   The page at the front of a document.

Hexadecimal   Base 16, a number base system used by computers.

Hierarchical organisation   An organisation type where there are lots of different levels of staff.

High-level language   Programming language where each instruction corresponds to several machine code instructions.

Human resources   Another name for personnel.

Hyphenation   The technique that a wordprocessor uses where long words at the end of a sentence have a dash inserted between them so that part of the word is on one line and the other part is on the next line.

Icons   Symbols displayed on a computer screen in the form of a menu.

IDE (integrated drive electronics)   A type of interface that is used to connect drives to a PC.

Immediate access store   Storage in the memory of the CPU.

Impact printer   Any printer that relies on a character pressing against an inked ribbon for its operation.

Import   Load a file into one package from another.

Indent   A section of text is moved slightly in compared with the normal position.

Index   A table which aids the manipulation of data in a database.

Information   Processed data.

Information flow diagrams   Diagrams used to show how information flows around an organisation.

Information retrieval   The process of recovering information after it has been stored.

Information system   A system used to convert data into information.

Ink-jet printer   A printer that works by spraying ink through nozzles onto the paper.

Input   Data fed into a computer for processing.

Input mask   A preset format for data being entered into a database. Brackets, dashes, etc. are added automatically, with the user having only to type in the variable data.

Input specification   Specification outlining the inputs needed to a system.

Integrity   The faith that the user has in the information gained from the system.

Interactive   A program or system that allows the user to respond immediately to questions from the computer, and vice versa.

Interface   The hardware and software used to enable devices to be connected together.

Interlacing   A special way in which the image on a monitor is built up.

Internet   Worldwide network of networks. The Internet forms the largest connected set of computers in the world.

Internet e-mail   E-mail sent over the Internet.

Interpreter   Program that converts a high-level language into machine code. The interpreter is different from a compiler because it translates each instruction and carries it out before moving on to the next instruction. Compilers translate the whole program first and then carries out each instruction.

Interrogation   The process of getting information from a file.

Interrupt   An instruction given to a computer telling it to stop whatever it is doing.

Intranet   A network used within an organisation which makes use of Internet technology.

Invoice   A bill or request for payment.

I/O controller   The link between the processor and its surrounding components.

I/O controller   The link between a microprocessor and its surrounding components.

ISA (industry standard architecture bus)   A type of bus used by older PCs.

ISDN (integrated services digital network)   A completely digital system which allows data to be sent along telephone lines at high speeds without the need for a modem.

ISP (Internet service provider)   A company that supplies a permanent connection to the Internet.

IT (information technology)   The application of computing, electronics and communications.

Iteration   A section of a program which is repeated a number of times.

Itinerary   A plan or record of a journey.

Janet (joint academic network)   A network serving the higher education funding council and the research councils and institutions, with links to international networks.

Java   A programming language.

Jaz drive   Backup storage device.

Joystick   An input device used instead of the cursor keys or mouse as a way of producing movement on the screen.

JSP (Jackson structured programming)   A method of using structure diagrams for analysing data and programs.

Jumper   Small switch used to configure expansion boards.

Just-in-time stock control   A method of stock control used by supermarkets to ensure that the goods arrive just in time for the customers to buy them.

Justification   Where text is aligned (i.e. flush) with the right and left margins, the text is justified right and left and this is called fully justified.

Key   One of the attributes of an entity on which an index has been created or a relation set.

Key-to-disk   A way of putting data directly onto disk using the keyboard.

Keyboard   The set of keys used to input data into a computer.

Keypad   A miniature keyboard or set of keys grouped in a certain way.

Lan (local area network)   A network of computers on one site.

Laptop   A computer small enough to fit on your lap. Laptop computers use rechargeable batteries.

Laser printer   A printer which uses a laser beam to form characters on the paper.

Library routine   Part of a program that is kept so that it can be copied and used in other programs, thus speeding up the writing of new programs.

Line spacing   Documents are normally produced using single line spacing. This may be altered to double, 1.5, etc.

Linear bus   Type of bus network where terminals are connected off a single cable called the bus.

Linked lists   A collection of data stored in a certain order, with a pointer which points to the next set of data.

Linux   An operating system.

Logical data model   A data model that is only conceptual.

Logical design   A design that looks only at the data itself and not at the physical implementation of the model.

Logical operator   An operator such as AND, OR and NOT.

Logo   A simple language which enables a 'turtle' to move according to the instructions given to it.

Lookup table   A table where a value is looked up.

Loop   A sequence of steps in a program which is repeated once or more.

Low-level language   A programming language very similar to the machine language of the computer. Each low-level instruction can be easily converted into a machine code instruction.

Machine code   The language a computer can understand without it being translated. Each type of computer has its own machine code.

Machine readable   Data that can be input directly into a computer without any data preparation, e.g. the magnetic ink characters on cheques.

Macro   A program written using applications software tools to automate a collection of keystrokes or events.

Magnetic media   Media such as tape and disk where the data is stored as a magnetic pattern.

Magnetic strip   A strip containing magnetically encoded information which can be read by a special reader. Cards such as credit cards have a magnetic strip.

Magnetic strip reader   Device that reads the data contained in magnetic strips.

Magneto-optical disk   Combines the technologies of magnetic media and CD-ROM to produce a disk that looks like a CD-ROM but which you can read and write to.

Mail merge   Combining a master file with a secondary file containing variable data such as names and addresses to produce multiple documents such as mail shots.

Mainframe   A large computer system with a number of dumb terminals attached to it.

Mainstore   Memory inside the CPU.

Many-to-many relationship   A type of relationship between two tables or entities.

MAR (memory address register)   The part of the memory responsible for giving the address of the memory location where data is either put into the memory or taken from it.

Margin   The blank edge on a page.

Marketing information   Information concerning the reasons why customers buy certain products.

Master file   The main source of information; the most important file.

Maths coprocessor   Incorporated as part of most chips to perform calculations at very high speed.

MB (megabyte)   Roughly one million bytes (1 048 576).

MBR (memory buffer register)   A place in the memory through which data must pass when it is being moved between memory locations, the data bus and the memory address register.

Medium   The material on which data can be stored such as magnetic disk, tape, etc.

Memo   A short message that is used internally.

Memory   Area of storage inside chips, consisting of ROM and RAM.

Menu selection interface   Interface which provides the user with a list of options to choose from.

Merge   To combine data from two different sources.

Mesh topology   Network arrangement where each computer has many wires connecting it to other computers.

Methodology   The method used when investigating, developing and documenting a system.

MHz (megahertz)   One million cycles per second. The speed of the internal clock that controls the speed of the pulses in a computer. Chip design and clock speed determine the overall performance of the CPU.

MICR (magnetic-ink character recognition)   Method of input that involves reading magnetic ink characters from a document.

Microcomputer   A cheap, relatively slow computer with limited memory that is only able to work on one program at a time; includes home computers as well as personal computers.

Milestones   Points in a project which mark the ends of logical stages in the project. Also known as mileposts.

Minutes   A detailed record of discussions, decisions, etc in a meeting.

MIS (management information system)   System, usually centred around a database, that is able to extract information in a way which enables management decisions to be made.

Mode   A particular way of running a program.

Model   A software representation of a real situation or system which can be used for analysing its operation and for investigating its behaviour under certain conditions.

Modem   Modulator/demodulator. A device which converts data from a computer into a form that can be passed along a telephone wire.

Modem response string   A string of bits given out by a modem which is used to tell a computer if the modem is on-line.

Modulated signal   The signal produced when a carrier signal is used to carry a data signal.

Motherboard   The main circuit board of a computer into which most of the devices are connected.

Mouse   An input device which is moved over the desk top to create equivalent movement of the cursor on a computer screen. Buttons on the mouse are pressed to make menu selections.

MS-DOS (microsoft disk operating system)   An operating system used by personal computers.

Multi-access system   System which allows many different users to gain access to the computer. Because of the high speed of the CPU, each user has the equivalent of sole access, even though the time is being shared between a number of users.

Multimedia   Software that combines more than one medium for presentation purposes, such as sound, graphics and video.

Multiplexer   Hardware device used to allow simultaneous transmission of multiple messages using a single communications channel.

Multitasking   The ability of a computer to run several different programs at the same time.

Multi-user processing   Processing which makes use of programs and operating systems that are able to support more than one user at a time.

Netscape navigator   A program which can be used for searching for information on the Internet.

Network   A group of computers that are able to communicate with each other.

Network administrator   The person responsible for the running and administration of a network. Also known as a network manager.

Network computer   Computer which is to be connected to a network.

Network management software   Software used to help the network Manager/administrator in their job of looking after a network.

Network operating system   A special operating system with all the features of an ordinary operating system but with additional facilities to deal with the problems encountered when communicating with other computers connected on a network; also known as networking software.

Newsletter   A document like a newspaper used within an organisation.

NIC (network interface card)   Card which slots into a motherboard and is used primarily to reduce the amount of cabling in a network.

Normalisation   The process of converting an invalid data model into a valid data model, ensuring data consistency and integrity of the data model.

Novell   A company that provide networking solutions.

Null modem   A cable used to allow two computers to communicate with each other by emulating a modem.

Object   The data, and the structure used to manipulate it, as a single, self-contained unit.

Object orientated language   A language which attaches data to the programming routines used to manipulate it.

OCR (optical character recognition)   A combination of software and a scanner which is able to read characters into a computer.

Off-line/on-line   Terms used to describe whether a device is under the control of a computer or not.

OLE (object linking and embedding)   Where one application can be linked directly to another. For example, a spreadsheet can be linked to a wordprocessing document, and if the figures are changed in the spreadsheet, the figures will be changed automatically in the document.

OMR (optical mark recognition)   Method of data input that involves detecting marks, usually shaded areas, on a piece of paper.

One-to-many relationship   A type of relationship between two entities/tables.

One-to-one relationship   A type of relationship between two entities/tables.

Operating system   The software that controls the hardware of a computer and runs the programs. The operating system controls the handling of input, output, interrupts, storage and file management; also known as systems software.

Operational information   Information used for the day-to-day of the organisation.

Operator   A symbol which represents an arithmetic operator ($+, -, /, \star$ etc.) or a relational operator ($=, <, >, >=, <=, <>$).

Optical mark reader   Device used for detecting marks on a sheet of paper and putting the data contained in them into a computer.

Optional relationship   A relationship that may or may not exist.

Oracle   Sophisticated database used for large networks that has its own programming language.

Order processing   The job of dealing with orders as they arrive in the organisation.

Organisation   A group of people all working towards a common purpose.

Orphan   The first line of a paragraph when separated from the rest of the paragraph by a page break.

OS/2   An operating system used with PCs.

Output   The results from processing data.

Outsourcing   Using an outside organisation for the development of new computing facilities.

Page layout   A term meaning the layout of the page that includes margins, justification, indents, tabulations, line spacing, fonts, page numbering, headers and footers, column layout etc.

Page orientation   The two ways of printing text on the page: portrait and landscape.

Pagination   The process of deciding where the page breaks are placed.

Paragraph numbering   A facility for the automatic numbering of paragraphs in a structured document.

Parallel port   An interface used for parallel communication.

Parallel processing   Processing which allows a computer to carry out more than one task at a time.

Parallel transmission   Transmission of bits side by side.

Parent form   The main form.

Parity bit   An extra bit added at the end of a group of bits to keep the number of binary digits in the transmission either even or odd.

PCI (peripheral component interconnect)   An interface developed by intel the chip manufacturer.

Peer-to-peer   Network arrangement in which each computer is of equal status.

Pentium   Computer based around the intel pentium or pentium pro chip.

Peripheral   A device connected to and under the control of the CPU.

Permanent storage   Place where permanent data such as boot programs is stored.

PERT (project evaluation and review technique)   A project management technique that involves charting out the time and other resources needed to complete a project.

Physical design   A design which can be physically implemented.

Piracy   The illegal copying and use of software.

Pixel   The smallest dot of light on a computer screen that can be individually controlled.

PKZIP   A file compression utility program.

Platform   The hardware used by an operating system, or the operating system used by an applications programs.

Plug and play   A system whereby an operating system is able to recognise the peripherals attached to it automatically. Windows 98, 2000 and NT make use of plug and play.

PNC (police national computer)   The central computer system used by the police which holds details of criminals, etc.

Point-to-point topology   Network arrangement where each terminal can send data to any other terminal on the network using a single link.

Polling   A system where each station on a network is invited to transmit data.

Port   An external connection point on a computer to which peripheral devices are attached.

POST   (Power On Self-Test). A systematic series of tests performed by the computer when it is first switched on.

Power management   A method where the power to different devices is lowered if the computer detects a period of inactivity.

Presentation graphics   Applications package for producing slides, graphics, etc.

Primary key   A field/attribute that uniquely defines a particular table/entity.

Primary storage   Storage area in ROM or RAM for holding data and instructions.

Printed report layout   The layout of a printout obtained from a database.

Printer server The computer which contains the printer server software and controls for a local area network (LAN) and which controls the printer queue.

Private wide area network A network which makes use of communications that are not shared by anyone else and are either owned or leased from a telephone company.

Problem statement An outline of the problem to which an it solution needs to be found.

Process Something that is done to raw data, e.g. calculating, sorting, etc.

Process control software The program which controls a certain process.

Processor Another name for the central processing unit.

Program A complete set of structured instructions given to a computer to tell it how to carry out a particular task.

Program generator A program which expands simple statements into program code, enabling inexperienced programmers to write their own programs.

Programmer A person who writes computer programs.

Project brief An outline of a project supplied by the senior managers or directors of an organisation.

Prolog A declarative programming language used primarily for expert system programming.

Proofread A way of reading carefully through what has been input to spot and correct errors.

Proofreading marks The standard marks made on a document during the proofreading process to mark up the alterations that need to be made.

Protecting cells The process of ensuring that cells are not altered by the user.

Protocol A set of standards that allow the transfer of data between computers on a network.

Prototyping software See fourth generation language.

Pseudocode A combination of English and programming language used to express the flow of a program.

PSTN (public switched telephone network) The ordinary telephone system.

Purchase information Information needed when purchasing goods or services from a supplier.

Qbasic A version of the basic programming language which can be interpreted or compiled.

QBE (query by example) A quick way of entering SQL commands in order to extract certain information from a database.

Query A request for specific information from a database. Action queries perform certain actions such as deleting or updating, and select queries find and extract certain data.

Query language Language used to construct queries to a database.

Queue A list of data in which the data is added at one end and taken away at the other.

Quickbasic A version of the basic programming language which can be interpreted or compiled.

Ram (random access memory) A fast temporary memory area where programs and data are stored while a computer is switched on.

Range check Data validation technique which checks that the data input into the computer is within a certain range.

RDMS (relational database management system) Database system where the data is held in tables with relationships established between them. The software is used to set up and hold the data as well as to extract and manipulate the stored data.

Readability This is a measure of the ease with which the writing can be read and understood.

Reading age An indication of the reading ability of a proficient schoolchild at that age.

Real time A real-time system accepts data and processes it immediately. The results have a direct effect on the next set of available data.

Record A set of related information. Records are subdivided into fields.

Refresh rate The number of times a complete picture is built up on a computer screen per second.

Relational database Database where data is stored in separate tables and the tables are related to each other. The use of tables cuts down on data duplication and enables access to the data to be more flexible.

Relational logic Using AND, OR and NOT in expressions.

Relationship The way in which entities are related to each other. Relationships can be one-to-one, one-to-many or many-to-many.

Relative cell reference A reference made to a cell in a formula which will change when the cell is copied to a new position.

Relative referencing When a cell is used in a formula and the formula is copied to a new address, the cell address changes to take account of the formula's new position.

Repeater Device used to increase the strength of a data signal sent through a wire over a large distance.

Repetition Repeating program statements until a certain condition is satisfied; also known as iteration.

Report Printed information extracted from a database.

Reservation Process whereby a terminal keeps exclusive use of a communication line while it is sending data to another computer or terminal in a network.

Resolution The number of dots per inch for a printer or the number of dots of light on a screen.

Retailer A person who sells goods to the general public.

Ring bus Type of network bus topology where terminals are connected in a circle or a loop.

Ring topology A type of network arrangement where the terminals are connected together with wires in a ring.

ROM (read only memory)   Computer memory that cannot be changed by a program.

Root directory   The first directory on a magnetic disk.

Router   Hardware device that is able to decide which path an individual packet of data should take so that it arrives in the shortest possible time.

RSI (repetitive strain injury)   A condition caused by typing at high speed with inadequate precautions.

Run-time error   An error detected during the running of a program (e.g. division by zero).

Sales-based ordering   Ordering new stock on the basis of what has been sold before.

Sales information   Information concerning the sales made by customers.

Scheduled task   A task which is to be carried out automatically on a certain date and time.

Screen input form   A form created on the screen to enable input into the system by the user.

Screen report   A report that is displayed on the screen rather than printed out.

Screendump, screenshot   A printout of what appears on the screen.

SCSI (small computer system interface)   A parallel interface used for attaching peripheral devices to a computer.

Search   To look for an item of data.

Search engine   Program which searches for specific information on the internet.

Secondary storage   Storage outside the CPU such as magnetic disk, CD-ROM, tape, etc.

Selection   Choosing between two or more alternatives.

Sensor   Device that measures physical quantities such as temperature, pressure, etc.

Sequence   Where all the steps are carried out in the order in which they are presented.

Serial access   Accessing data in sequence; the time it takes to locate an item depends on its position.

Serial port   An interface used for serial communication.

Serial transmission   Transmission of bits simultaneously.

Service   An organised system of labour rather than products, provided by organisations to supply the needs of companies.

Service provider   A company which provides a service enabling connection to the Internet.

Shading   Making a certain area darker or patterned.

Share level security   Security system provided by many network operating systems which gives each resource on a network its own password.

Shell   The name given to the program used to control the operation of a computer.

Simulation   An imitation of a system or phenomenon produced using computer software, e.g. a flight simulator used to train airline pilots.

Single-user processing   A system with single user processing can only carry out processing for one user at a time. It is used by stand-alone machines.

Slide show   A series of screens that are put together in a certain order to explain something.

Software   The programs used by a computer.

Software package   Software that can be bought off-the-shelf to solve a business problem.

Sorting   Putting data into ascending or descending order.

Sound card   An add-on card used in multimedia applications to output high quality sound through speakers.

Sound sensor   A sensor used to detect a certain level of sound.

Source document   The original document from which data is taken.

Special fields   Form fields designed to contain special data such as date, time, page number or calculations.

Spell checker   A program, usually included with wordprocessing software, which checks the spelling in a document and suggests corrections.

Spreadsheet   A software package, often used to produce financial predictions, which consists of a grid used to contain text, numbers or formulae.

SQL (structured query language)   A special language used for extracting specific information from a database.

Ssdam (structured systems analysis and design methodology)   The most popular way of approaching systems analysis.

Stand-alone computer   A computer system which is self-contained, e.g. a PC which is not connected to a network.

Star topology   Network arrangement where the cabling looks like a star with the terminals at the points of the star.

Stock control   The system used to ensure that there is enough stock to satisfy orders.

Structure diagram   A diagram showing how a job can be broken down into smaller tasks.

Structured english   A shorthand form of English used to define the elements of a computer program.

Sub form   A smaller form that is placed inside the main form called a parent form.

Subdirectory   A directory within or under the main directory.

Subscript   A number or letter placed below another number or letter.

Superscript   A number or letter placed above another number or letter.

Supplier information   Information needed about a supplier when ordering goods or services from them.

Syntax error   An error reported by a computer due to the incorrect use of the rules governing the structure of the language. Similar to the rules governing grammar in English.

System   Hardware and software working together to do a job.

System specification   The data and information used by analysts and programmers to produce a system which meets the users' requirements.

Systems analysis   Investigating and analysing the requirements for an information system.

Systems analyst   A person who studies the overall organisation and implementation of a system.

Systems design   The process in systems analysis where the new system is designed.

Systems development life cycle   The cyclical development of a system.

Systems software   Another name for the operating system.

TA (terminal adapter)   A piece of equipment needed for ISDN.

Tab   Space created by pressing the Tab key; useful for lining up text and/or numbers

Table   Used in a relational model to represent an entity type. The columns represent attributes of the entity and each row corresponds to an entity occurrence.

Tailor-made software   Software specifically written to solve a particular problem.

Tape   Magnetic media used to store data.

Task analysis   Breaking a project down into a series of smaller tasks to estimate the amount of time and resources to be allocated to the project.

TE (terminal equipment)   A piece of equipment needed for ISDN.

Team building   The process of enhancing the ability of team members to work together in a coherent and positive way, rather than as a collection of individuals.

Telecommunications   The field of technology concerned with communicating at a distance (e.g. telephones, radio, cable, etc.)

Teleconferencing   System enabling two or more people to hold a conference using the telephone system.

Teletext   Information sent by means of a television signal and displayed on a specially equipped television set.

Template   An electronic file which holds a standardised document layout or screen format.

Temporary storage   Storage on a medium which allows data or files to be deleted.

Terabyte   Approximately one trillion bytes.

Terminal   A computer on a network, or a keyboard and VDU connected to a mini- or mainframe computer.

Terminal emulation   The type of terminal selected will set the character and control codes used in transmission and reception.

Test data   Data used to test a program or flow chart for logical errors.

Text concatenation   Joining text together.

Text orientation   The way the text is positioned on the screen.

Text/picture box   A box placed around a section of text or around a picture.

Thesaurus   Software which suggests words with similar meanings to the word highlighted in a document; usually included with wordprocessing documents.

Timesharing   A system whereby the processing time of the CPU is divided among many terminals.

Time-slice   The time given to a particular terminal; also known as a time slot.

Token-ring   A network where a token (a signal which gives workstations permission to transmit data) passes around a ring network.

Toolbar   A bar containing a series of icons and pull-down menus that may be selected.

Top-down approach   Problem-solving strategy whereby an overall problem is broken down into progressively smaller and more manageable problems.

Topology   The way a particular network is arranged.

Trace   A method of following what is going on in a complex spreadsheet to aid troubleshooting.

Transaction file   A file on which all the transactions (bits of business) over a certain period of time are kept. It is used to update a master file.

Transaction processing   A real-time system where transactions are processed as soon as they are received.

Translator   A program used to convert a program written in a high- or low-level language into machine code.

Transmission modes   Configuration settings used with networks.

Transmission rate   The speed of data flow in bits per second (bps) through a medium.

Tuple   A row of data in a table of a relational database.

Turbo pascal   A declarative high-level programming language.

Turnaround document   A document produced by a computer which is subsequently filled in manually and used as the input to the computer.

Uninstall   A utility program to remove programs.

Unix   An operating system used in multi-user computing.

Update   Changing details which have become out of date.

USB (universal serial bus)   A fast external bus that supports fast transfer of data.

User interface system   A system used to improve the ease with which users can interact with the system.

User level security   Level of security provided by a network operating system where each user is given a user name and a password in order to gain access to the network.

Utility program   Software which helps the user perform tasks such as virus checking, file management, etc.

Validate   A check performed by a computer program to make sure that data is allowable.

VDU (visual display unit)   The screen on which data is displayed; also known as a monitor.

Vector graphics   Graphics which are defined using co-ordinate geometry. They are easy to scale without any loss of resolution. Each part of a vector graphic can be manipulated individually.

Verbal information   Spoken information.

Verification   Checking the accuracy of data entry.

Video card   Another name for a graphics card. Gives the computer its display characteristics.

Video compression card   Used to compress the large amount of data needed to store a video sequence so that it can be stored on disk or sent over a network.

Video conferencing   System enabling meetings to be conducted without leaving the office, using video cameras and special computer software.

Viewdata   A computer-based information retrieval system such as teletext which uses screen messages to display information.

Virtual reality   Computer technology which creates a simulated multi-dimensional environment for the user.

Virus   A program that has been created to do damage to a computer system.

Virus checking   The process of using software to scan for computer viruses.

Visual basic   A version of basic which makes use of the Windows environment.

Voice recognition   The ability of a computer to 'understand' spoken words by comparing them with stored data.

WAN (wide area network)   A network where the terminals are remote from each other and telecommunications are used to communicate between them.

Web site   A site containing information on the Internet.

Wholesaler   A person who supplies products in bulk from manufacturers to shops which in turn sell the products to the general public.

Widow   The last line of a paragraph when separated from the rest of the paragraph by a page break.

WIMP (windows icons menus pointing devices)   Using a graphic user interface (GUI) rather than typing in commands at the command line.

Windows   A graphic user interface which provides a common way of using programs.

Windows NT   An operating system used mainly for networking.

Windows 2000   An operating system used mainly for networking which is replacing windows NT.

Wizards   Programs within Microsoft Office which can be run to speed up certain tasks.

Wordprocessor   A program that allows text to be entered, styled, displayed on a VDU and edited before being printed out.

Worksheet   The working area on a spreadsheet.

Workstation   A computer or terminal at which a user works.

Wrap   The process by which a computer starts a new line automatically.

Writing style   A style of writing such as formal, business, social etc.

Wysiwyg (what you see is what you get)   What appears on the computer screen is exactly what is printed out.

Xenix   A version of the Unix operating system.

Zip drive   Backup storage device.

Index